Chinese Medicine for Lower Body Pain

Chinese Medicine for Lower Body Pain

Benjamin Apichai

CRC Press
Taylor & Francis Group
Boca Raton London New York

CRC Press is an imprint of the
Taylor & Francis Group, an **informa** business

First edition published 2021
by CRC Press
6000 Broken Sound Parkway NW, Suite 300, Boca Raton, FL 33487-2742

and by CRC Press
2 Park Square, Milton Park, Abingdon, Oxon, OX14 4RN

CRC Press is an imprint of Taylor & Francis Group, LLC

Library of Congress Cataloging-in-Publication Data

Library of Congress Control Number: 2020948725

ISBN: 978-0-367-50795-4 (hbk)
ISBN: 978-0-367-23585-7 (pbk)
ISBN: 978-0-429-28059-7 (ebk)

Typeset in Times
by Deanta Global Publishing Services, Chennai, India

Thank you

To my dearest wife Agnes Wong for making numerous meals, painstakingly taking care of my family, and supporting me throughout the months that I sat in front of the computer writing instead of helping her with housework and taking her to appreciate the beautiful Seattle summer weather.

To my most important teacher Professor Dr. Chongyun Liu of Bastyr University, who taught me and mentored me.

To my dear son Dominic Apichai who edited and helped me with the photography.

Contents

Preface

While I was working in China, we referred postoperative patients who suffered from acute or chronic pain to the department of Chinese medicine—one of the departments in hospitals in China which includes the services of acupuncture medicine, herbal medicine, and Tui Na rehabilitation. Located within the hospital, the department is very convenient for postoperative patients, and there is no need for traveling since the physicians attend the patients minutes after a referral is sent.

Since I came to the US in 1994, I've noticed that patients who suffer from both acute and chronic pain, including postoperative pain, are prescribed hydrocodone, methadone, oxycodone, fentanyl, and many other narcotic analgesic medications. These pain medications attach to receptors on nerves in the brain that increase the threshold of pain—the level of pain stimulus at which we start to notice it. Then these medicines reduce the perception of pain.[1] When taken as directed, these medications can relieve both acute and chronic pain. However, narcotic pain medications have a high potential for misuse, abuse, and addiction. They can have serious side effects including constipation, headache, trouble breathing, abnormal heartbeats, cardiac arrest, and death.[2]

With the opioid crisis in America still at record levels, it is time we educate communities that pain medications, especially narcotics, are not the only options for pain management. Acupuncture, Chinese herbal medicine, and Tui Na treatments work effectively to arrest pain and help patients with acute and chronic conditions feel more comfortable and improve their daily function. Acupuncture has no side effects. It activates the body's own self-healing mechanism, works through repairing the source of pain, and stimulates the body to self-recovery.

Even more commendable, acupuncture treatments also work for opioid dependence.

An acupuncture needle is just a piece of metal; there is no opioid medication nor any other chemical pain relievers on it. Thanks to the US FDA, on December 6, 1996, and revised as of April 1, 2019, the FDA Title 21, Volume 8 classified acupuncture needles as class II medical therapeutic devices.[3]

I am a clinical practitioner. I am also an educator. My intention in writing this book not only is to educate the patients who are looking for a solution for managing their pain, but also to address several treatment modalities in Chinese medicine for acupuncturists.

For patients, this book helps you to easily understand your pain, as well as providing simple but effective self-care in the Patient Advisory section at the end of each topic. It advises you on self-management of your pain and also gives you ideas for prevention and dietary suggestions.

For acupuncturists, this book explains the mechanism of each disease, including the concise regional anatomy related to the pain, western and Chinese medicine pathology and etiology, and the most selected treatment plans.

NOTES

1. https://www.medicinenet.com/narcotic_analgesics-oral/article.htm#what_are_the_side_effects_of_pain_medications
2. https://www.medicinenet.com/narcotic_analgesics-oral/article.htm#what_are_the_side_effects_of_pain_medications
3. https://www.accessdata.fda.gov/scripts/cdrh/cfdocs/cfcfr/cfrsearch.cfm?fr=880.5580

About the Author

Dr. Apichai is a clinical supervisor at Bastyr Center for Natural Health and a core faculty member in the Department of Acupuncture and East Asian Medicine at Bastyr University, Seattle, Washington. He also has a private practice in Seattle, Washington. Dr. Apichai serves as an advisor for the Traditional Chinese Medicine (TCM) Sports Medicine Club. Dr. Apichai specializes in musculoskeletal concerns such as pain associated with the neck, back, and legs and in headaches. He earned his Doctor of Medicine (MD) degree from Jinan University, Guangzhou, The People's Republic of China in 1992, where he also completed several medical internships and research projects and joined the Tui Na Rehab Department at First Branch of Guangzhou TCM Hospital in 1993.

1 Mechanisms of Pain

There are three types of pain:

- Nociceptive pain
- Neuropathic pain
- Psychogenic pain

NOCICEPTIVE PAIN

Nociceptive pain is caused by damage to body tissues. The noxious stimuli that cause tissue damage activate the nociceptors, which are the pain receptors located on the surface of the body or in the musculoskeletal tissues. A signal is then sent from the neuron via the spinal cord to the medulla of the brain, which determines the location and intensity of the stimulus.

There are two main types of nociceptor:

- A-fibers are thinly myelinated and associated with fast conduction and response, such as first acute pain and heat.
- C-fibers are unmyelinated and associated with longer-lasting and dull pain such as reactions to chemicals and thermal and mechanical stimuli.[1]

Nociceptive pain can be further classified:

- Somatic nociceptive pain originates from the nociceptors located on the surface of the body or in the musculoskeletal tissues. It is associated with body tissue injury or inflammation. Most acute sport injuries result in this type of pain. It is usually localized, sharp or dull, sore, stabbing, or aching in nature. It is often exacerbated with activity, coughing, laughing, deeply breathing, or changing wound dressings and alleviated upon rest (Robertson 2018). This type of pain is commonly seen in musculoskeletal pain, arthritic pain, and postsurgical pain. It responds well to the treatments of acupuncture and Tui Na.
- Visceral pain originates from pain receptors located in the internal organs. This type of pain is usually described as cramping, deep, aching, squeezing, constant, and pressure-like, but it is not localized because it may refer, as with appendicitis or gallstones.[2]

NEUROPATHIC PAIN

Neuropathic pain is caused by injury to the nerves that send the wrong pain signals to the brain. This type of pain is typically described as tingling, numb, tight, prickling, itching or like the sensation of needles, electric shock, burning, or shooting that is often intense and chronic. The pain can be intermittent or continuous. Examples include stroke, diabetes, postherpetic neuroglia, vitamin B deficiency, phantom limb pain, and carpal tunnel syndrome.

PSYCHOGENIC PAIN

This is a term used to describe pain that is associated with some degree of psychological disturbance. Examples include stress, anxiety, depression, hypochondriac pain, headache, stomach pain, abdominal pain, and back pain. The pain is induced by emotional stress, worry, and grief.[3] This type of pain responds well to the treatments of acupuncture and Chinese herbal medicine.

NOTES

1. https://www.tocris.com/research-area/nociception
2. https://www.painmanagement.org.au/2014-09-11-13-34-03/2014-09-11-13-35-16/130-visceral-pain.html
3. https://en.wikipedia.org/wiki/Psychogenic_pain

2 Classification of Pain

CLASSIFICATION UNDER THE CHINESE ETIOLOGY

The meridians in traditional Chinese medicine are a network in the body, not the nerves as many people believe, but Qi in the meridian network. Each meridian is connected with a specific organ which is called Zang-Fu in Chinese medicine, and meridians are connected to themselves through certain acupuncture points. Blockage of the meridian can affect the Zang-Fu; conversely, a Zang-Fu's deterioration may manifest on the meridian as well. Thus, Chinese etiology will have to consider the conditions of both Zang-Fu and meridians. It will not just focus on the regional anatomy; the Zang-Fu's etiology and pathology will have to be considered as well.

Even though there are many factors that cause pain, they can be divided into these categories:

1) The meridian blockage can be caused by the Six Pathogenic factors 六淫: Wind, Dampness, Cold, Dryness, Heat, and Summer Heat. The pain can occur from the invasion of one or more of these climatic factors. For example, joint pain or musculoskeletal disease may manifest as more intense pain during the autumn and winter seasons when the climate is cold and rainy.
2) The meridian blockage can be affected by Seven Emotions: joy 喜, anger怒, anxiety憂, pensiveness思, grief悲, fear驚, and fright恐. Joy and anger are two main factors that cause chronic pain.
   a) Joy in Chinese medicine refers to two states.
      (i) Su Wen素問 chapter 39 (Unschuld and Tessenow 2011) states "喜則氣和志達 When one is joyous, then the Qi is in harmony and the mind is unimpeded. 榮衛通利The Ying Qi (Nutrient Qi) and the Wei Qi (Defensive Qi) pass freely."
      (ii) Su Wen素問 chapter 2 (Unschuld and Tessenow 2011) states "心志為喜， 喜傷心 The Heart controls joy, joy injures the Heart. 故氣緩矣Hence, the movement Qi flow is slow and relaxed." Su Wen 素問 chapter 5 (Unschuld and Tessenow 2011) states "violent joy harms the Yang 48." In the case of extreme joy, the Qi slows down excessively and reaches a state of dispersion and unable to be kept together.
   b) Anger is a harmful emotion that creates physical, emotional, and social problems that prevent healing. Research has shown that anger is significantly correlated with increased chronic pain levels. Anger is the emotion associated with the Liver.

      Su Wen素問 chapter 2 (Unschuld and Tessenow 2011) states "肝在志為怒。怒傷肝 The Liver controls anger, anger injures the Liver."

Su Wen 素問 chapter 39 (Unschuld and Tessenow 2011) states "怒則氣逆 When one is angry, then the Qi moves contrary [to its regular course]. 故氣上矣 Hence, the Qi rises. The Liver Qi normally effuses downwards. Hence, in the case of anger the Liver Qi rises, resulting in headache and hypertension."

3) Meridian blockage can be caused by Blood stasis, stagnant food, calculus. For example, a sprained ankle causes a bruise. Bruising in Chinese medicine is "Blood stasis," and Blood stasis in meridians manifests as pain in the affected ankle. Eating too fast and too much in the emotional state under the Seven Emotions causes stomach pain; a kidney stone causes lumbar pain.
4) Deficiency in the Zang-Fu and the meridian can also manifest pain. The deficiency includes Blood deficiency, Yin deficiency, and Yang deficiency. According to traditional Chinese medicine, the Liver governs the Tendons and the Kidneys control the bones. When there is insufficient Liver Blood, the Tendons will not be adequately nourished. As such, the patient may experience trembling hands or feet, numbness of the limbs, or even difficulty in flexing and extending the limbs. When Kidney Yin and Yang are deficient, the patient may have symptoms related to the skeletal system such as cold hands, knees, feet, lumbar, and painful joints[1] (Unschuld and Tessenow 2011).

CLASSIFICATION UNDER THE SYMPTOMS IN CHINESE MEDICINE

Inquiry to learn the patient's history of pain is helpful; the sensation of pain is also helpful in finding out the etiology and pathology of pain. In Chinese medicine, pain can be caused by the blockage of external pathogenic factors, Phlegm Damp, Blood stasis, Blood deficiency, food, parasites, calculus, Yin deficiency, and Yang deficiency.

According to Chinese medicine, the factors of pain are categorized into excess and deficiency. Pain occurs when there is a blockage in meridians is under excessive condition; a deficient pain is primarily a result of deficiency of Qi causing insufficient nourishment in Zang-Fu.

Pain under Excessive Conditions

1) *Distended pain*: The sensation of pain is accompanied by a sensation of distention. Distention presented in the head and eyes indicates Liver Fire or Liver Yang uprising. Distention presented in the chest and hypochondriac regions indicates Liver Qi stagnation.
2) *Stabbing pain*: The pain is similar to needle pricking. The pain indicates Blood stasis.
3) *Burning pain*: The sensation of pain is accompanied by a hot sensation. The patient prefers cold and dislikes heat. It is caused by hyperactivity of Yang or deficiency of Yin.

4) *Colic pain*: The pain is intense, just like a knife cutting. It is caused by tangible evil Qi blocking the meridians and resulting in stagnation of Qi and Blood.
5) *Heavy pain*: The pain is accompanied by heavy sensation. It is caused by dampness blocking the meridians, resulting in Qi stagnation.
6) *Fixed pain*: The location of the pain is fixed. If the pain is located in the chest and abdomen, it indicates Blood stasis fixed Bi syndrome. If the pain is located in the joints of the extremities, it indicates Cold Damp Bi syndrome.
7) *Wandering pain*: The location of the pain is not fixed; it may change places from one side of the limb to the other side. It indicates Wind Bi syndrome. Or it is in the chest, hypochondriac regions, or abdomen, which indicates Qi stagnation.

Pain under Deficient Conditions

1) *Cold pain*: The sensation of pain is accompanied by a sensation of cold, alleviated by warmth. It indicates pain from cold pathogen blocking meridians. It is caused by Yang deficiency.
2) *Achy pain*: The pain is mild, intermittent. It indicates Yin and Blood insufficiency; the patient is undernourished, or Yang deficiency causes the patient's body to lose functions of Kidney Yang.
3) *Empty pain*: The pain is accompanied by an empty sensation. It is caused by Qi and Blood insufficiency; the body lacks essence and is undernourished.
4) *Sore pain*: The pain is sore and the region is weak. If the whole body is sore, it indicates superficial disease or Damp Phlegm accumulation. If the soreness is on the lumbar and knees, it indicates Kidney deficiency.
5) *Referred pain*: The pain radiates from one place to another. It is caused by the blockage of evil Qi leading sluggish Qi flow in meridians, or it is caused by insufficiency of Yin and essence, resulting in undernourishment of tendons.

CLASSIFICATION UNDER THE DEGREE OF PAIN

Patients with the concern of pain usually have a difficult time describing their pain. Health care providers usually will obtain the information related to pain through the below history.

- *The anatomic location*: The location tells the possible involvement of the anatomic structures. For example, if a patient comes for low back pain, the patient will not say "My L3 transverse process has been causing my right lumbar pain." Per the health care providers, the severity of the injury at the right side L3 transverse process will not cause sciatic pain.
- *History of pain and injury*: Even though pain intensity is higher in acute onset of pain, the healing is sooner and the prognosis is better. Acute pain

may be able to be completely healed. Compared to acute pain, chronic pain may have lower pain intensity, but due to its longer duration, the healing takes longer, and the prognosis may not be as good as for acute pain. The history of injury in terms of the type of injury provides an idea of the treatment plan and prognosis. Chinese medicine will not only stop pain, it will also prevent future pain reoccurrence. For example, when a patient suffers from sciatic pain due to prolapse of the lumbar intervertebral disc, Chinese medicine will strengthen the affected disc and the surrounding soft tissues by tonifying Liver and Kidney and then advise the patient on dietary, weight management, exercise, and postures.

- *Verbal descriptors*: The health care providers will need to know the quality of pain and whether there is any referring pain. The quality of pain usually gives an idea of how to obtain the clinical information in a more straightforward way. For example, say it is a cold pain at the right lumbar; the pain is less when applying warmth pad. The practitioners will obtain information regarding Kidney Yang deficiency. If it is a stabbing pain, then the information that is possibly needed will relate to traumatic injury, such as an acute lumbar sprain. If the low back pain radiates to the lateral side of the tibia and to the foot and toes, then the possible diagnosis is the sciatic nerve-related pain such as prolapse of the lumbar intervertebral disc, lumbar spinal stenosis, or piriformis syndrome. The degrees of pain give an idea of the prognosis in terms of the quality of pain. Stabbing pain usually is an acute onset of trauma, cold pain usually indicates that it is an acute onset underlined Kidney Yang deficiency, and the concerns may not be just in the lumbar regions; there might also have knee pain.
- *Time course*: The pain schedule narrates the degree of pain. For example, if the low back pain is worse at night, in the morning, or after sitting a long period of time and better with movements, such information provides us with the degree of the pain when the temperature is cooler by the end of the day. The lumbar pain or high intensity of pain occurs when the weather is cold and rainy.
- *Aggravating factors*: Both physical and mental factors can result in an increased degree of pain. When the patient continues to have the same physical activity which is the cause of the pain, the degree of the pain may increase. For example, a patient suffers from an acute left ankle sprained from running on an uneven surface, and instead of resting until fully recovered, the patient continues to practice running on the same uneven surface. Stress is a cause of certain joint pain; patients who suffer from the injury need mental relaxation during healing. For example, a patient suffers from an acute subluxation of the sacroiliac joint; mental stress may be the cause. Thus, the joint treatment needs mental rest; conversely, the patient continues to be under the same mental stress resulting in pain aggravation.
- *Alleviating factors*: In Chinese medicine, patients are asked to describe their preferable temperature of food and drink and their body sensation. The answers help to ensure accurate diagnosis and to learn the alleviating

factors. For example, a chronic low back pain patient feels cold in the lumbar and knees and prefers warm water; from these two symptoms, we learn that the alleviating factor for this patient is applying a heating pad on the lumber. Health care practitioners also want to obtain information that could decrease the pain, including pain medications and nonpharmacologic approaches such as acupuncture, massage, rest, movements, relaxation, and heating pad or cold pack.

- *Changes in functional status caused by pain*: Some physical activities may be limited by pain. Besides, physical activities pain may affect sleep, appetite, energy, relationships, and mood. For example, sciatic pain patients may not be able to walk a distance, may not be able to stay asleep, may suffer from constipation from taking pain medications, and may be fatigued all day long.
- *Severity of pain*: Severity of pain helps the health care practitioner narrow the differential or presumptive diagnosis and prognosis. A high intensity pain may be easier to heal than a low intensity pain. For example, a patient suffering from an acute lumbar sprain that manifests as a high intensity sharp stabbing pain may take a shorter period or fewer office visits to heal the pain than a diabetic patient who develops peripheral neuropathies that manifest as a low intensity tingling sensation on all toes. To ensure an accurate diagnosis in the initial visit and to evaluate the effectiveness of the prior acupuncture treatments, practitioners need to provide an easy understanding and a simple method of access for patients to describe their level of intensity of pain.
  a) The Visual Analogue Scale (VAS) consists of a 10–15 cm straight horizontal line that is empty of words, figures, or vertical lines in the middle of the line. The line has the words "No pain" at the far left and "Pain as bad as it could be" at the far right. This allows patients to mark the line(s) representing the level of their pain intensity over the past 24 hours. Patients also may write some words following their marks such as "morning waking up with" or "after taking pain medications," as so many factors may influence pain perception (Haefeli and Elfering 2006). This tool was first used in psychology by Freyd in 1923.
  b) The Graphic Rating Scale (GRS) is a modified VAS that adds three descriptive terms "mild," "moderate," and "severe" to the Visual Analogue Scale. Patients are instructed to mark the line at a point representing their pain intensity over the past 24 hours. The GRS can be administered in person or by telephone.
  c) The Numeric Pain Rating Scale (NPRS) is a unidimensional measure of pain intensity. It is a segmented numeric version of the Visual Analogue Scale (VAS). Commonly, clinics use a horizontal bar with 11 whole numbers ranging from 0–10 integers. Some clinics may use a scale ranging from 0 to 100 with 0 indicating no pain and 100 indicating the worst pain.[2] Patients are instructed to mark a number that best represents their pain intensity over the past 24 hours. The NRS

can be administered verbally. 0 represents no pain. 1 represents a little pain, but it doesn't impact daily living, and patients feel no need to see medical attention. 2 represents mild pain; patients notice the pain but can easily ignore it, as doesn't restrict movements. 3 represents moderate pain. Patients notice the pain but can easily ignore and forget it. It is uncomfortable but tolerable. It slightly inhibits movement but the patient is able to continue daily activities, and the pain doesn't impact the quality of life. 4 represents uncomfortable pain. It is constant and starts to radiate to other places, like low back pain that starts to radiate to the hip, groin, leg, and foot. It impacts some degrees of normal life. 5 represents distracting pain. It is constantly noticeable pain that the patient cannot ignore. It impacts daily activities and quality of life. 6 represents distressing pain. The patient becomes more psychologically emotional about the ability to recover the pain and starts seeking treatments. 7 represents severe pain that completely controls the mindset on pain recovery. The pain impedes the ability to operate daily life and cognitive thinking. Patients rely on medical attention to manage the pain. 8 represents unmanageable severe pain. The extreme discomfort impedes the ability to be mobile and active. 9 represents severe pain that is intolerable and excruciating. Patients may seek stronger pain medications or surgery. 10 represents such debilitating pain that patients may go unconscious. This level of pain usually is incurred during a severe accident, and emergency treatment is needed.[3]

d) Verbal Rating Scales (VRSs) are another modified VAS. Both ends of the line still have the words "No pain" at the far left and "Pain as bad as it could be" at the far right. Adjectives are then added to describe the levels of the pain. Different VRS are used in clinical practices, ranging from four-five to fifteen-point. The objective words in a four-point VRS scale are no pain, mild, moderate, severe; or no pain at all, some pain, considerable pain, and Pain that could not be more severe. The objective words in five-point VRS are none, mild, moderate, severe, and very severe. The objective words in fifteen-point VRS are extremely weak, very weak, weak, very mild, mild, very moderate, slightly moderate, moderate, barely strong, slightly intense, strong, intense, very strong, very intense, and extremely intense.

e) The Descriptor Differential Scale of Pain Intensity (DDS-I) is a vertical scale with a sub-scale in each objective word. The words include faint, moderate, barely strong, intense, weak, strong, very mild, extremely intense, extremely intense, very weak, slightly intense, very intense, and mild.[4]

f) The Modified Graphic Rating Scale (C-GRS) is a modified Numeric Pain Rating Scale (NPRS). Patients are instructed to mark the line at a point representing their pain intensity. It is a horizontal bar with the integers ranging from 10 at the left endpoint to 0 at the right endpoint. The left endpoint position 10 indicates "no pain"; the right endpoint position 10 indicates "worst pain ever" or increased pain intensity. Patients are instructed to mark the line from the 0 to the left if the pain is reduced and to the right if the pain is increased.

CLASSIFICATION UNDER THE ONSET, DEVELOPMENT, AND DURATION

Acute Pain

Acute pain usually comes on suddenly. It is a temporary response that alerts the body. The pain can be greatly in intensity. Acute pain usually has a cause, and the pain doesn't last longer than three months (Robertson 2018). After acute pain subsides, a person can go on with life as usual.

The common causes of acute pain include:

- Surgery
- Injury
- Fracture
- Dental work
- Burns
- Cuts

Chronic Pain

The definition of chronic pain is persistent or recurrent pain lasting longer than three months.

- *Chronic pain associated with emotion*

  According to traditional Chinese medicine, the Liver is the Zang-Fu organ associated with emotional stress, anxiety, and depression. The signs and symptoms of Liver Qi stagnation include throat pain that feels like there is a plum seed stuck there, headache, migraines, neck pain, feeling oppression on the chest, hypochondriac pain, breast distension, tendon tightness, stiffness and spasms, abdominal cramps and pain, painful menstruation, low back pain, and knee pain.

  Stress, anger, and depression are the leading causes of Liver Qi stagnation. Meanwhile, Liver Qi stagnation manifests stress, anger, and depression. These three emotional factors insidiously debilitate joint tendons by deteriorating the functions of "Liver stores Blood" and "Liver governs Tendon," resulting in chronic pain.
- *Chronic pain associated with cancer*

  Chronic cancer pain includes pain caused by the tumor itself, such as brain tumor headaches, brain tumor trigeminal nerve tooth pain,[5] and the pain from cancer treatments such as surgery, chemotherapy, or radiotherapy.

- *Chronic pain associated with neuropathy*
  Neuropathic pain that lasts longer than three months. Examples include multiple sclerosis or diabetes that causes numbness, pain, or burning and stinging sensations on limbs and fingers.
- *Chronic pain after surgery*
  Surgery is recognized as one of the most frequent causes of chronic pain in patients attending pain clinics (Reddi and Curran 2014). Central nervous system changes may be the leading cause of chronic pain following surgical trauma and nerve injury (Reddi and Curran 2014).
- *Chronic pain after trauma*
  Chronic pain that is associated with trauma can be divided into two types:
  a) Chronic pain caused by an initial injury such as a shoulder injury that becomes frozen shoulder.
  b) Chronic pain without any prior injury such as fibromyalgia or temporomandibular joint dysfunction (TMJ).
- *Chronic headache*
  Chronic headache is commonly defined as headaches occurring on 15 or more days in a month for at least three months.
  Factors causing chronic headaches include overuse of pain medications. The factors in traditional Chinese medicine include:
  a) Obesity from Spleen Qi deficiency causes empty headache accompanied by persistent heavy sensation, poor appetite, and loose stool.[6]
  b) Patients who are Blood and Yin deficient from severe Blood loss in menstruation may experience dull headache, palpitations, and discomfort in hypochondriac regions.
  c) Head injury causes Blood stasis.
  d) Liver Qi stagnation causes migraine headache.
  e) Hyperactivity of Liver Yang causes headache with dizziness, irritability, insomnia, and flushed face.
  f) Cold and Damp accumulation causes headache with heaviness sensation, cold limbs, distention in the chest, pain in upper abdomen, nausea or vomiting with excessive sputum, poor appetite, and fatigue.
- *Chronic orofacial pain*
  Trigeminal neuropathic pain (TNP) and pain related to temporomandibular joint disorders (TMD) are the common chronic orofacial pain seen in acupuncture clinics.
- *Chronic visceral pain*
  Visceral pain is the pain derived from internal organs such as stomach pain from gastritis.
- *Chronic musculoskeletal pain*
  Chronic musculoskeletal pain is caused by an initial injury to the bones, joints, muscles, tendons, ligaments, or nerves, and the period of pain lasts longer than three months. Chronic musculoskeletal pain can also be caused by overuse, as is the case with osteoarthritis and inflammatory arthritis.

CLASSIFICATION UNDER THE REGION OF PAIN

Lumbar Pain

- *Low back pain caused by disorders of the soft tissues surrounding the facet joints*: Includes lumbar muscle strain, myofibritis, and local injuries caused by strains, sprains resulting bleeding, bruises, edema, adhesions, and muscle spasms. Patients are usually able to describe the causative factor such as shoveling snow, bending the low back to pick up a big box, or lifting a heavy object a few hours or a day before the pain occurs.[7]
- *Low back pain caused by disorders in the vertebrae*: Includes rheumatoid spondylitis, osteoporosis, hyperostosis, tuberculous spondylitis, spinal trauma, and disc herniation.
- *Low back pain caused by nerve and spinal cord*: Includes spinal cord compression, degenerative disc disease, acute transverse myelitis, radiculitis, and spinal cord tumors.
- *Low back pain caused by internal organ disorders*: Includes nephritis, urinary tract infections, acute and chronic pyelonephritis manifested as lumbar pain (and severe cases radiate along the ureter to the perineum), urinary stones (urolithiasis), cholecystitis, gallbladder stones, pancreatitis, gastric and duodenal ulcers, prostatitis, etc. Tumors can also cause lumbar pain.
- *Low back pain in female patients*: Female patients often suffer from back pain caused by endometritis, adnexitis, salpingitis, retroverted uterus, pelvic inflammatory disease, uterine fibroids, cervical cancer, and ovarian cysts. Low back pain is also common during pregnancy as the baby develops.
- *Low back pain caused by mental factors*: Patients with hysteria may also complain of lumbar disease, but when there is no objective evidence or objective examination and subjective narration that cannot be explained by physiological anatomy and pathological knowledge, this kind of back pain is often a manifestation of hysteria.
- *Low back pain caused by kidney deficiency*: The low back is the "mansion of the kidneys." The back pain is not clearly caused by any specific factor or disease, abnormality, or injury of the spine. There is no objective finding nor objective examination. Low back pain isn't accompanied by redness, swelling, or heat. And the patients are not mentally ill nor suffering from hysteria. The pain is low intensity, aggravated by fatigue, and alleviated by bed rest.

Hip Pain

- *Hip pain caused by arthritis*: This is commonly seen in older patients. Arthritis in the hip joint causes inflammation. Osteoarthritis (OA) and rheumatoid arthritis (RA) are common arthritis disorders in the hip.
- *Hip pain caused by fractures*: A fall, especially a sideways fall, often causes the femur fracture. Certain medical conditions such as cancer and osteoporosis may make patients more susceptible to fractures.

- *Hip pain caused by cancer*: Bone cancer can spread to the bone in the hip, resulting in hip pain.
- *Hip pain caused by sprain*: Athletes are susceptible to hip sprain. Insufficient warm-up prior to training and overstretching are common causes.
- *Hip pain caused by muscle strain*: The injury is usually related to exercise or repetitive use of certain muscles, tendons, and ligaments that support the hips. Symptoms can be swelling, tenderness, and stiffness of the hip, but often patients cannot pinpoint the exact location of the hip pain because the pain is rarely felt in just one area.[8] Pain is usually manifested with movements. Iliotibial Band Syndrome is another common cause of hip pain.

Knee Pain

- *Knee pain caused by sprained or strained knee ligaments and/or muscles*: The common cause of this injury is a blow to the knee or a twist while moving. Examples include lateral collateral ligament injury, medial collateral ligament injury, anterior cruciate ligament injury, and posterior cruciate ligament injury.
- *Knee pain caused by torn cartilage*: Both medial and lateral menisci can be torn with a sprain of the knee, resulting in knee pain.
- *Knee pain caused by tendon overuse*: Inflammation of the tendons or called tendonitis can cause knee pain. Examples include patella tendonitis and quadriceps tendon rupture.
- *Knee pain caused by arthritis*: Knee arthritis is seen in rheumatoid arthritis and osteoarthritis.
  a) Rheumatoid arthritis often affects younger patients. It is knee joint inflammation and destroyed cartilage, resulting in pain. It is symmetrical.
  b) Osteoarthritis often affects older patients. It is a degenerative change to the knee where the cartilage has worn out. It is more commonly seen than rheumatoid arthritis.
- *Knee pain caused by knee fractures*: Knee fractures can be from injury or osteoporosis, resulting in knee pain. Examples include patellar fracture, femoral condyle fracture, and tibial eminence fracture.
- *Knee pain caused by knee bursitis*: It is the result of overuse or sustained pressure such as kneeling on hard surfaces or from bacterial infection of the bursa, resulting in knee pain.
- *Knee pain caused by mechanical issues*: The common mechanical causes are iliotibial band syndrome, and displacement of patella.

Ankle Pain

- *Ankle pain caused by a sprain*: It is an injury to the ligaments. The anterior talofibular ligament (ATFL) spraining is the most commonly seen. The injury results in pain, swelling, and bruising.

- *Ankle pain caused by a fracture*: The fracture commonly occurs on the fibula. Severe pain immediately will be located at the fracture area accompanied with ankle swelling.
- *Ankle pain caused by a dislocation*: Severe ankle pain occurs immediately at the dislocated ankle joint accompanied with a deformed look to the ankle and ankle swelling.
- *Ankle pain caused by overuse*: Overusing or injury to the gastrocnemius muscle and Achilles tendon can result in ankle pain. It is more common in runners.
- *Ankle pain from uncommon causes*: Includes arthritis, gout, and infection.

NOTES

1. https://www.asiaone.com/health/how-tcm-can-help-numbness-fingers-and-foot
2. http://ratologydisabled.blogspot.com/2009/06/self-report-scales-and-procedures-for.html
3. https://www.painscale.com/assets/pdf/PainScale_lowerback.pdf
4. https://www.verywellhealth.com/pain-scales-assessment-tools-4020329
5. https://www.pennmedicine.org/updates/blogs/neuroscience-blog/2016/june/signs-of-a-brain-tumor
6. http://www.shen-nong.com/eng/exam/headaches_chinese_medicine_understands.html
7. https://www.uptodate.com/contents/low-back-pain-in-adults-beyond-the-basics
8. https://www.beaconortho.com/blog/common-hip-flexor-injuries-and-treatments/

3 Low Back Pain

PROLAPSE OF THE LUMBAR INTERVERTEBRAL DISC 腰椎間盤突出癥

Introduction

- It is commonly called sciatica or sciatic pain, and it is also called slipped disc, ruptured disc, and lumbar disc herniation.
- Lumbar radicular pain is defined as low back pain radiating to the leg below the knee level, while sciatica is defined as lumbar radicular pain along with clinical findings suggestive of nerve root compression.
- Research suggests that up to 90% of sciatica is caused by a herniated lumbar intervertebral disc (Kumar 2011, 2).
- A study reveals that the intervertebral discs of middle-aged (35–50 years of age) human donors already show degeneration (Martins et al. 2018; Smith 2018).
- Approximately 30% of people show evidence of disc degeneration at one or more levels. By the age of 60, more than 90% of people will show evidence of some disc degeneration (The Spine Hospital 2020).
- Degeneration itself does not necessarily cause pain. Painless degeneration is just called degeneration. The term "degenerative disc disease" describes disc degeneration that causes pain and other symptoms.

Anatomy

- The vertebral column of lumbar is made up of five vertebrae, the sacrum is made up of five sacral vertebrae which are fused together, and the coccyx region is made up of four fused pieces of bone.
- The sciatic nerve is the largest and longest nerve in the human body. Every person has two sciatic nerves, the right and left nerves. The sciatic nerve is formed by the combination of five nerves in the lower lumbar and sacral spine—L4, L5, S1, S2, and S3.
- The intervertebral discs lie between the vertebral bodies and consist of two regions, with the central, more gelatinous nucleus pulposus surrounded by a fibrous ring, the annulus fibrosus. The discs are separated from the adjacent vertebrae by a thin layer of hyaline cartilaginous tissue, the cartilage endplates.
- The intervertebral disc is the largest avascular tissue in the body. The cells in the center of an adult lumbar disc are approximately 8 mm away from the nearest blood supply (Choi 2009).

Western Medicine Etiology and Pathology

Sciatica is a medical condition affecting the lumbar spine in which a tear in the fibrous rings, or annulus fibrosus, of an intervertebral disc allows the nucleus pulposus, which is the gelatinous central portion of the intervertebral disc, to bulge out beyond the weakened damaged outer fibrous rings, resulting in back pain and nerve root irritation. It most commonly arises in middle-aged patients.

Although all discs are the same age, discs at the lower lumbar segments, which undergo mechanical loading, are more vulnerable to degenerative changes than the upper lumbar segments. This suggests that mechanical loading, rather than simple aging, is a causative factor.

As aging progresses, with more degenerative changes, the intervertebral discs fail to bear hydrostatic pressure, and tearing of the annulus clefts can occur easily, as the axial load is then transferred directly to the vulnerable annulus fibrosus. The load will cause a loss of height, leading to bulging of the discs when there is less resistance against tension under the load. Therefore, degenerated discs no longer function hydrostatically under the load, resulting in the concentration of stress on the endplate and the annulus, predisposing the disc to injury (Choi 2009).

Internal Pressure

1) Improper lifting

    The trauma occurs when using the back muscles to lift heavy objects.

    When the spine is straight, such as in standing or lying supine, the internal pressure is equalized on all parts of the discs. The internal pressure on a disc can increase from 17 psi in lying postures to over 300 psi on lifting with a rounded back.[1] Herniation of the disc into the spinal canal often occurs when the anterior side of the disc is compressed while sitting or during lumbar flexion, and the nucleus pulposus becomes pressed against the tightly stretched annulus fibrosus on the posterior side of the disc. Increased internal pressure results in the rupture of the confining membrane. The nucleus pulposus, which is the jelly-like contents of the disc, then moves into the spinal canal, pressing against the spinal nerve roots and thus, producing intense low back pain and referred pain symptoms.

2) Overweight

    Being overweight puts added pressure on the discs in the low back.

    Studies reported that obesity is associated with a higher prevalence of low back pain, which is more pronounced in women. However, the increase of and the greatest risk for back pain were highest among obese men.[2]

3) Repetitive activities that strain the spine

    - Athletes or physical workers who often have to perform repetitive lifting, twisting, and bending or who are frequently put in awkward positions may damage the intervertebral discs. Lifting and carrying heavy objects, such as weight lifting in a gym and working for a job as a furniture moving worker, predisposes to segmental structure failure.
    - Studies have found associations between vibration exposure and back pain. For example, for a person who operates a jackhammer as part of their daily job, the risk of sciatica is higher. And the risk is even higher if the operation is in a sitting position (Bongers et al. 1990, 1007–1026).

4) Lack of exercise

Prolonged sitting in an awkward posture, such as office jobs that involve using computers or long-distance driving, may increase the risk of sciatica. The intervertebral discs can become stuck, and blood supply can become poor, so that the bones and discs of the spine start to degenerate.

Earlier studies have found sitting to be associated with the emergence of low back disorders and a prevalence of sciatic pain in occupations in which sitting is a major requirement additionally co-exposed to awkward postures. Intradiscal pressure seems to increase while sitting; the pressure increases by almost 50% while sitting with hips and knees flexed and almost doubles if the torso is flexed while sitting.[3]

5) Smoking

It has been suggested that smoking can cause sciatica by interfering with the nutrition of the intervertebral discs. As aging progresses, the nutritional deficiency of the disc cells is accelerated by smoking. And it is believed that smoking reduces oxygen supply to the discs and causes more rapid degeneration.

Aging

Degeneration of the intervertebral disc usually starts around the age of 30. Consequently, the risk of developing sciatica begins around then, and approximately 30% of people will show evidence of disc degeneration at one or more levels. By the age of 60, more than 90% of people will show evidence of some disc degeneration (The Spine Hospital 2020).

Unlike other tissues of the body, the intervertebral discs have a very low blood supply. Daily physical activities and sports may cause damage to the disc, and it cannot repair itself once it is injured.

Overall, because of work and social and sports activities, this age group between 30 and 50 tends to be very active in comparison with older age groups, which increases the likelihood of physical injury. Additionally, the discs themselves have begun their descent into vulnerability.

Not everyone at that age has back pain, however. Spinal stenosis, arthritic changes in the spine such as bone spurs, usually first crops up in people older than 50 and can cause sciatica.

Traditional Chinese Medicine Etiology, Pathology, and Manifestations

Wind, Damp, Cold Invasion

Externally, wind, an external pathogenic factor of mysterious nature, is thought to carry the pathogenic influences into the body and contribute to the variable nature of the pain.

Internally, a deficiency of body constitution under post-traumatic circumstances, poor blood circulation of the intervertebral discs, and exposure to a wet, windy, and cooler environment cause Bi syndrome. Bi syndromes manifest as pain, soreness, aches, numbness or heaviness of muscles, sinews, and joints, and/or swelling and burning pain. In general, the causative factors of Bi syndrome are due to the blockage of meridians and collaterals and invasion—by wind, cold, damp, and heat.

For example, when a person who previously suffered from lumbar stiffness or trauma is standing on a rainy and cool day, the cold and rain affect the body and stay in the lumbar region, where the blood circulation is already poor; this external pathogen will stay and block the flow of Qi and blood at the lumbar joints; therefore, the external wind may not be the main factor causing the pain, although due to the unique character of constant movement of wind, it can still combine with cold and dampness and stir them up to exacerbate the underlying condition and move on to the legs.

Symptoms and Signs

This type of sciatica will not have an obvious history of injury; the causative factor in western medicine is not clear; patients have the sensation of a heavy leg and difficulty in turning the low back. Palpation by a medical provider will find tender spots along the lumbar spine. The pain is aggravated by change of weather.

- Tongue: White greasy coating
- Pulse: Deep and slow

Kidney and Liver Deficiency

Liver and kidney deficiency are also a main cause of low back pain and sciatica.

Liver governs tendons; it directly associates with gall bladder meridians and is associated with emotions. Liver depression due to stress may cause sciatica. When liver Qi is stagnant, the tendons are susceptible to injury. The radiating pain onto the leg is on the gall bladder meridians.

The liver function is to store blood. Due to old age, there is less blood returning to the liver, and liver blood deficiency may delay the self-healing of trauma in the lumbar region.

The lumbar region is the home of kidneys. When kidneys are strong, a low back injury does not remain for long. Many factors can deplete kidney Qi: they include constant fear, excessive physical activity, prolonged periods of standing, excessive sexual activity, drug abuse, and aging.

Wind, damp, and cold can settle in the low back when kidneys are weak. These exterior factors block meridians, resulting in pain.

Symptoms of Liver Blood Deficiency

- Tremor of legs. Patients may report that their legs twitch or tremble spontaneously at night while they are sleeping and may waken them from sleep. Sometimes, instead of shaking or trembling, the leg muscles may cramp as well while sleeping.
- In women, menses are scanty and light-colored, with a long cycle, or they may have no periods at all.
- The moods of anxiety, lack of confidence, and poor short-term memory may be manifested.

- The complexion looks pale, and the lips are pale.
- The body tends to be underweight.
- Tongue: Pale or pink, especially at the sides. Dry. Tongue coating may be thin or normal.
- Pulse: Thready and weak.

Symptoms of Kidney Yang Deficiency

- Patients generally are weak in constitution and/or suffering from chronic illnesses.
- Low back pain and sore, weak, the pain is on and off for years. Fatigue, cold limbs, pale face, impotence, 5 am diarrhea, dripping urination.
- Tongue: Flabby, pale tongue body, thick white greasy coating.
- Pulse: Deep, weak, thin or slippery, and slow.

Symptoms of Kidney Yin Deficiency

- Patients are generally weak in constitution and/or suffering from chronic illnesses.
- Low back pain and sore, weak, the pain is on and off for years; dizziness, blurry vision, tinnitus, loss of hearing; red face, afternoon fever, night sweats, insomnia, dry mouth, and five-center heat.
- Tongue: Small body size, red body, peeled coating.
- Pulse: Deep, weak, thin, short, and rapid.

Qi and Blood Stagnation

Patients may have an obvious history of injury.

Internal factor: Liver Qi stagnation is the main cause of this type. Emotional stress, anger, and anxiety are the causes of the blockages in meridians; because Qi is the commander of blood, the stagnation of Qi leads to stasis of blood. When Qi and blood are stagnant, the soft tissues lack nourishment, manifesting in spasm, low back pain, and radiating pain.

External factor: Improper use, overuse, or trauma of the lumbar region results in microscopic tears in ligaments, tendons, and/or muscles of the low back; this low back pain is called lumbar strain and is the most common cause of low back pain. The pain usually is classified as acute if it has only been present for days to weeks; it is referred to as chronic if the strain lasts longer than 3 months.

After the strain or injury to the soft tissues of muscles, tendons, or ligaments in the spine, the back muscles try to protect themselves from the lumbar strain by going into spasm. This type of soft tissue injury typically heals to a degree within a week or two, and the muscle spasms stop.

If the back spasm does not get better in 1 to 2 weeks, or it comes and goes over time in the same area of the back, it may indicate an underlying anatomical problem in the spine. Some examples of underlying issues that could cause the back to spasm include facet joint osteoarthritis, degenerative disc disease, spinal stenosis, and herniated disc.

When these or other underlying anatomical problems are present in the spine, muscle spasms are likely to keep recurring due to ongoing inflammation or instability.

The muscles stay in spasm in response to an underlying anatomical problem; this affects the blood and Qi circulation to the disc, eventually causing sciatica.

Symptoms and Signs

Patients may have an obvious history of injury. The pain is unbearable; the patient is afraid to move; the lumbar spine is deviated. There are tender spots along the spine at the regions of L4, L5, and S1. There is a radiating pain on the leg, aggravated by cough. There is pain and numbness or muscle atrophy of the leg at later stages.

- Tongue: Purple
- Pulse: Wiry, choppy, and rapid

Physical Examination

- Inspection:
  - Visual inspection is not able to diagnose the condition of the disc.
- Palpation:
  - Tender spots along the lumbar spine may be present in some patients.
- Range of motion:
  - Check the lumbar flexion, extension, and lateral flexion from side to side.
- Extra tests:
  - Straight leg raise test is positive.

Techniques:

- The patient lies supine with the legs extended.
- The therapist grasps the patient's ankle and places another hand on the patella to maintain the ipsilateral knee in full extension.
- Slowly raise the leg until the patient complains of pain or the leg is raised to the maximal range at 60–120 degrees.
- Note the degree of elevation when pain presents and the quality and distribution of pain.
- Carefully return the leg to the table.
- The result is positive if there is reproduction of pain along the distribution of the sciatic nerve dermatome. Note that changing the back pain is not a positive test.
- The result is negative if there is no pain felt upon maximal flexion of the leg.

Differential Diagnosis

Lumbosacral Strain

Both diseases cause low back pain. Lumbosacral strain is negative in the straight leg raise test.

Treatment

Treatment under the onset development and duration.

Herbal Treatments

Qi and Blood Stagnation

Treatment principles:

- Early or acute stage—moves blood and Qi, stops pain.

Herbal formula:

- Shen Tong Zhu Yu Tang 身痛逐瘀湯.

Formula ingredients: Dang Gui 當歸 9 g, Chuan Xiong 川芎 9 g, Tao Ren 桃仁 9 g, Hong Hua 紅花 9 g, Mo Yao 沒藥 6 g, Wu Ling Zhi 五靈脂 6 g, Qin Jiao 秦艽 9 g, Qiang Huo 羌活 9 g, Xiang Fu 香附 3 g, Chuan Niu Xi 川牛膝 9 g, Di Long 地龍 6 g, Gan Cao 甘草 6 g.

Ingredient explanation:

- Dang Gui 當歸, Chuan Xiong 川芎, Tao Ren 桃仁, and Hong Hua 紅花 move blood, disperse bruising, and stop pain due to blood stasis.
- Qin Jiao 秦艽 and Qiang Huo 羌活 expel wind, cold, and dampness, unblock painful obstruction, and alleviate pain.
- Wu Ling Zhi 五靈脂, Mo Yao 沒藥, and Xiang Fu 香附 move Qi and blood and stop pain due to Qi and blood stagnation.
- Chuan Niu Xi 川牛膝 and Di Long 地龍 unblock and promote movement in the channels and collaterals; stop spasms and convulsions.
- Gan Cao 甘草 harmonizes the herbs.

Formula actions:

- Moves Qi and blood.
- Unblocks painful obstruction in meridians.
- Stops pain.

Contraindications:

- Contraindicated during pregnancy.

Modifications:

- For heat symptoms, add Chai Hu 柴胡 and Huang Bai 黃柏.
- For Qi deficiency, add Dang Shen 黨參 and Huang Qi 黃芪.
- For pain in lumbar region and leg, add Xu Duan 續斷, Du Zhong 杜仲, and Sang Ji Sheng 桑寄生.

- For severe pain, add Quan Xie 全蠍, Wu Gong 蜈蚣.
- For cold symptoms, omit Qin Jiao 秦艽 and add Zhi Chuan Wu 制川烏.
- For cold symptoms and severe pain, add Gui Zhi 桂枝 and Xi Sin 細辛.
- For wind traveling pain, add Fang Feng 防風.
- For radiating pain to the leg, omit Wu Ling Zhi 五靈脂 and add Shen Jin Cao 伸筋草.
- For damp heat joint edema, add Cang Zhu 蒼朮, Huang Bai 黃柏, and Fang Ji 防己.
- For stiffness in joints, add Shen Jin Cao 伸筋草, Quan Xie 全蠍, and Wu Gong 蜈蚣.
- For mild joint edema and more damp Bi pain, add Fang Ji 防己, Yi Yi Ren 薏苡仁, and Cang Zhu 蒼朮.

Sub-Acute Stage with Kidney Deficiency

Treatment principles:

- Tonify kidney and liver.

Herbal formula:

- Bu Shen Huo XueHuoH Tang 補肾活血湯.

Formula ingredients: Shu Di Huang 熟地黃 10 g, Bu Gu Zhi 補骨脂 10 g, Tu Si Zi 菟絲子 10 g, Du Zhong 杜仲 3 g, Gou Qi Zi 枸杞子 3 g, Dang Gui 當歸尾 3 g, Shan Zhu Yu 山茱萸 3 g, Rou Cong Rong 肉蓯蓉 3 g, Mo Yao 没藥 3 g, Du Huo 獨活 3 g, Hong Hua 紅花 2 g.

Ingredient explanation:

- Shu Di Huang 熟地黃, Bu Gu Zhi 補骨脂, Tu Si Zi 菟絲子, Du Zhong 杜仲, Gou Qi Zi 枸杞子, Shan Zhu Yu 山茱萸, and Rou Cong Rong 肉蓯蓉 tonify liver and kidney, strengthen bone, and benefit essence and blood.
- Dang Gui 當歸尾, Mo Yao 没藥, Du Huo 獨活, and Hong Hua 紅花 move blood, disperse bruising, unblock stagnation, open meridians, and stop pain.

Formula actions:

1. Moves blood.
2. Nourishes liver and kidney.
3. Strengthens the lumbar region.
4. Stops pain.

Contraindications:

- Contraindicated during pregnancy.

Modifications:

- For bruising, add Tu Bie Cong 土鱉蟲 and Di Long 地龍.
- For damp heat, add Cang Zhu 蒼朮, Huang Bai 黃柏, and Yi Yi Ren 薏苡仁.
- For severe pain in leg, add Niu Xi 牛膝 and Wu Jia Pi 五加皮.
- For severe pain in lumbar region, add Xi Xian Cao 豨莶草 and Sang Ji Sheng 桑寄生.
- For Qi deficiency, add Huang Qi 黃芪 and Dang Shen 黨參.
- For Yang deficiency, omit Shu Di 熟地 and Dan Pi 丹皮, and add Xian Mao 仙茅 and Xian Ling Pi 仙靈脾.
- For blood deficiency, add Dan Shen 丹參, Ji Xue Teng 雞血藤, and He Shou Wu 何首烏.
- For Yin deficiency, add Nu Zhen Zi 女貞子 and Han Lian Cao 旱蓮草.

Intermediate or Chronic Stage

Treatment principles:

- Expels wind, damp, and cold, tonifies kidney and liver, nourishes blood, and stops Bi pain.

Herbal formula:

- Du Huo Ji Sheng Tang 獨活寄生湯.

Formula ingredients:

Du Huo 獨活 9 g, Sang Ji Sheng 桑寄生 6 g, Du Zhong 杜仲 6 g, Niu Xi 牛膝 6 g, Xi Xing 细辛 3 g, Qin Jiao 秦艽 6 g, Fu Ling 茯苓 6 g, Rou Gui 肉桂 6 g, Fang Feng 防風 6 g, Chuan Xiong 川芎 6 g, Dang Shen 党参 6 g, Zhi Gan Cao 甘草 3 g, Dang Gui 當歸 6 g, Bai Shao 白芍 6 g, Shu Di Huang 熟地黄 6 g.

Ingredient explanation:

1. Du Huo 獨活 expels wind, damp, and cold in lower Jiao and joints; removes chronic Bi stagnation in lower limb.
2. Xi Xing 细辛 enters Shao Yin meridian to expel wind, damp, and cold.
3. Qin Jiao 秦艽 expels wind and damp, opens meridians, and benefits joints.
4. Rou Gui 肉桂 warms meridians, expels cold, and opens meridians to promote blood circulation.
5. Fang Feng 防風 expels wind and drains damp.
6. Sang Ji Sheng 桑寄生, Du Zhong 杜仲, and Niu Xi 牛膝 are for deficiency of Qi and blood in liver and kidney, and strengthen bone and tendons; Sang Ji Sheng 桑寄生 is able to expel wind and drain damp; Niu Xi 牛膝 is able to move blood and opens meridians to benefit joints and tendons.
7. Dang Gui 當歸, Chuan Xiong 川芎, Shu Di Huang 熟地黄, and Bai Shao 白芍 are to nourish and harmonize blood.
8. Dang Shen 党参, Fu Ling 茯苓, and Gan Cao 甘草 are for tonifying spleen Qi.

Formula actions:

- Dispels wind, cold, and damp.
- Relieves painful obstruction.
- Supplements Qi and blood.
- Tonifies the liver and kidney.

Contraindications:

- Damp heat excessive Bi syndrome.

Cautions:

- Avoid food that is raw, cold, or oily.
- Keep warm during windy and cold weather.

Modifications:

- With severe pain, add Zhi Chuan Wu 制川烏, Zhi Cao Wu 制草烏, Bai Hua She 白花蛇, Xu Duan 川斷, Gou Ji 狗脊, Tu Si Zi 菟絲子, Hong Hua 紅花, and Di Long 地龍.
- With more cold symptoms, add Zhi Fu Zi 附子 and Gan Jiang 乾薑.
- With more damp symptoms, omit Shu Di Huang 熟地黃and add Fang Ji 防己, Yi Yi Ren 薏苡仁, and Cang Zhu 蒼朮.

Acupuncture Treatments

According to traditional Chinese medicine, sciatica is caused by blockages in mainly two meridians:

1) The gall bladder meridian of the foot—Shaoyang 足少陽經 (GB). The pain radiates from the lumbar→region posterior compartment of the thigh→posterior compartment of the leg.
2) The urinary bladder meridian of the foot—Taiyang 足太陽經 (UB). The pain radiates from the lumbar region→posterior compartment of the thigh→lateral compartment of the leg.

Acupuncture point formula:

- GB34 陽陵泉 Yanglingquan and GB30 環跳 Huantiao are two key "acupuncture points" for treating sciatica.

Local points:

1) Pain on the gall bladder meridian of the foot—Shaoyang 足少陽經(GB), select GB34 陽陵泉 Yanglingquan, GB35 陽交, Yangjiao, GB39 懸鐘 Xuanzhong.

2) Pain on the urinary bladder meridian of the foot—Taiyang 足太陽經(UB), select UB37 殷門 Yinmen, UB40 委中 Weizhong, UB57 承山 Chengshan.

Points for consideration:

- Tun Zhong 臀中—this is an extra point located at the center of the buttock, 3.5 cun lateral to the posterior midline, at the level of the fourth sacral foramen. The needle depth needs to be at least 2 cun because it is going to touch the sciatic nerve, so there is a need for a sensation radiating downward along the dermatome of the sciatic nerve.
- GB30 環跳 Huantiao—this is another good point you must select for sciatic pain. The needle depth also needs to be at least 2 cun, because it is also going to touch the sciatic nerve, so there is a need for a sensation radiating downward along the dermatome of the sciatic nerve. The patient needs to be lying on the opposite side with the knee flexed. If the needle sensation does not reach the ankle, add a needle to GB34 陽陵泉 Yanglingquan to extend the sensation further to the ankle or the foot. This point is also an excellent point for adding moxa to the needle (warming needle technique) for the sciatic pain with kidney Yang deficiency. The moxa needs to burn for the duration of 3–5 cones.
- GB31 風市 Fengshi—select this point when the sciatic pain is on the gall bladder meridian of the foot—Shaoyang 足少陽經. This point is a good point for sciatic pain with the invasion of wind-cold Bi syndrome 風寒痺症.
- GB34 陽陵泉 Yanglingquan—this point is a local point for the sciatic pain passing by when L5–S1 nerve roots are involved. It is also a point to warm the meridians for wind-cold Bi syndrome 風寒痺症. This point also benefits the sinews and tendons.

Distal points:

- UB40 委中 Wei Zhong—this is a command point for any acute low back pain. The command points are individual points that have strong and broad effects governing particular parts of the body. This point is better for acute low back pain than for chronic cases. This point is for cooling, sedating, reducing, and opening the blockage in meridians instead of tonifying the deficiency conditions. Therefore, it is a frequently chosen point for Qi and blood stagnation sciatica; it is not the best point for the cases of liver-kidney deficiency sciatica, except when the sciatica condition is of acute onset with an underlying deficiency.
- UB57 承山 Chengshan—this point is for low back pain and sciatica with underlying liver-kidney deficiency. It is for intermediate, subacute, or even chronic low back pain.
- KD4 大鐘 Dazhong—this point is good for chronic low back pain and sciatic pain with kidney deficiency.

- UB60 崑崙 Kunlun—this is for chronic low back pain and sciatica.
- GB40 丘墟 Qiuxu—this is a good point for pain at the L4 dermatomal region.
- UB58 飛陽 Feiyang—this is a good point for sciatic pain from a herniated disc between L4 and L5.
- UB62 申脈 Shenmai—this is a good point for sciatic pain from a herniated disc between L5 and S1.

Treatment under the Differentiations

1) Sciatic pain worse on rainy days, heavy sensation and stiffness, cold sensation in the lumbar region, limitation of movement in the lower back, and low back movement limitation.
   - Tongue: White greasy coating.
   - Pulse: Superficial and slippery.

   Treatment principles:
   - Drains damp, warms meridians.

   Herbal formula:
   - Qiang Huo Sheng Shi Tang 羌活勝濕湯.

   Formula ingredients:
   - Qiang Huo 羌活 3 g, Du Huo 獨活 3 g, Gao Ben 藁本 1.5 g, Fang Feng 防風 1.5 g, Chuan Xiong 川芎 1.5 g, Man Jing Zi 蔓荊子 0.9 g, Zhi Gan Cao 炙甘草 1.5 g.

   Ingredient explanation:
   - Qiang Huo 羌活 and Du Huo 獨活 enter Taiyang meridians to drain damp and expel wind in the meridians. They benefit joints and tendons. Qiang Huo 羌活 drains the upper body damp, while Du Huo 獨活 drains the lower body damp.
   - Fang Feng 防風 and Gao Ben 藁本 expel wind and drain damp in Tai Yang meridians.
   - Chuan Xiong 川芎 moves blood, expels wind, and stops pain.
   - Man Jing Zi 蔓荊子 expels wind and stops pain.
   - Gan Cao 甘草 harmonizes the herbs.

   Modifications:
   - With more wind damp symptoms, add Fang Ji 防己, Qin Jiao 秦艽, Zhi Fu Zi 附子, Wu Tou 烏頭.
   - With effusion in lower limb joints, add Niu Xi 牛膝.
   - With Qi stagnation, add Chen Pi 陳皮and Mu Xiang 木香.
   - With blood deficiency, add Dang Gui 當歸 and Bai Shao 白芍.

   Acupuncture treatments:
   - LI4 合谷 He Gu, GB20 風池 Fengchi, ST40 豐隆 Fenglong, SI3 後溪 Houxi and UB62 申脈, UB36 承扶 Chengfu, UB58 飛陽 Feiyangand, UB63 金門 Jinmen, GB29 居髎 Juliao, GB36 外丘 Waiqiuand, GB40 丘墟 Qiuxu, ST30 氣沖 Qichong, ST34 梁丘 Liangqiu, ST40 豐隆

Fenglong and ST42 沖陽 Chongyang, LIV3 太沖 Taichong, LIV5 蠡溝 Ligou, LIV6 中都 Zhongdu and LIV12 急脈 Jimai, UB32 次髎 Ciliao and UB53 胞肓 Baohuang, SP4 公孫 Gongsun, UB20 脾俞 Pishuand, ST36 足三里 Zusanli.

2) Sciatic pain worsened by emotional states such as anger, anxiety, or depression. Patients also have pain in the hypochondriac or genital region with chest distension and irritability.
   - Tongue: Red tongue body with thin yellow coating
   - Pulse: Wiry and/or rapid

   Treatment principles:
   - Moves liver Qi, stops pain.

   Herbal formula:
   - Chai Hu Shu Gan Wan 柴胡疏肝散.

   Formula ingredients:

   Chen Pi 陳皮 6 g, Chai Hu 柴胡 6 g, Chuan Xiong 川芎 4.5 g, Zhi Ke 枳殼 4.5 g, Bai Shao 芍藥 4.5 g, Zhi Gan Cao 炙甘草 1.5 g, and Xiang Fu 香附 4.5 g.

   Ingredient explanation:
   - Chai Hu 柴胡 moves liver Qi and releases stagnation.
   - Xiang Fu 香附 moves liver Qi and stops pain.
   - Chuan Xiong 川芎 moves blood and Qi, and stops pain.
   - Chen Pi 陳皮 and Zhi Ke 枳殼 regulate Qi and remove blockage.
   - Bai Shao 芍藥and Gan Cao 甘草 nourish blood, soothe liver, and stop pain.

   Modifications:
   - With pain in the hypochondriac region. add Yu Jin 鬱金, Qing Pi 青皮, Dang Gui 當歸, and Wu Yao 烏藥.
   - With liver fire, add Shan Zhi Zi 山梔子 and Huang Qin 黃芪.

   Cautions:
   - This formula is warm and dry. It may impair Yin Qi. It should not be taken long term.

   Acupuncture treatments:

   Acupuncture point formula:
   - LIV3 太沖 Taichong, P6 內關 Neiguan, P6 內關 Neiguan and REN17 膻中 Danzhong, LIV2 行間 Xingjian, HT3 少海 Shaohai, HT7 神門 Shenmen, and 安眠 Anmian.

3) Sciatic pain worse with fatigue and alleviated by bed rest. The low back pain is dull and sore, accompanied by lassitude and weak knees. Patients also might have the symptoms of dizziness, poor memory, and poor concentration.
   - If patients are kidney Yang deficient, the low back pain is also accompanied with cold sensation in the lumbar region and abdomen, cold

sensation on knees, impotency, nocturia, cold hands and feet, and aversion to cold temperature.
  - Tongue: Pale tongue body with wet white coating, teeth marks
  - Pulse: Deep, thready, and slow
- If patients are kidney Yin deficient, the low back pain is also accompanied with five-center heat (heat manifesting on the Yin surfaces of the body, which are the palms, the bottom of the feet, and the chest, especially in the afternoon and evening). Dizziness, malar flush, ringing in the ears.
  - Tongue: Red tongue with little or no coating or geographic maps
  - Pulse: Floating and empty pulse

Treatment principle:
- Nourish kidney, stop pain.

Herbal formula:
- Pattern herbs for kidney Yang deficiency: 金匱腎氣丸 Jin Gui Shen Qi Wan.
- Pattern herbs for kidney Yin deficiency: 左歸丸 Zuo Gui Wan.

Acupuncture treatments:
- Tonify kidney, strengthen the lumbar region and bones: UB23 腎俞 Shenshu, KD3 太溪 Taixi, KD6 照海 Zhaohai and KD10 陰谷 Yingu.
- Promote Qi and blood circulation and stop pain: UB40 委中 Weizhong—this is one of four commanding points.
- Strengthen the tendons and bones: GB34 陽陵泉 Yanglingquan and GB39 懸鐘 Xuanzhong.
- Harmonize the meridians and stop pain: UB25 大腸俞 Dachangshu, UB57 承山 Chengshan, UB58 飛陽 Feiyang and UB60 崑崙 Kunlun.
- Nourish liver, spleen, and kidney to promote Qi and blood circulation: SP6 三陰交 Sanyinjiao.
- Nourish blood: LIV8 曲泉 Ququan.
- Tonify kidney Yang: Moxa DU4 命門 Mingmen with a moxa box or warming needle techniques.

Moxibustion or Moxa:
- Moxa is used to open the blockage by warming up the cold and tonifying kidney Yang when the blockage is caused by the wind, damp, and cold invading the meridians in the lumbar region.
- It is extremely beneficial for sciatica.
- Damp-heat in the urinary bladder meridians or gall bladder meridians is contraindicated.

The following acupuncture points are commonly used for adding moxibustion:
- UB23 腎俞 Shenshu—this is to warm kidney Yang.

- 臀中 Tun Zhong, GB30 環跳 Huangtiao, GB34 陽陵泉 Yanglingquan, and GB39 懸鐘 Xuanzhong are excellent local points for moxibustion.
- A cold-chasing method is good to consider as well. Gently heat up the urinary bladder meridians from UB36 承扶 Chengfu all the way down to UB40 委中 Weizhong until a red line appears.
- For pain and a cold sensation on the posterior superior iliac spine (PSIS) or sacral region, add a moxa box on the affected regions to eliminate cold and relieve pain.

Gua Sha treatment:

- Apply the Gua Sha tool on the urinary bladder meridian on the lower back, the gall bladder meridian on the leg, and the urinary bladder meridian on the leg until the skin becomes red or shows spots, or even to the point of bruised skin for Qi and blood stagnation type sciatica.

Seven-star needle treatment:

- UB23 腎俞 Shenshu: cupping after bleeding.
- Point selected is UB23 腎俞 Shenshu.

Procedures:

- Prior to the procedure, the tools should be prepared and ready to be used in the working area.
- Set up a clean field with paper towels.
- Place these tools in the clean field, including dry sterile cotton balls, medical alcohol wipes or cotton swabs soaked in alcohol, sterile gloves, a biohazard sharps container, a biohazard trash container, goggles and facemask, and paper towels. Also, a medical lancet or a traditional three-edge needle or the seven-star needle is a bleeding tool; the seven-star needle is selected for bleeding UB23 腎俞 Shenshu in this case.
- The patient lies prone with the legs extended, UB2 3 腎俞 Shenshu facing upward.
- The room should be well lit.
- The therapist washes their hands thoroughly and wipes their hands with a single-use napkin. Put on sterile gloves.
- Wipe the acupuncture point UB23 腎俞 Shenshu with a medical alcohol wipe or a cotton swab soaked in alcohol. Move from the center of the acupuncture point to the periphery.
- Place the seven-star needle at a perpendicular angle to the skin and hold the end of the handle.
- Flick the flexible handle to tap the needles onto the acupuncture point lightly and superficially to allow the skin to develop a mild rash. The flexible handle is flicked at a fast speed, so it actually takes only a short time to get the rash result.
- Lastly, put on a cup to release more bleeding.

- Dispose of the tools in the proper containers.
- Wash the hands thoroughly.

Cupping treatment:

- UB23 腎俞 Shenshu with a needle in the cup.
- UB40 委中 Weizhong with bleeding technique from seven-star needle treatment.

Tui Na Treatments

Tui Na techniques:

- Kneading, pressing, pulling, shaking, grasping, rolling, stretching

Tui Na procedures:

- The patient is in the prone position. The Tui Na therapist stands beside the patient.
- Apply kneading and pressing techniques with the heel of a hand on the lower back to promote local blood circulation and relax the spasming muscles.

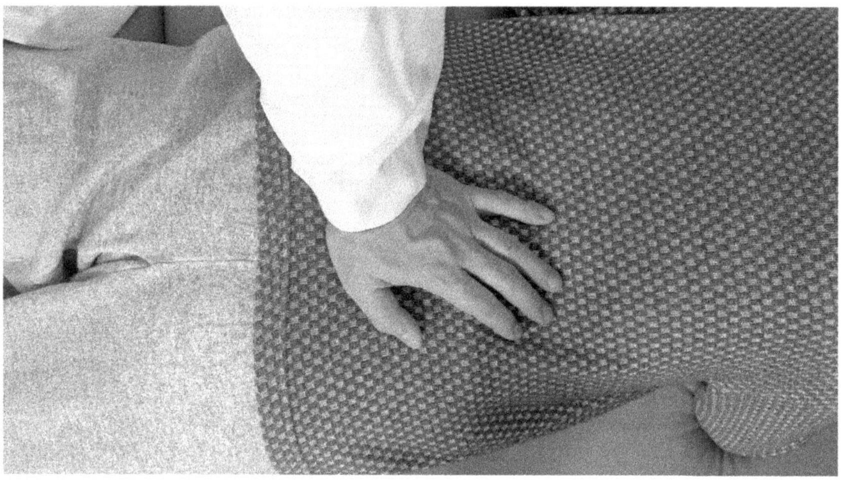

- Apply pulling and shaking on the ankle for the lower back vertebrae. The therapist stands near the feet and with both hands, grasps the ankles and lifts up the legs and then shakes the legs to send the vibration waves to the lumbar region.

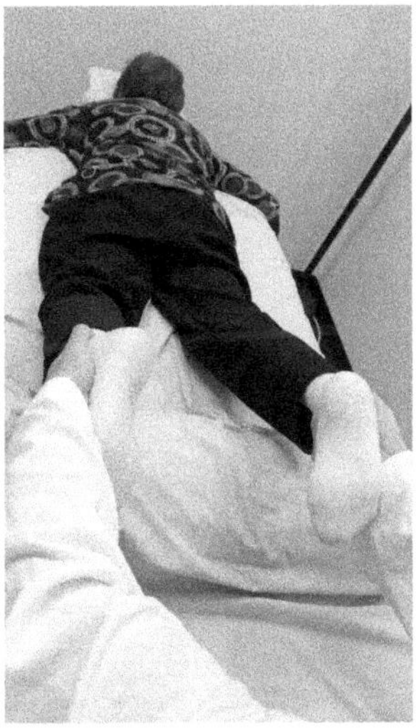
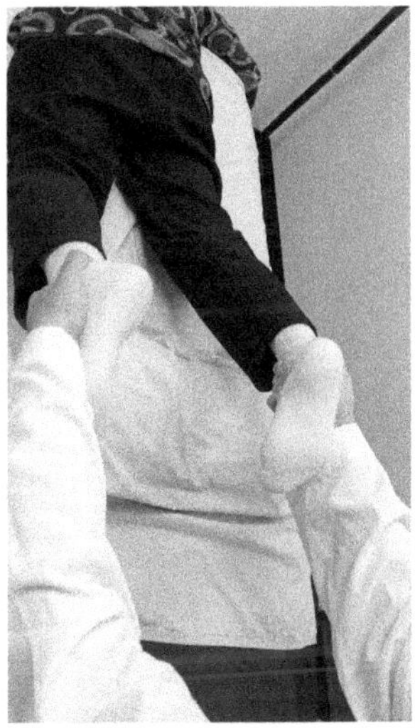

- Apply the grasping technique on the patient's lower back with all fingers for 10 to 20 minutes. The therapist stands next to the lower back, places both thumbs together on the acupuncture points UB23 腎俞Shenshu, and places all the remaining fingers on UB23 腎俞Shenshu located on the other side of the lumbar region, bending the distal phalangeal joints to press the finger tips to UB23 腎俞Shenshu.

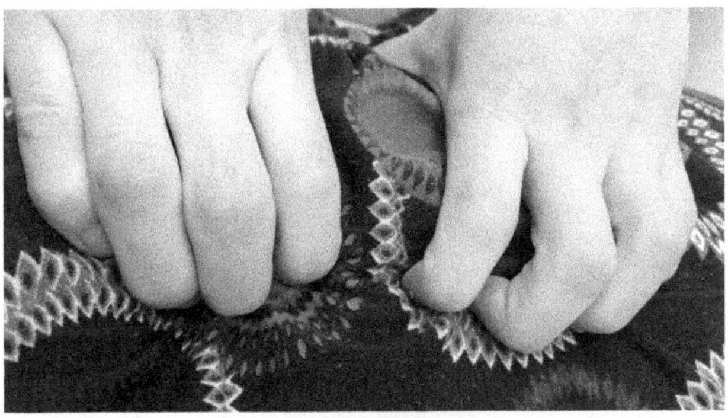

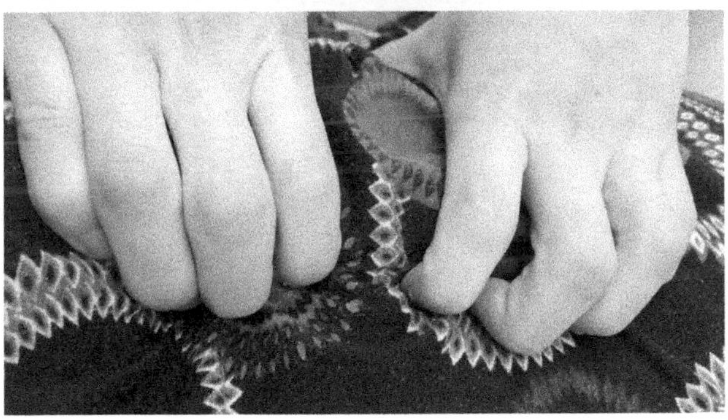

- Apply the rolling technique on the muscles beside the spinal column with both fists. The therapist clenches the fists, places the knuckles on both sides of the lumbar urinary bladder meridians, and then rolls the fists forward and backward.

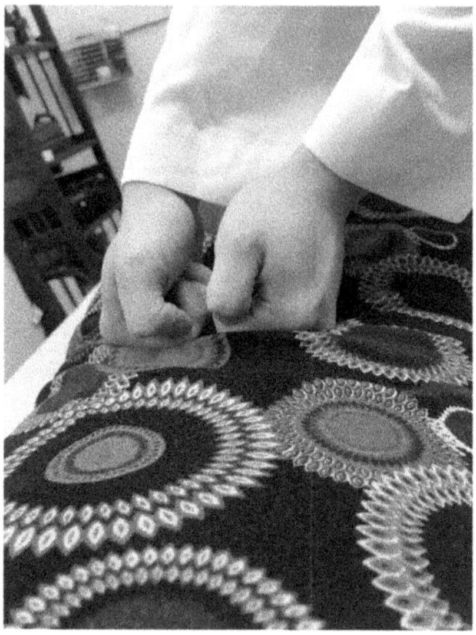
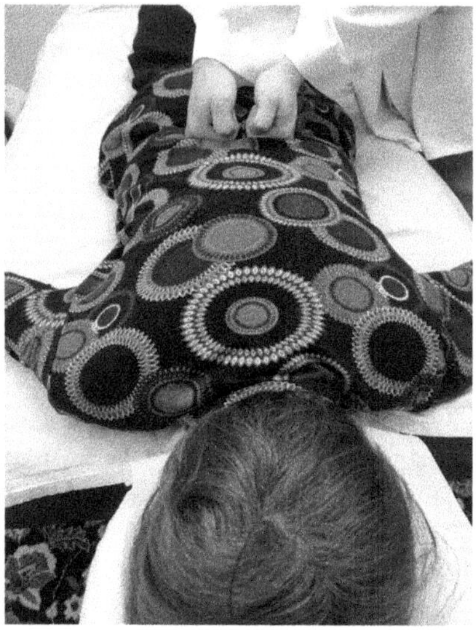
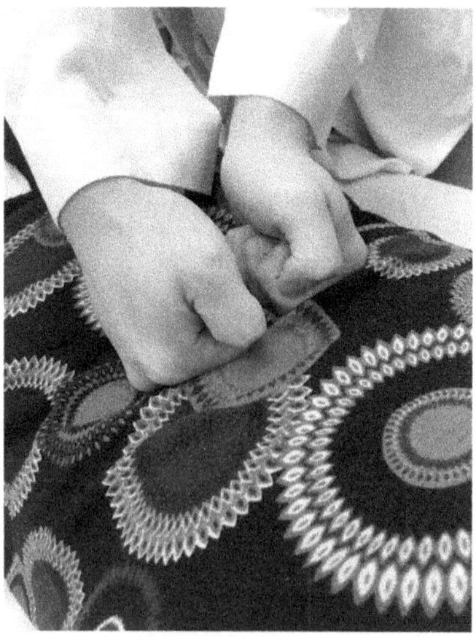
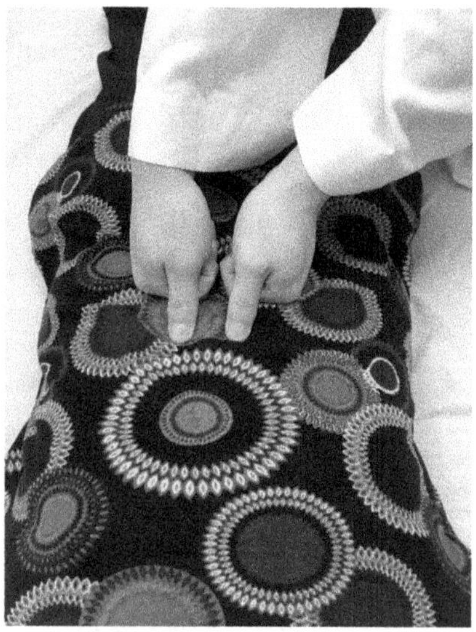

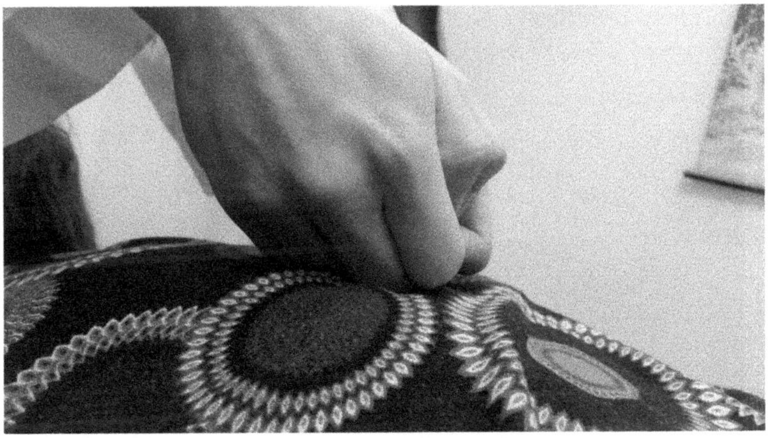

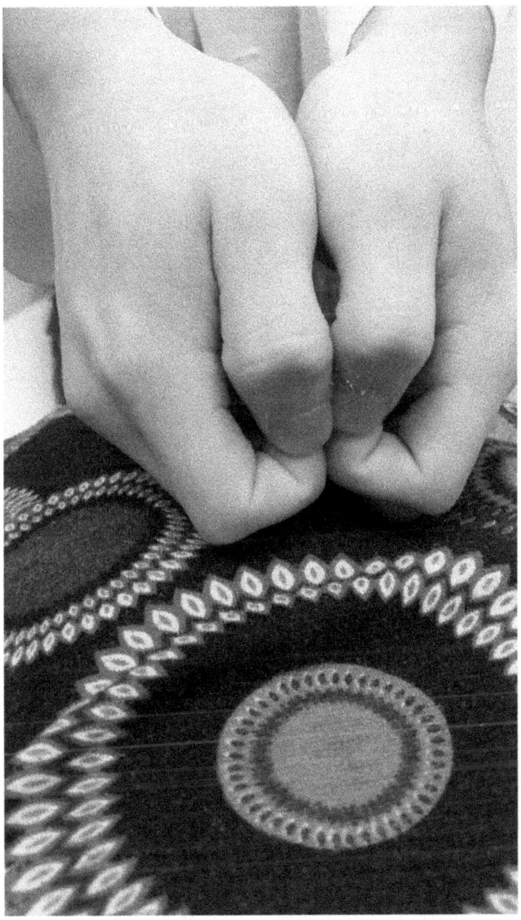

- Apply elbow pressing on the hips. The therapist leans back, places an elbow olecranon on GB30 環跳 Huan Tiao, and presses it.

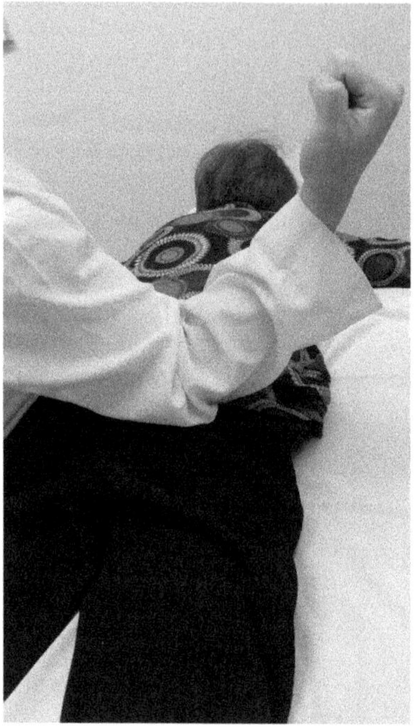

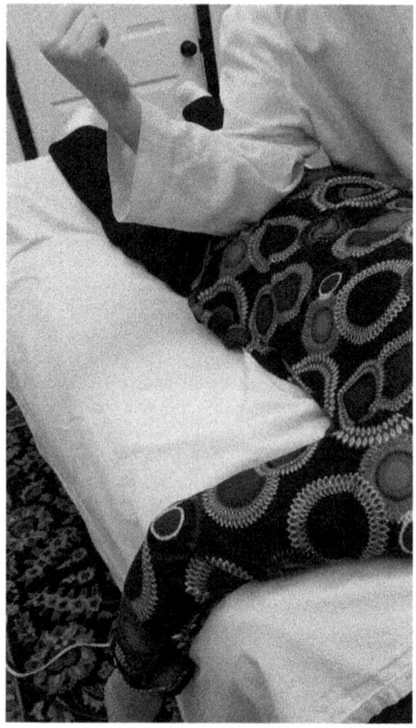

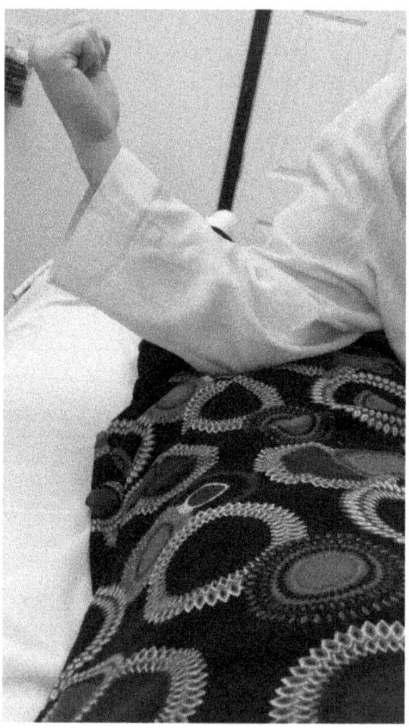

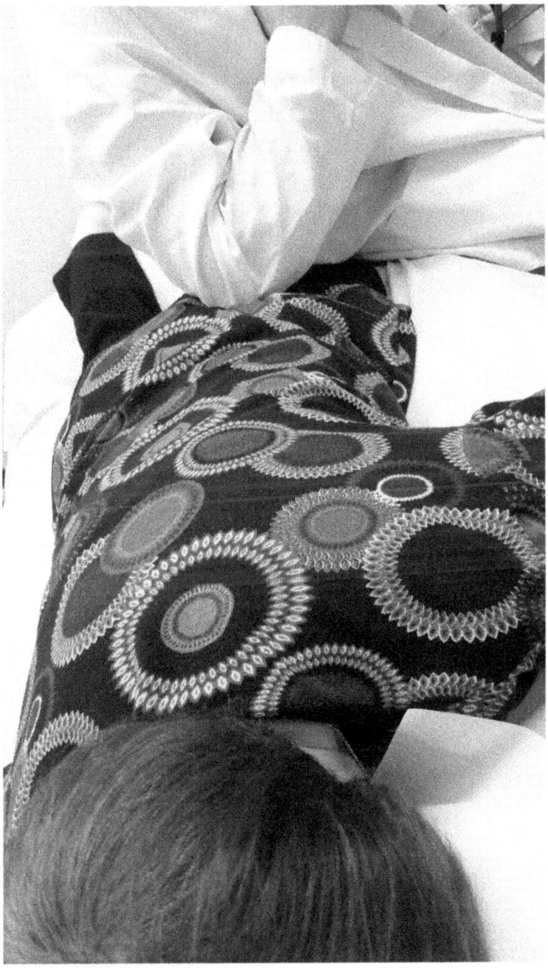

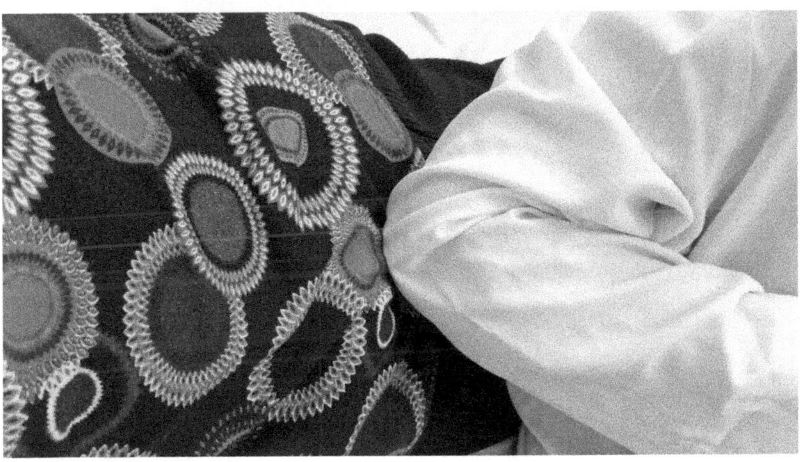

- Apply stretching to the leg. The therapist places a hand on the lumbar region and presses the lumbar region while at the same time lifting up the leg that is on the opposite side and then moving it to cross the midline.

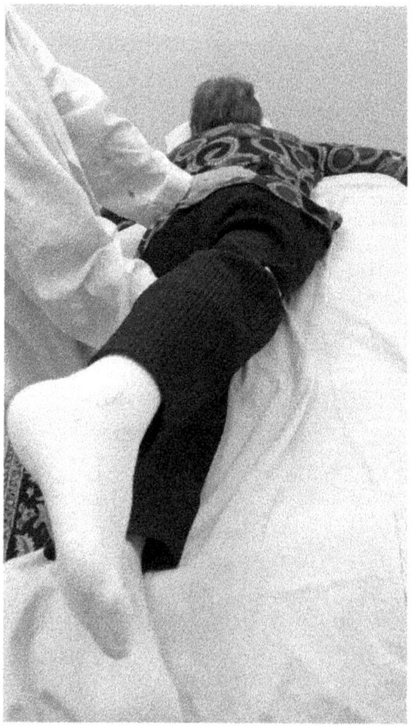
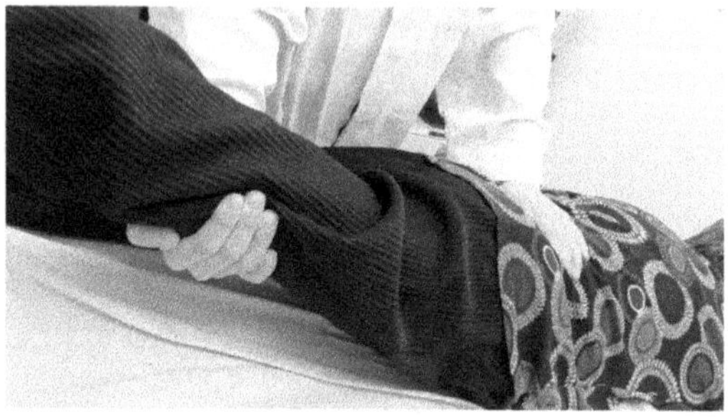
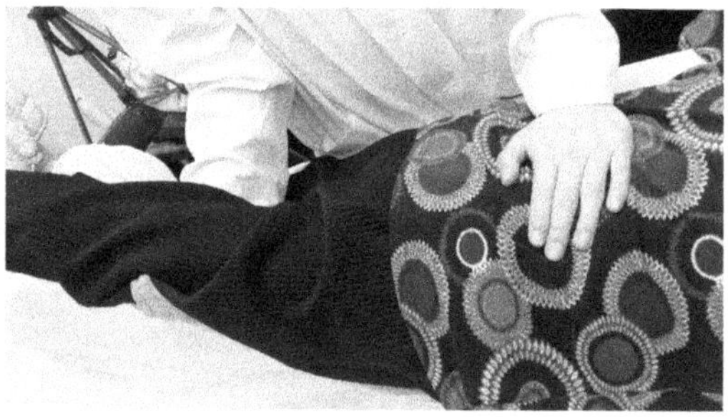
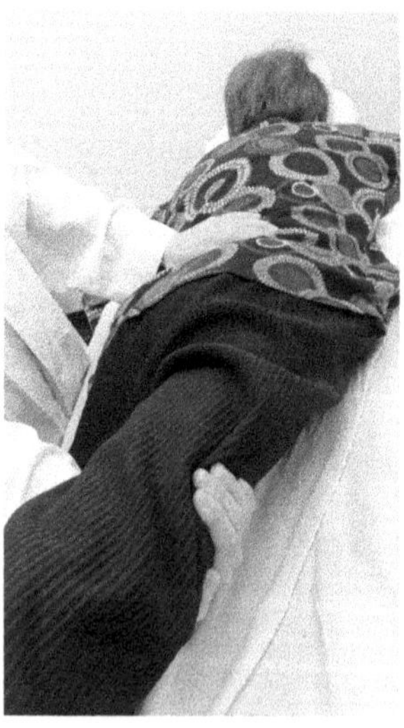
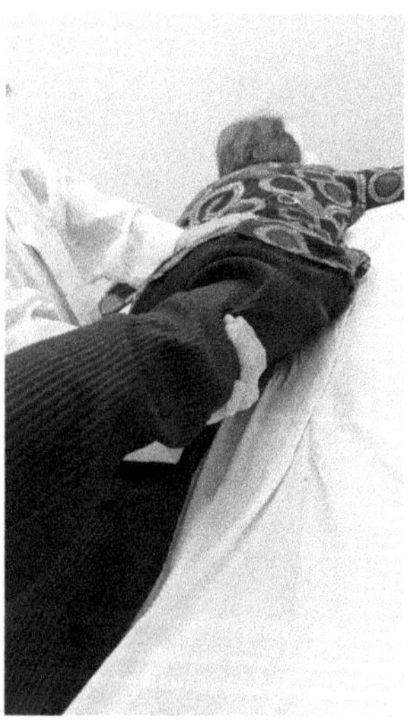
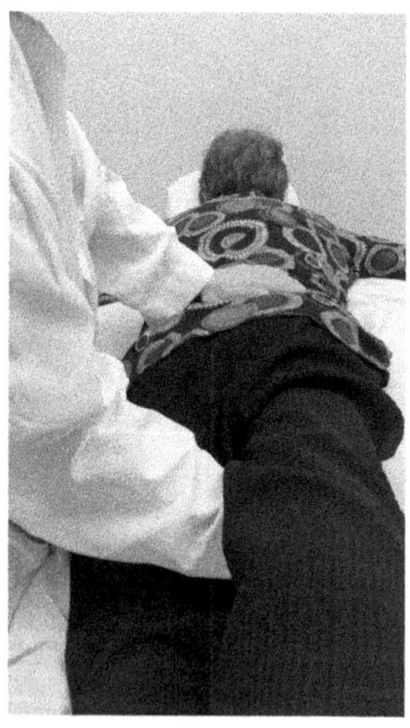

- Apply stretching to the lower back. The patient lies on the lateral side facing the therapist; the leg on the top is flexed, while the other one is extended. The therapist places an elbow on the hip and places a hand on the shoulder. Both hands press toward the opposite directions.

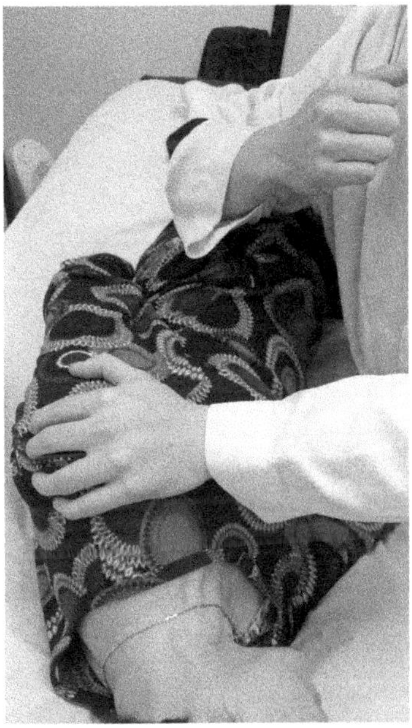
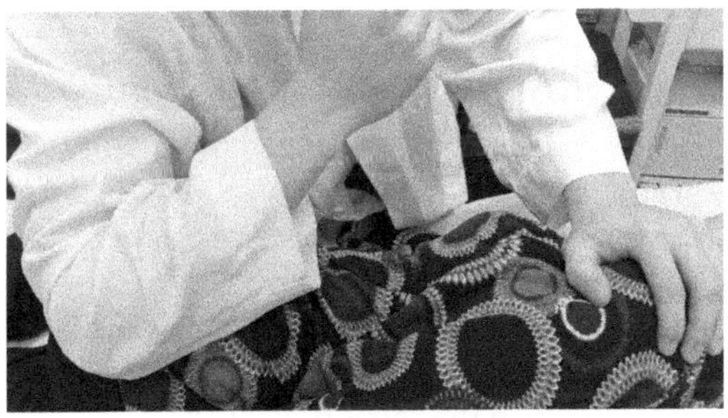
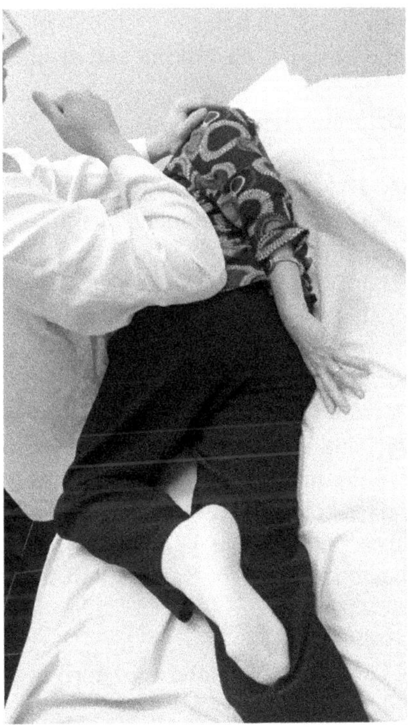
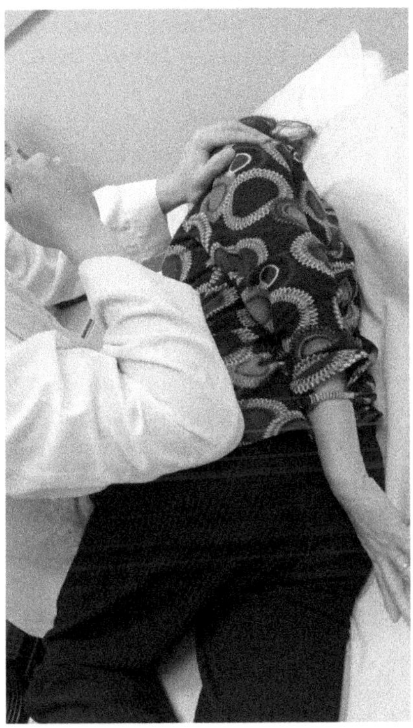

- Apply rolling, kneading, and rubbing on UB23 腎俞 Shenshu, GB30 環跳 Huan Tiao, and urinary bladder meridians on the legs.

Caution:

- Never apply Tui Na on the central type of prolapse of the lumbar intervertebral disc.
- Differentiate the central type of prolapse of the lumbar intervertebral disc from other ailments.

Patient Advisory

1) Sleeping on a firm bed is recommended. Avoid sleeping on a cold surface.
2) The diet should be light in taste with more green vegetables; drink water regularly, about 8–10 cups daily; increase fresh fruits. Increase wheat, sunflower seeds, pork, and peanuts.
3) Avoid exercise in the acute stage. Bed rest is highly recommended.
4) Appropriate routine stretching exercise is recommended after the acute stage; the following lower body stretching exercise is helpful for sciatic pain recovery:
   - Lie supine on a firm pad or carpet with both legs extended.
   - Flex the affected knee and hip; place both hands on the patella.
   - Continue to flex the leg and move the knee gently toward the opposite shoulder.
   - Once the limit before feeling pain is reached, hold for 30 seconds. Be mindful; gentle stretching is better than overstretching. Repeat 5–10 times.
   - Practice the same procedure on the other leg 5–10 times.
   - Practice the same procedure on both legs 5–10 times.
   - The frequency is twice daily, in the morning on waking and at night before sleep.
5) Keep good posture so that Qi and blood flow can be free.
   - Good sitting posture:
     a) Be sure to straighten the back and avoid leaning forward. Place a cushion behind the lower back to support the normal lumbar curvature.
     b) Flex the hips and knees at about 90 degrees.
     c) Keep the legs shoulder width apart.
     d) Keep the feet flat on the floor. If the feet cannot reach the floor, place the feet on a footrest.
     e) Do not remain sitting for longer than an hour. It is recommended to place a cup of tea within reach, with a few chrysanthemum flowers and goji berries with hot water, and to finish each cup within an hour. Goji berries help with eye tiredness, chrysanthemum flowers help to clear heat in the eye, and tea helps to clear the heat inside the body. With 250 cc of water, this is enough to make someone get up and walk to the bathroom. Someone sitting for 8 hours will finish drinking eight cups of water.
   - Good standing posture:
     a) Be sure to stand with erect head and back. Keep the head upright on the midline between the shoulders.

b) Stand with both legs slightly apart and with both knees slightly flexed.
c) Keep the weight on the front third of the feet.
d) Be mindful not to lean your body weight to either side.

- Good sleeping posture:
  a) Sleeping on a relatively firm bed is good for sciatic pain recovery.
  b) Supine is the appropriate sleeping posture; both sides of the body receive equal pressure.
  c) Place a cushion or pillow below the knees to alleviate lumbar pressure.
  d) In the acute stage, lift the lower legs higher by placing more pillows below the tibia.
  e) The normal lumbar curvature also can be supported by placing a thin cushion or a rolled towel under the back.

LUMBAR FACET JOINT DISORDER 腰椎後關節紊亂症

INTRODUCTION

This is a low back injury. It may be an acute low back pain, unusually intense low back pain occurring immediately after the injury, or it may be a chronic repetitive microtrauma and eventually become osteoarthritis in the chronic state. It happens to all ages and all genders.

It is often misdiagnosed as acute psoas sprain, because the symptoms are similar to acute lumbar sprain; the exact diagnosis should be dysfunction of the posterior joints of the lumbar vertebrae or the lumbar facet joint disorder or the lumbar facet syndrome.

This condition often occurs in the lumbosacral intervertebral joints, accounting for 50% to 60% of acute soft tissue injuries of the lumbar region,[4] and the chronic lumbosacral facet joint syndrome is reported to be the most common source of mechanical low back pain in 15–45% of patients with chronic low back pain (Physiopedia, 2020).

ANATOMY

Each level of the human spine has a three-joint complex to provide for the joint functions; these are two facet joints on the back side and a large disc that sits between the bodies of the two vertebrae on the front side.

The facet joints are composed of the lower articular processes of the superior vertebrae and the superior articular processes of the lower vertebrae.

The facet joints are the synovial joints. The articular surface is contained in a cartilaginous sheath over an encapsulated fibrous capsule. Each facet joint includes a small joint cavity, which is capable of accommodating between 1 and 1.5 ml of synovial fluid to facilitate joint movement, a synovial membrane, and hyaline cartilage surfaces and is surrounded by a joint fibrous cyst, which is only 1 mm in thickness.

The lumbar facet joint surface is arranged in a semi-frontal and semi-sagittal position; its transverse section is approximately curved and is flexible for flexion, lateral flexion, and rotation. Because the range of lumbosacral movements is large, the small joints are also slack.

Western Medicine Etiology and Pathology

Lumbar facet syndrome is a dysfunction at the posterior facet joint of the spine. The term "dysfunction" means that the lumbar facet joints are not functioning normally. Because the facet joints together with the disc form the intervertebral joint, changes to the anatomic structure at the posterior facet joints can influence the intervertebral disc, resulting in low back pain. Facet joint degenerative osteoarthritis is the most frequent form of facet joint pain; loss of joint space, narrowing, loss of synovial fluid, and loss of cartilage and bony overgrowth are the common changes due to osteoarthritis.

- The most common cause is repetitive microtrauma. In daily living, work, and sports that involve repetitive excessive hyperextension or arching of the lumbar spine, or excessive hyperextension with rotation of the back, swelling of tissues from the surrounding structures of facet joints arising from local inflammation caused by the repetitive microtrauma irritates the nerve roots, resulting in pain.
- Overloading to the facet joints may damage the intervertebral disc and change the biomechanics of the joint.
- Physical trauma to the lumbar region may tear the capsule, leading to a subluxation of the facet joint.
- A hypertonic contraction of the surrounding muscles may occur due to a self-protection mechanism, resulting in fibrosis and osteophyte formation or spur.

Chinese Medicine Etiology and Pathology

Qi and Blood Stagnation

A direct blow or cut to the medial, lateral aspect of the knee or to the quadriceps tendon and the patellar tendon causes blockages to the flow of the local Qi and blood in meridians almost immediately after the violence.

Liver and Kidney Deficiency

Ligament injury is considered to be "Jin Shang 筋傷" or "traumatic injury to tendons" in its meaning. The definition of tendon is not just limited to tendons; traditionally, "tendon" in Chinese medicine includes muscle and all of its sounding soft tissues. Therefore, injury to the tendons is trauma to muscle, ligament, cartilage, meniscus, the intervertebral disc and its outer fibrous ring, blood vessel, nerve, tendon and its sheath, bursa, and joint capsule. *Treatise on the Engenderment of the Five Viscera*《素問·五藏生成篇 第十》states: "All tendons belong to the joints 諸筋者, 皆屬於節" (Unschuld and Tessenow, 2011).

In the theory of traditional Chinese medicine (TCM), the liver governs the tendons. Treatise on the *Engenderment of the Five Viscera*, Chapter 10 《素問·五藏生成篇 第十》, states:

> The Liver stores the Blood and the Heart moves it. When a person is active, Blood moves through all the meridians to Tendons; when the body is in a state of rest, the Blood returns and stores in the Liver. Why is this so? Because the Liver governs the sea of Blood. 肝藏血，心行之，人動則血運于諸經，人靜則血歸于肝藏。何者？肝主血海故也。…When a person is at the state of rest, the Blood returns to the liver. The liver receives the Blood and the person is able to see. The feet receive the Blood and are able to walk. The palms receive the Blood and are able to grasp. The fingers receive the Blood and are able to hold 人臥血歸於肝，肝受血而能視，足受血而能步，掌受血而能握，指受血而能攝。(Unschuld and Tessenow 2011).

When injury occurs to tendons, it consumes liver blood. As the trauma becomes chronic, the blood is insufficient. Liver's element is wood, which needs water nourishment from kidney. As deficiency of liver continually demands sustenance from kidney, kidney becomes deficient.

Manifestations

- Usually, patients do not have an obvious history of injury; the pain occurs suddenly.
- Pain on palpation or pressure pain is at the affected facet joint, but some patients may have a hard time pointing out the exact location, because the area of pain is wide.
- Local pressure pain is found at the multifidus muscle and erector spinae muscles.
- The episodes are intermittent; they may occur a few times per month or year.
- The pain occurs when rising up with a flexed torso, such as standing up from a sitting position.
- The pain is greater in lateral flexion, hyperextension, and y-axis rotation in extension.
- The sensation of pain is very stiff due to hypertonia and sensibility.
- There is referred pain from the upper lumbar facet joints extending into the flank, hip, and upper lateral thigh but not further than the knee; referred pain from the lower lumbar facet joints can penetrate deep into the thigh, laterally and/or posteriorly; L4–L5 and L5–S1 facet joints can refer pain extending into the distal lateral leg, and in rare instances, to the foot; and both the lumbar facet joints are able to produce pain that can radiate to the groin.
- Pain intensity is different between patients, and the intensity is also different at different lumbar positions within the same patient.
- Sciatic nerve dermatomal distribution on the leg is negative unless the patient has nerve root impingement from a bone spur or a synovial cyst, or unless L4–L5 and L5–S1 facet joint hypertrophy encroaches on the intervertebral foramen.
- Patients prefer a prone sleeping posture, and they prefer having a cushion under the abdomen and refuse to be touched because of the severe pain.[5]

- Other patients prefer side lying with both legs flexed toward the chest and are afraid of being touched or moved.
- Spinal movements, coughing, and sneezing aggravate the pain.
- When standing, the patients need to semi-squat the hips and knees; they prefer placing the hands on the knees to lift the lumbar region.

Physical Examination

- Begin by obtaining a medical history.
  - General health conditions such as chronic low back pain; patients are usually lacking regular lumbar exercise and flexibility.
  - History of injury to the facet joint may not be obvious even if the pain occurs suddenly.
  - Careers that may be vulnerable to injury, such as jobs that require repetitive flexing, extension, and lateral flexing.
  - Sport that requires repetitive excessive hyperextension with rotation or arching of the back, such as golf, football, gymnastics, tennis, and other contact sports.[6]
- Inspection:
  - Inspect the spine for increased or decreased lordosis, the tightness of the paraspinal muscles; posture asymmetry of both sides of the lumbar region and signs of muscle atrophy may be commonly seen in chronic facet joint disorder of the lumbar vertebrae; flattening of the lumbar lordosis may be seen in the chronic state as well.
- Palpation:
  - Pain on palpation of the facet joints is located about 2–3 centimeters lateral to the spine; the facet joints are not really palpable, so palpation should be on the spinous process and along the paravertebral regions, and then directly over the transverse processes. The purpose of this palpation is to attempt to locate and narrow down the exact spot of pain for the nerve block or Ashi acupoint needling. The palpation will reproduce tenderness, which is usually present in the facet joint, and in some cases, the pain may radiate to the gluteal or posterior thigh region.
- Range of motion:
  - The pain is aggravated by prolonged standing, sitting, and walking; lumbar extension, lateral flexion towards the affected side, and axial rotation; and coughing or sneezing.
  - The pain is improved by short periods of sitting or walking; lumbar flexion,[7] lumbar lateral flexing toward the unaffected side, supine sleeping with knees flexed; medication (Physiopedia 2020) such as nonsteroidal anti-inflammatory drugs, acetaminophen, muscle relaxants, antidepressants, and oral steroids during acute flares (Alexander 2020).
- Extra tests:
  - Springing test is positive.
  - Kemp's test is positive.

- Straight leg raise test is negative.
- Skin temperature is unchanged.
- Since the clinical symptoms of lumbar facet joint disorder are so similar to those of other low back pain, the manifestations cannot be reliably clinically diagnosed (Jackson, 1992). The most frequently used systems to diagnose this syndrome are an X-ray, a computed tomography (CT) scan of the spine, or a magnetic resonance imaging (MRI) scan (Physiopedia, 2020).
  - According to Christopher Alexander and Mathew Varacallo, "In patients with chronic low back pain, degenerative changes on CT scan are cited to range from 40% to 85%. Studies have shown MRI to be more than 90% sensitive and specific in visualizing facet degeneration" (Alexander 2020).

Differential Diagnosis

Myofascial pain:

- The same manifestations between lumbar facet joint disorder and myofascial pain:
  - Localized areas of muscle pain.
  - Gets worse with muscle stretching.
  - Muscle stiff.
  - Reduced range of motion.
- The different manifestations between lumbar facet joint disorder and myofascial pain:
  - Myofascial pain has trigger points are that are nodules on the lumbar muscle, while facet joint pain has tender spots only near the spine.

Prolapse of the lumbar intervertebral disc:

- The same:
  - Pressure pain near the spinous process.
- The difference:
  - Leg pain on the dermatomal distribution.
  - Straight leg raise test is positive.

Acute sprain of waist:

- The same:
  - The pain may occur suddenly.
  - Severe pain.
- The difference:
  - The painful region is ipsilateral or bilateral.
  - Pressure pain is wide.

TCM Diagnosis

a) Qi and blood stagnation
b) Bi syndrome

Treatments

Chinese Herbal Methods

Acute State

Treatment principle:

- Moves blood, disperses bruise, moves Qi, and stops pain.

Herbal formula:

- Shun Qi Huo Xue Tang -順氣活血湯

Formula ingredients:

Su Geng 蘇梗 3 g, Hou Po 厚朴 3 g, Zhi Ke 枳殼 3 g, Sha Ren 砂仁 1.5 g, Dang Gui Wei 歸尾 6 g, Hong Hua 紅花 1.5 g, Mu Xiang 木香 1.2 g, Chi Shao 炒赤芍 3 g, Su Mu 蘇木 6 g, Tao Ren 桃仁 9 g, Xiang Fu 香附 3 g.

Formula explanations:

- Su Geng 蘇梗, Hou Po 厚朴, Zhi Ke 枳殼, Sha Ren 砂仁, Mu Xiang 木香, Xiang Fu 香附 move Qi, remove stagnation.
- Tao Ren 桃仁, 紅花, Dang Gui Wei 歸尾, Chi Shao 赤芍, and Hong Hua 紅花 move blood and disperse bruising.

Modifications:

- For soothing liver Qi, add Chai Hu 柴胡.
- For clearing heat, add Huang Qin 黄芩.
- For helping rebellion,[8] add Ban Xia 半夏.
- For dispersing lung, add Xing Ren 杏仁.
- For sedating the excess, add Da Huang 大黄.

Formula actions:

- Moves Qi and blood.
- Eliminates stagnation.
- Breaks up stubborn blood stasis.

Intermediate State

Treatment principle:

- Moves blood, disperses bruising, moves Qi, stops pain, tonifies kidney, strengthens bone.

Herbal formula:
He Ying Zhi Tong Tang - 和營止痛湯.

- Xu Duan 續斷 12 g, Dang Gui 當歸 9 g, Chi Shao 赤芍 9 g, Chuan Xiong 川芎 9 g, Su Mu 蘇木 9 g, Chen Pi 陳皮 9 g, Tao Ren 桃仁 9 g, Wu Yao 烏藥 9 g, Mo Yao 沒藥 6 g, Ru Xiang 乳香 6 g, Gan Cao 甘草 6 g.

Ingredient explanation:

- Dang Gui 當歸 nourishes blood, moves blood.
- Chi Shao 赤芍, Chuan Xiong 川芎 9 g, Su Mu 蘇木, Tao Ren 桃仁, Mo Yao 沒藥, and Ru Xiang 乳香 move blood, disperse bruising, and stop pain.
- Wu Yao 烏藥 and Chen Pi 陳皮 move Qi and stop pain.
- Xu Duan 續斷 tonifies kidney and strengthens bone.
- Gan Cao 甘草 harmonizes the ingredients.

Modifications:

- To make the formula stronger, add Shu Di Huang 熟地, Huang Qi 黃芪, Bu Gu Zhi 補骨脂, and Gu Sui Bu 骨碎補.

Formula actions:

- Moves the blood.
- Breaks up stagnation.

Acupuncture Treatment Methods

Acupuncture Point Formula

Distal points:
Actions: Opens Du meridian, promotes Qi and blood, stops pain.

Acupuncture points:

- SI6 養老 Yanglao
- SI3 後溪 Houxi
- DU26 人中 Renzhong

Local points:
Acupuncture points:

- UB23 腎俞 Shenshu, UB25 大腸俞 Dachangshu, UB40 委中 Weizhong, Ashi points, and Hua Tuo Jia Ji points 華佗夾脊.

Select Hua Tuo Jia Ji Points 華佗夾脊 on the affected lumbar facet joints. The method is located 0.5 cun lateral to the spinous processes above and below the affected one.

Insert an acupuncture needle size 28, length 1.5–2 cun to each selected Hua Tuo Jia Ji Point with needle tips at the angle 15–30 degrees pointing toward the spine and needle depth at 1–2 cun, and then manipulate the needles to De Qi as the patient feels a sensation like electric currents at the acupuncture points or meridians. Connect the needles with an acupuncture electro machine; set the rhythm as continuous speed and the level of current to the degree that the patient is able to accept. The retention period is 20–30 minutes.

Treatment: 1 treatment per day, 10 times as 1 course. Rest 1 day and continue with the second course of treatment.

Cupping Treatment

- UB23 腎俞 Shenshu, UB25 大腸俞 Dachangshu.

Moxibustion Treatment

- Warming needling techniques on UB23 腎俞 Shenshu, UB25 大腸俞 Dachangshu.

Bleeding Treatments

- UB40 委中 Weizhong.
- Point selected is UB40 委中 Weizhong.

Procedures:

- Prior to the procedure, the tools should be prepared and ready to be used in the working area.
- Set up a clean field with paper towels.
- Place these tools in the clean field, including dry sterile cotton balls, medical alcohol wipes or cotton swabs soaked in alcohol, sterile gloves, a biohazard sharps container, a biohazard trash container, goggles and face-mask, and paper towels. Also, a medical lancet or a traditional three-edge needle or the seven-star needle is a bleeding tool; a medical lancet or a traditional three-edge needle is selected for bleeding UB40 委中 Weizhong in this case.
- The patient lies prone with the legs extended, UB40 委中 Weizhong facing upward.
- The room should be well lit.
- The therapist washes their hands thoroughly and wipes their hands with a single-use napkin. Put on sterile gloves.
- Wipe the acupuncture point UB40 委中 Weizhong with a medical alcohol wipe or a cotton swab soaked in alcohol. Move from the center of the acupuncture point to the periphery.

- Place the medical lancet or the traditional three-edge needle at an angle of 90 degrees to the skin.
- In one quick move, prick the needle into the acupuncture point about 0.1 cun deep, eliciting a few drops of blood.
- If more drops of blood are needed, absorb the blood with a sterile cotton ball.
- Lastly, the point is pressed with a sterile cotton ball until the bleeding ceases.
- Dispose of the tools in the proper containers.
- Wash the hands thoroughly.

Electro-Acupuncture Techniques

- UB23 腎俞 Shenshu ~UB25 大腸俞 Dachangshu 2/100 Hz for 20 minutes daily.[9]

Tui Na Treatments

Tui Na techniques:

- Pressing, kneading, stretching, rotating, back-to-back "carrying on the tiger"

Tui Na procedures:

- The patient is in a prone position; the therapist stands at the side of the lumbar region next to the patient.

Pressing techniques: These acupuncture points are selected for pressing with thumbs.
  - Points: Hua Tuo Jia Ji 華佗夾脊 on the affected region, UB25 大腸俞 Dachangshu, UB28 膀胱俞 Pangguangshu, UB40 委中 Weizhong, UB57 承山 Chengshan, and Ashi acupoints 阿是穴.
  - Gradually increase pressure while pressing to the level that is tolerable to the patient and with the sensation of soreness.
  - Duration is 30 seconds on each acupuncture point.
  - For a better result, low back exercise after acupressure treatment is recommended. The recommended exercise is to repeatedly direct the patient to change position between sitting down and standing up. This promotes the flow of Qi in the meridians.

- The patient is in supine position; the therapist stands at the patient's right side near the lumbar region. The procedure explained here assumes that the painful facet joint is on the right side.

Rotating techniques on the lumbar region:

- The therapist's right hand grasps the patient's right ankle; the therapist then places the left hand on the patient's right patella.

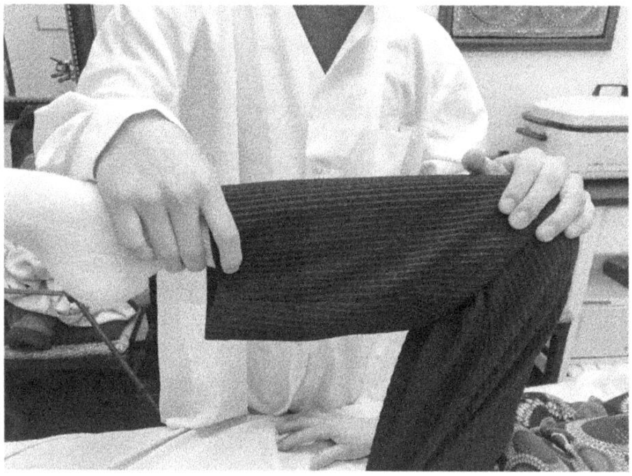

- Flex the patient's right knee toward the right shoulder by the right hand pushing the knee upward. The speed is slow.
- Extend the patient's right leg. The speed is also slow.
- Repeat the second and third steps 10–15 times. The speed gradually increases.

- The patient is in prone position; the therapist stands at the patient's left side next to the lumbar region.

Kneading techniques on the lumbar region:

- Apply the palms and the heels of the hands to both sides of the lumbar region with kneading techniques until warm.

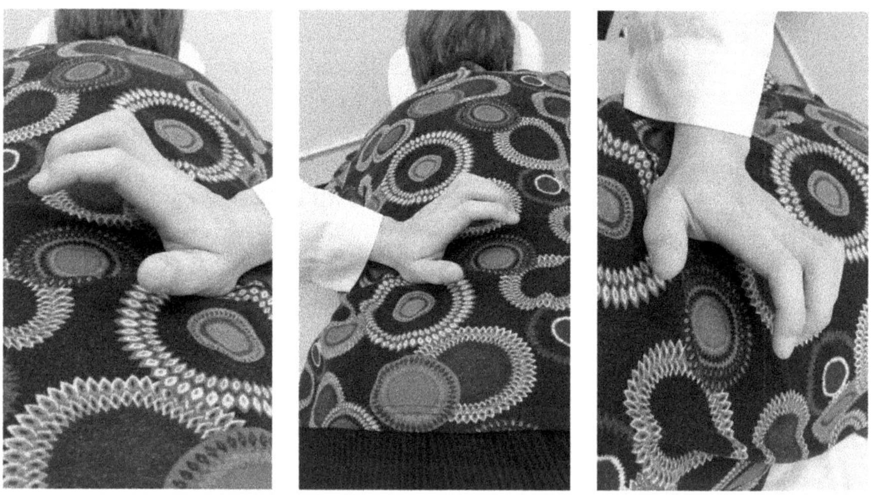

- Apply grasping techniques on the affected lumbar region to soften the spasming tissues.

Stretching techniques on the lumbar region:

- The patient is lying on the left side (assuming the affected facet joint is on the right side) facing the therapist. The left leg is fully extended and the right leg flexed.
- The therapist places the right hand or the right elbow on the patient's right shoulder and presses the crista iliaca with the left hand or the left elbow. Both hands or elbows press both the shoulder and the crista iliaca to the end of their range of motion and then give them a quick but short press to the opposite direction.

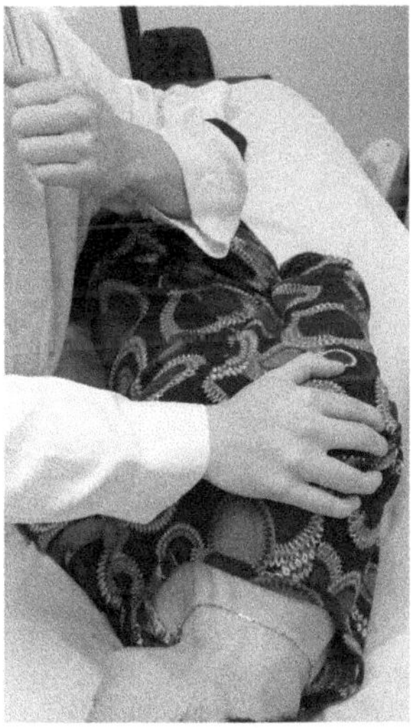

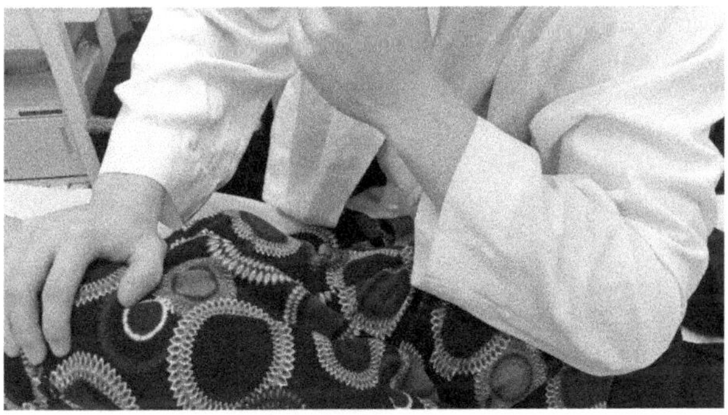

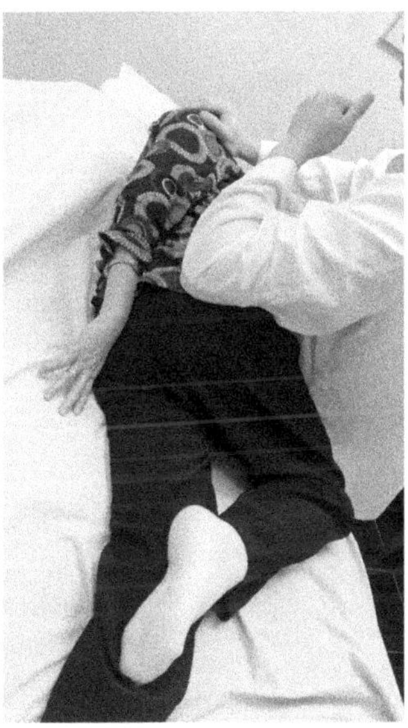

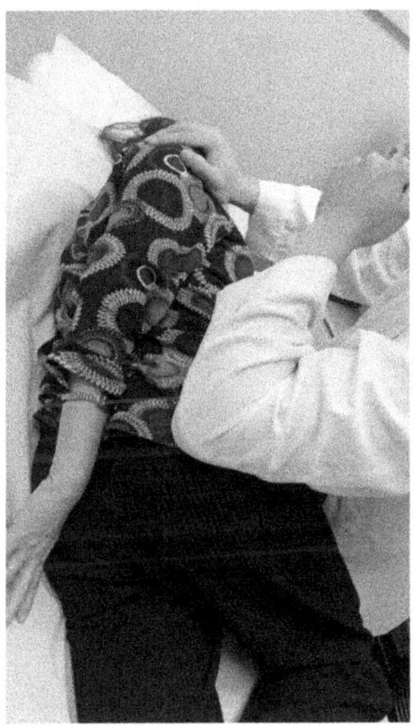

Back-to-back "carrying on the tiger":

- Standing back-to-back with the patient, the therapist places the hands on the anterior aspects of the patient's shoulders by going through the armpits.
- Carry the patient on the back, and then shake the patient upward and downward and leftward and rightward two to five times, ending the treatment.

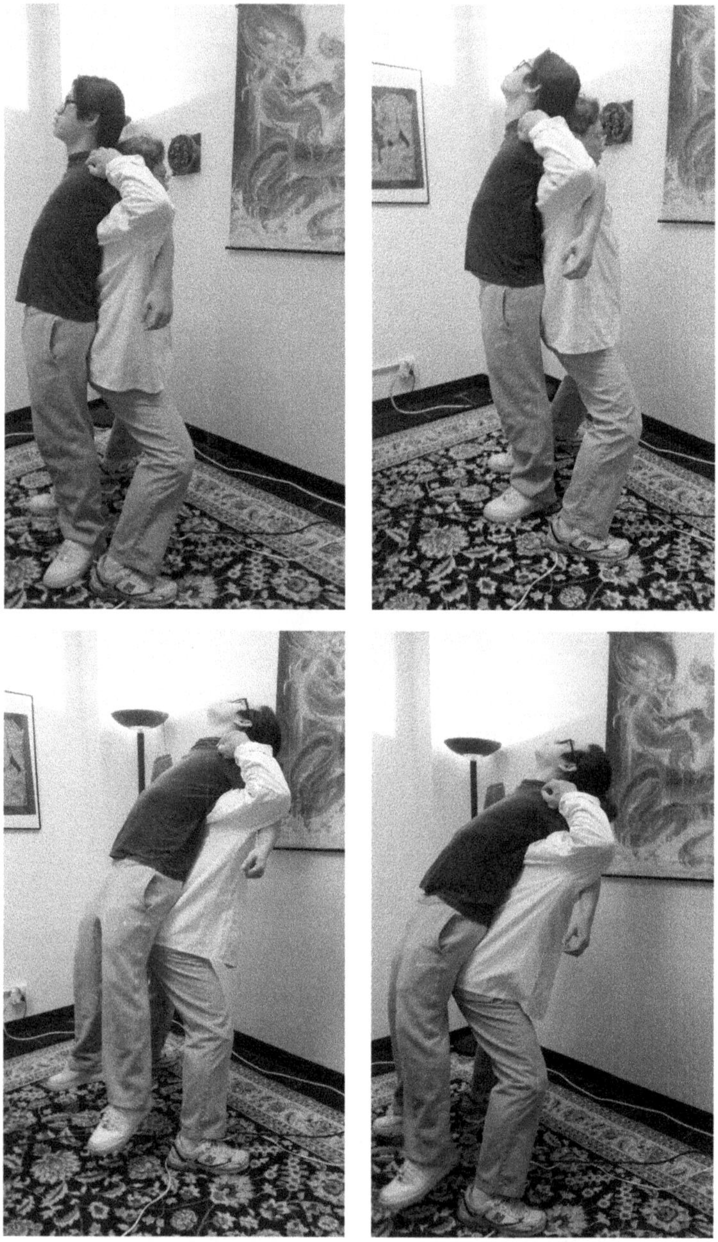

- The patient is prone, and the therapist stands next to the patient's lumbar region.
- Apply rubbing techniques with palms and the heels of hands on the lumbar region along the Urinary Bladder meridians on both sides until the skin is warm.

Patient Advisory:

- Bedrest is recommended for 2–3 days after the initial treatment.
- Do not rotate the lumbar region for at least 1 week.
- Keep the lumbar region warm.
- The diet should be light in taste with more green vegetables; drink water regularly, about 8–10 cups daily; increase fresh fruits. Increase wheat, sunflower seeds, pork, and peanuts.
- Avoid exercise in the acute stage. The patient should rest in bed.
- Keep good posture so that Qi and blood flow can be free.

- Good sitting posture:
  a) Be sure to straighten the back and avoid leaning forward or to one side with the spine bent. Avoid crossing the legs or ankles. Place a cushion behind the lower back to support the normal lumbar curvature.
  b) Flex the hips and knees at about 90 degrees and try to keep the knees at the same height as the hips.
  c) Keep the legs shoulder width apart.
  d) Keep the feet flat on the floor and in front of the knees. If the feet cannot reach the floor, place the feet on a footrest.
  e) Do not remain sitting for longer than an hour. It is recommended to place a cup of tea within reach, with a few chrysanthemum flowers and goji berries with hot water, and to finish each cup within an hour. Goji berries help with eye tiredness, chrysanthemum flowers help to clear heat in the eye, and tea helps to clear the heat inside the body. With 250 cc of water, this is enough to make someone get up and walk to the bathroom. Someone sitting for 8 hours will finish drinking eight cups of water.
- Good standing posture:
  a) Be sure to stand with erect head and back. Keep the head upright on the midline between the shoulders.
  b) Stand with both legs slightly apart and with both knees slightly flexed.
  c) Keep the weight on the front third of the feet.
  d) Be mindful not to lean your body weight to either side.
- Good sleeping posture:
  a) Sleeping on a relatively firm bed is good for sciatic pain recovery.
  b) Supine is the appropriate sleeping posture; both sides of the body receive equal pressure.
  c) Place a cushion or pillow below the knees to alleviate lumbar pressure.

d) In the acute stage, lift the lower legs higher by placing more pillows below the tibia.
e) The normal lumbar curvature also can be supported by placing a thin cushion or a rolled towel under the back.

- Exercise to strengthen the lumbar muscles:
  a) Lie prone on the floor with legs straight and slightly apart.
  b) Place both hands on the floor in front of the shoulders.
  c) Lift up the upper trunk with the arms slowly as high as possible without feeling low back pain.
  d) Be sure that only the upper trunk is lifted up, and the hips and legs are still on the floor.
  e) Hold the position for 5 to 30 seconds depending on the lumbar discomfort. The exercise is practiced breathing naturally and free from pain.
  f) Be mindful that the lumbar muscles should remain fully relaxed.
  g) To return to the beginning, slowly flex the elbows to lower the upper trunk until the chest reaches the floor
  h) Repeat the procedure three to five times.
  i) The pace of the movements should be slow; the lumbar muscles work harder when moving slowly.

ACUTE LUMBAR SPRAIN 急性腰肌扭傷

Introduction

Acute lumbar sprain is commonly known as the Flash Waist. About 10% to 25% of low back pain patients experience pain from the sacroiliac joint (Raj 2019). Those sports that require leaning or bending the lumbar region, such as lifting heavy furniture from the side, can lead to a lumbar sprain. In addition, sports that require sudden twisting of the lower back, such as tennis, basketball, baseball, and golf, can lead to this injury. Certain risk factors, for example, excessive lower back curvature, forward-tilted pelvis, weak lumbar region or weak abdominal muscles, tight hamstrings, and falling down, can increase the risk for this injury. Flash Waist is a very common cause of lumbago.

Anatomy

Acute spraining of the waist may affect many anatomic structures in the lumbar region.

1) The sacroiliac joint, commonly called the SI joint, is the joint between the lateral articulating surfaces of the sacrum and the articulating surfaces of the ilium bones of the pelvis. There are two joints, one on each side of the sacrum. The joint is a synovial articulation between the articular surface of the sacrum and ilium on either side. In humans, when standing (Hayat 2020),[10] the sacrum

and the ilium support the spine, and the sacroiliac joint serves to transfer weight from the upper body to the lower limbs. This acts as a shock absorber and reduces the pressure on the spine.[11] The joint is stabilized by strong ligaments and muscles. Even though they are joints, the movement of the joints is minimal; it is limited to only 2 to 4 mm in any direction (Raj 2019). When either of the joints comes out of alignment, it can be painful.[12]

2) Sacrospinalis muscles[13] are also known as erector spinae muscles. These muscles run essentially vertically on either side of the vertebral column spinous processes and extend throughout the lumbar, thoracic, and cervical regions. They are large, roughly one hand's width from the spinous processes. They stabilize the entire vertebral column.
3) The lumbosacral joint is a joint between the last lumbar vertebra (L5) and the first of the auricular surfaces of the sacrum (S1).[14] This joint has an intervertebral disc.
4) The ligaments connecting the L5 and the S1 are the continuation downward of the anterior and posterior longitudinal ligaments, the ligamenta flava, and the intervertebral fibrocartilage.

Western Medicine Etiology and Pathology

The traumatic changes are usually in the waist, the lumbosacral region, the sacroiliac region, and both sides of the sacrospinous muscle.

An improper posture of leaning or bending during physical exercise or during physical labor, such as bending the body to one side while lifting heavy furniture, may cause tearing of muscles, ligaments, and fascia or dislocation of the lumbosacral joint, the sacroiliac joint, and minor joints of the intervertebral disc. The injury causes local hemorrhage and edema, leading to local traumatic inflammation that results in pain and spasm of the related muscles, ligaments, and fascia.

The posterior sacral ramus innervates the sacroiliac joint, and it is a source of pain when the joint is compressed or inflamed. Overall, obese patients are more commonly affected by sacroiliac joint disorder. The pain is seen in all genders and people of all races (Raj 2019).

Chinese Medicine Etiology and Pathology

Qi and Blood Stagnation

According to Chinese medicine, blood is stagnant when Qi is stagnant; the injury is impacted by external violence or chronic indulgence and long-term emotional imbalance, resulting in Qi stagnation. In general, Qi stagnation leads to blood stasis, because Qi is a "commander" of blood. Qi and blood stagnation cause local blockages, and pain is the result of the blockages.

Liver and Kidney Deficiency

In the theory of traditional Chinese medicine (TCM), the liver governs tendons.[15] Tendons are soft tissues in Chinese medicine.

Traditional Chinese medicine holds that chronic traumatic change to soft tissues causes liver blood and Yin deficiency; bone injury causes kidney Qi, Yin and Yang, or essence deficiency. When physical trauma occurs and becomes chronic, liver and kidney are affected, resulting in deficiency; this is because liver governs tendons and kidney governs bone. When soft tissues are injured or inflamed, liver provides blood and Yin to repair the trauma; kidney provides Qi, Yin and Yang, or even essence to assist in healing the bones. When the injury lesion takes too long to recover and it becomes chronic, or the patient has been suffering from liver and kidney deficiency, or due to old age, both liver and kidney are deficient and lacking in substance to provide for healing or repairing.

To achieve more rapid healing and a relatively permanent curative effect, and to strengthen the soft tissues and bones, tonifying both liver and kidney is necessary and efficacious.

Manifestations

- Patients may have an obvious history of injury; the pain is sudden in onset with a history of trauma.
- Pain is in the buttocks and the legs, often in the posterior aspect of the thigh.
- The patient is unable to move or even turn the body in bed. The pain is persistent. Sometimes, the patient can ignore the pain and continue working but finds that next morning, the pain is unbearable and movement is limited. The pain is aggravated by activity, cough, defecation, speaking loudly, sneezing, or even deep breathing.
- The pain is not eliminated by rest.
- It is more common in young adults and obese patients.

Physical Examination

- Begin by obtaining a medical history.
  - Patients usually have a noticeable history of lumbar injury.
  - Careers that often flex the lumbar region in different ways.
  - Sports that require sudden lumbar twisting.
- Inspection:
  - Spinal deviation may not be found in the acute condition; it may be found in the later stage.
  - Swelling of the lumbar muscles may be visible.
- Palpation:

  There are superficial tender points on the lumbar region.

  1) If the trauma is on the sacrospinal muscle or the lumbodorsal fascia, the tender points are located on both sides of the vertebrae.
  2) If the trauma is on the supraspinal ligament or the interspinous ligament, the tender points are located on the mid-spinal line.
  3) Spasm of muscles may be found on the sacrospinal muscle and the gluteus maximus muscle.

4) The affected muscles are extremely sensitive to the touch.

- Range of motion:
  - The patient is in a compulsive painful position[15] with spasm of muscles in the lumbar region.

Differential Diagnosis

- Acute spraining of waist needs to be differentiated from sciatica, because they both have very similar symptoms, including pain in the lumbar region; pain is aggravated by activity, coughing, defecation, speaking loudly, sneezing, or even deep breathing.
- Lumbar pain is constant on acute spraining of waist, and bed rest does not relieve the pain.
- Acute spraining of waist may also be associated with pain radiating into the buttocks but does not affect the legs, while sciatica will follow the dermatomal pattern, which radiates into the legs, passes the knee to the sides of the tibia, and may go to the foot and the toes as well. Acute spraining of waist may have a positive straight leg raise test, but it will be negative on the dorsiflexion of the foot.

Treatments

Treatment principle:

- Relax and prevent adhesion of the tendons, move blood, and stop pain.

Herbal Treatments

Early Stage

Early stage—move blood, clear stasis, drain edema, and stop pain.

Herbal formula:

- He Ying Zhi Tong Tang 和營止痛湯.

Ingredients:

Xu Duan 續斷 12 g, Dang Gui 當歸 9 g, Chi Shao 赤芍 9 g, Chuan Xiong 川芎 9 g, Su Mu 蘇木 9 g, Chen Pi 陳皮 9 g, Tao Ren 桃仁 9 g, Wu Yao 烏藥 9 g, Mo Yao 沒藥 6 g, Ru Xiang 乳香 6 g, Gan Cao 甘草 6 g.

Ingredient explanation:

- Dang Gui 當歸nourishes blood and moves blood.
- Chi Shao 赤芍, Chuan Xiong 川芎 9 g, Su Mu 蘇木, Tao Ren 桃仁, Mo Yao 沒藥, and Ru Xiang 乳香 move blood, disperse bruising, and stop pain.
- Wu Yao 烏藥 and Chen Pi 陳皮 move Qi and stop pain.

- Xu Duan 續斷 tonifies kidney and strengthens bone.
- Gan Cao 甘草 harmonizes the ingredients.

Modifications:

- To make the formula stronger, add Shu Di Huang 熟地, Huang Qi 黃芪, Bu Gu Zhi 補骨脂, and Gu Sui Bu 骨碎補.

Formula actions:

- Moves the blood.
- Breaks up stagnation.

External Wash Formula

Herbal formula:

- Xiao Yu Zhi Tong ointment 消瘀止痛藥膏.

Formula ingredients:

- Mu Gua 木瓜 60 g, Zhi Zi 梔子 30 g, Da Huang 大黃 150 g, Pu Gong Ying 蒲公英 60 g, Di Bie Chong 地鱉虫 30 g, Ru Xiang 乳香 30 g, Mo Yao 没藥 30 g.

Ingredient explanation:

- Mu Gua 木瓜 releases spasm, increases Qi and blood circulation.
- Zhi Zi 梔子 reduces swelling, opens meridians.
- Da Huang 大黃 moves blood and disperses bruising.
- Pu Gong Ying 蒲公英 clears heat in blood, reduces swelling.
- Di Bie Chong 地鱉蟲 moves blood, disperses bruising.
- Ru Xiang 乳香 moves blood, stops pain.
- Mo Yao 沒藥 moves blood, stops pain.

Direction:

- Apply the herb with maltose on the affected area.

Intermediate Stage

The pain is less; it is time to loosen the tendons, move blood, and strengthen tendons.

Herbal formula:

- Yao Shang Er Fang 腰傷二方.

Formula ingredients:

- Gou Teng 鉤藤 12 g, Xu Duan 續斷 12 g, Du Zhong 杜仲 12 g, Shu Di Huang 熟地黃 12 g, Dang Gui 當歸 12 g, Du Huo 獨活 10 g, Niu Xi 牛膝

10 g, Wei Ling Xian 威靈仙 10 g, Bai Shao 白芍 5 g, Zhi Gan Cao 炙甘草 6 g, Sang Ji Sheng 炙桑寄生 30 g.

Ingredient explanation:

- Gou Teng 鉤藤 expels wind, releases spasm.
- Xu Duan 續斷 nourishes liver and kidney, strengthens bones and tendons.
- Du Zhong 杜仲 tonifies the kidneys and liver, strengthens the sinews and bones, and lowers blood pressure.
- Shu Di Huang 熟地黃 for heat symptoms due to Yin deficiency and devastated blood.[17]
- Dang Gui 當歸 tonifies blood, stops pain.
- Du Huo 獨活 expels wind and damp, stops pain.
- Niu Xi 牛膝 moves blood, releases bruising.
- Wei Ling Xian 威靈仙 expels wind and damp, stops pain.
- Bai Shao 白芍 nourishes blood, astringes Yin, softens tendons, stops pain.
- Zhi Gan Cao 炙甘草 tonifies Qi, harmonizes other herbs
- Sang Ji Sheng 炙桑寄生 expels wind and damp, tonifies liver and kidney Yin.

Modifications:

- Add Qin Jiao 秦艽 to expel wind damp.
- Add E Jiao 阿膠 to tonify blood and nourish Yin; stops bleeding.
- Add Sang Zhi 桑枝 to expel wind and damp; opens meridians, releases spasm.
- Add Xu Duan 川續斷 to tonify liver and kidney; strengthens bones and tendons.

External Wash Formula

Herbal formula:

- Shu Jin Huo Luo ointment 舒筋活络膏.

Formula ingredients:

Chi Shao 赤芍 1 g, Hong Hua 紅花 1 g, Tian Nan Xing 天南星 3 g, Pu Huang 蒲黃 1.5 g, Zhi Zi 梔子 2 g, Xuan Fu Hua 旋覆花 1.5 g, Su Mu 蘇木 1.5 g, Zhi Cao Wu 制草烏 2 g, Zhi Chuan Wu 制川烏 2 g, Qiang Huo 羌活 2 g, Du Huo 獨活 2 g, Ban Xia 半夏 2 g, Da Huang 大黃 2 g, Mu Gua 木瓜 2 g, Lu Lu Tong 路路通 2 g.

Ingredient explanation:

- Chi Shao 赤芍 moves blood, reduces swelling, stops pain.

- Hong Hua 紅花 moves blood, releases bruising, opens meridians, stops pain.
- Tian Nan Xing 天南星 dries damp, expels phlegm, releases spasm.
- Pu Huang 蒲黃 stops bleeding, moves blood.
- Zhi Zi 梔子 clears heat, cools blood, resolves damp, reduces swelling.
- Xuan Fu Hua 旋覆花 expels phlegm.
- Su Mu 蘇木 moves blood, reduces swelling, stops bleeding.
- Zhi Cao Wu 制草烏 reduces swelling, stops pain.
- Zhi Chuan Wu 制川烏 expels wind and damp, warms meridians, stops pain.
- Qiang Huo 羌活 expels wind, cold, and damp, stops pain.
- Du Huo 獨活 expels wind and damp, stops pain.
- Ban Xia 半夏 transforms phlegm.
- Da Huang 大黃 moves blood and disperses bruising.
- Mu Gua 木瓜 releases spasm, increases Qi and blood circulation.
- Lu Lu Tong 路路通 moves blood, opens meridians, expels wind, promotes water metabolism.

Directions: Apply the herb with maltose on the affected area.

Late Stage

Qi and blood deficiency, weak constitution.

Herbal formula:

- Bu Shen Zhuang Jin Tang 補肾壯筋湯.

Formula ingredients:

Shu Di 熟地 12 g, Dang Gui 當歸 12 g, Niu Xi 牛膝 10 g, Shan Zhu Yu 山茱萸 12 g, Fu Ling 茯苓 12 g, Xu Duan 續斷 12 g, Du Zhong 杜仲 10 g, Bai Shao 白芍 10 g, Qing Pi 青皮 5 g, Wu Jia Pi 五加皮 10 g.

Ingredient explanation:

- Shu Di 熟地 nourishes blood; nourishes liver and kidney Yin.
- Dang Gui 當歸 tonifies blood.
- Niu Xi 牛膝 expels wind dampness, tonifies liver and kidneys, directs herbs downward.
- Shan Zhu Yu 山茱萸 nourishes liver and kidney, strengthens kidney Yang.
- Fu Ling 茯苓 drains damp, tonifies spleen.
- Xu Duan 續斷 nourishes liver and kidney, strengthens bones and tendons.
- Du Zhong 杜仲 nourishes liver and kidney.
- Bai Shao 白芍 nourishes blood, constricts Yin, softens tendons, stops pain.
- Qing Pi 青皮 disperses liver Qi, drains damp, transforms phlegm.
- Wu Jia Pi 五加皮 expels wind and damp, nourishes liver and kidney, strengthens bones and tendons.

External Wash Formula

Herbal formula:

- Shu Jin Huo Luo Bathing 舒筋活絡澡.

Formula ingredients:

Shen Jin Cao 伸筋草 9 g, Hai Tong Pi 海桐皮 9 g, Qin Jiao 秦艽 9 g, Du Huo 獨活 9 g, Dang Gui 當歸 9 g, Gou Teng 鉤藤 9 g, Ru Xiang 乳香 6 g, Mo Yao 沒藥 6 g, Hong Hua 紅花 6 g.

Ingredient explanation:

- Shen Jin Cao 伸筋草 expels wind, drains damp, relaxes tendons.
- Hai Tong Pi 海桐皮 expels wind, drains damp, reduces swelling.
- Qin Jiao 秦艽 expels wind and damp.
- Du Huo 獨活 expels wind and damp, stops pain.
- Dang Gui 當歸 nourishes blood, reduces edema, and stops pain.
- Gou Teng 鉤藤 expels wind, releases spasm.
- Ru Xiang 乳香 moves blood, stops pain.
- Mo Yao 沒藥 moves blood, stops pain.
- Hong Hua 紅花 moves blood, releases bruising, opens meridians, stops pain.

Modification:

- To move blood, add Tao Ren 桃仁 and Chuan Xiong 川芎.

Directions: Wash the affected area with this herbal soup.

Acupuncture Treatments

Acupuncture point formula:

Local points:

- UB20 脾俞 Pishu, UB23 腎俞 Shenshu, UB24 氣海俞 Qihaishu, UB25 大腸俞 Dachangshu, Huo Tuo Jia Ji 華佗夾脊 L1–L5 (select the affected side).

Distal points:

- LI4 合谷He Gu, LIV3 太沖Taichong, SI3 後溪Houxi (select the opposite side and insert the needle direction toward LI4), UB62 申脈Shenmai, UB40 委中Weizhong (apply bleeding techniques and draw 10 drops of blood), UB53 胞肓 Baohuang, UB57 承山 Chengshan, UB60 崑崙 Kunlun, GB39 懸鐘 Xuanzhong.

Extra points:

- Yao Yan Xue 腰痛穴 (select the affected side); Ashi points 阿是穴 on lower back and buttocks (insert the needles slowly and avoid puncturing the vessels).

Cupping Treatment

- UB23 腎俞 Shenshu, UB25 大腸俞 Dachangshu.

Moxibustion Treatment

- Warming needling techniques on UB23 腎俞 Shenshu, UB25 大腸俞 Dachangshu.

Bleeding Treatment

- UB40 委中 Weizhong.

Point Selected is UB40 委中 Weizhong

Procedures:

- Prior to the procedure, the tools should be prepared and ready to be used in the working area.
- Set up a clean field with paper towels.
- Place these tools in the clean field, including dry sterile cotton balls, medical alcohol wipes or cotton swabs soaked in alcohol, sterile gloves, a biohazard sharps container, a biohazard trash container, goggles and facemask, and paper towels. Also, a medical lancet or a traditional three-edge needle or the seven-star needle is a bleeding tool; a medical lancet or a traditional three-edge needle is selected for bleeding UB40 委中Weizhong in this case.
- The patient lies prone with the legs extended and UB40 委中 Weizhong facing upward.
- The room should be well lit.
- The therapist washes their hands thoroughly and wipes their hands with a single-use napkin. Put on sterile gloves.
- Wipe the acupuncture point UB40 委中 Weizhong with a medical alcohol wipe or a cotton swab soaked in alcohol. Move from the center of the acupuncture point to the periphery.
- Place the medical lancet or the traditional three-edge needle at an angle of 90 degrees to the skin.
- In one quick move, prick the needle into the acupuncture point about 0.1 cun deep, eliciting a few drops of blood.

- If more drops of blood are needed, absorb the blood with a sterile cotton ball.
- Lastly, the point is pressed with a sterile cotton ball until the bleeding ceases.
- Dispose of the tools in the proper containers.
- Wash the hands thoroughly.

Electro-Acupuncture Techniques

UB23 腎俞Shenshu ~UB25 大腸俞 Dachangshu 2/100 Hz for 20 minutes daily.

Tui Na Treatments

Tui Na techniques:

- Pressing, kneading

Tui Na procedures:

- The patient is in prone position; the therapist stands near the patient's lumbar region.
- Apply pressing techniques with thumb tip on a Tui Na extra point Shanyao Xue 閃腰穴 three to five times; then continue to apply kneading with the thumb pad for another 3–5 minutes. Shanyao Xue 閃腰穴 is located at the border between upper third and middle third on the line between UB57 承山 Chengshan and UB60 崑崙 Kunlun.
- Apply kneading techniques with the heel of the hand on the affected lumbar region; kneading should start with light pressure and gradually increase to heavy pressure for the deeper tissues. Treatment duration is about 5 minutes.

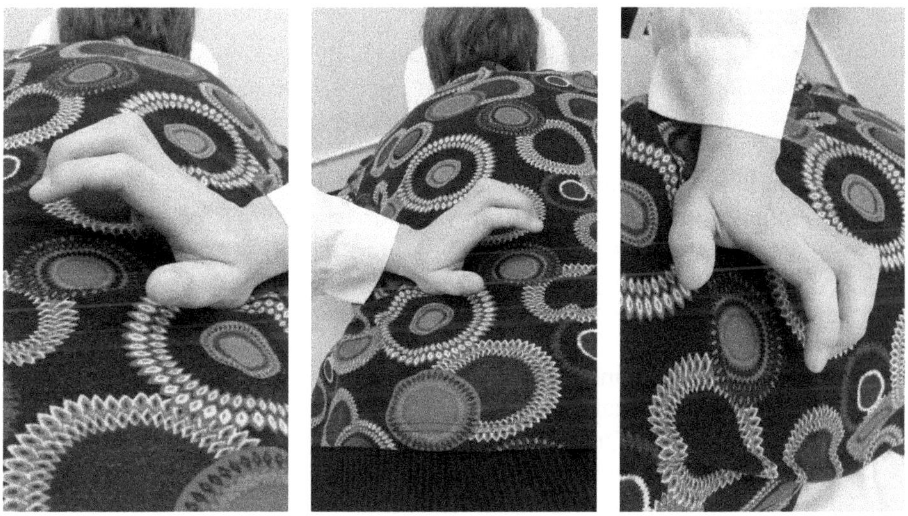

- Apply pressing techniques with the thumb on UB23 腎俞 Shenshu, UB40 委中 Weizhong, and DU3 腰陽關 Yaoyangquan. Avoid rough and hard pressing at the beginning; the pressure should start gentle, light, and superficial and then gradually increase but remain gentle to the deeper tissues. Treatment duration is about 5 minutes.

Patient Advisory

1) Avoid further injury by avoiding the movements or postures that cause pain. When pain occurs during any physical activity, stop as soon as possible and seek treatment.
2) Bed rest on a relatively firm bed for the first 3–7 days.
3) Apply a cold pack to the affected region on the first day; may start applying a heating pad 24 hours after the injury.
4) Keep the lumbar region warm when not using an ice pad.
5) Learn how to lift and bend properly.
6) Avoid standing for long periods of time. If standing for work, stand with correct posture. Take a break every hour. Do not lift heavy objects after a long period of standing.
7) Avoid sitting for long periods of time. If sitting for work or driving long distances, sit with correct posture. Take a break every hour. If the chair has a back rest, avoid leaning, and use a cushion or pillow to support the lumbar region. To rest the lower back, raise the knees higher than the hips by using a stool under the feet while sitting. Do not lift heavy objects after a long period of sitting.
8) Lose weight if overweight.
9) Do not wear high heels.
10) Quit smoking.
11) After the acute stage, routine lower body stretching exercise is helpful for waist pain recovery and improving flexibility. Try exercise to strengthen the abdominal muscles and lumbar muscles, as follows:
    - Lie supine on a firm pad or carpet with both legs extended.
    - Flex the affected knee and hip, and place both hands on the patella.
    - Continue to flex the leg and move the knee gently toward the opposite shoulder.
    - Once the limit before feeling pain is reached, hold for 5 to 30 seconds. Be mindful; gentle stretching is better than overstretching. Repeat 5–10 times.
    - Practice the same procedures on the other leg 5–10 times.
    - Practice the same procedures on both legs 5–10 times.
    - The frequency is twice daily, in the morning on waking and at night before sleep.
    - Keep good posture so that Qi and blood flow can be free.
    - Keep good posture generally when sitting, standing, sleeping, and walking so that Qi and blood flow can be free.
12) Learn to relax mentally and physically by doing relaxation exercise such as Tai Chi Chuan, Yoga, swimming, or dancing.

INTERSPINOUS LIGAMENT INJURY 棘間韌帶損傷

Introduction

Traumatic injury to the interspinous ligament causes low back pain and referred pain in the lower limbs. The straining of the ligament is commonly seen in the activities of arm lifting an object in an awkward position, twisting the lumbar region, or falling in an awkward manner and landing with a lumbar twisted posture.

Pain from injury to the interspinous ligament is involved in both acute and chronic injuries, from a single trauma to cumulative microtrauma over time.[18]

Anatomy

Interspinous Ligaments

The interspinous ligaments are composed of thin and membranous ligaments. They are anterior to the supraspinous ligaments, and connecting the spinous processes of the vertebrae in the spine from C1 to S1, they extend from the roots to the apexes of each spinous process. Anteriorly, their fibers connect with the ligamentum flavum.

The interspinous ligaments are narrow and elongated in the thoracic region and broader, thicker, and quadrilateral in form in the lumbar region.[19]

Function

The function of the interspinous ligament is to limit flexion of the vertebral column.[20]

Western Medicine Etiology and Pathology

During a long period of lumbar flexion without periodically changing position, the interspinal ligaments are constantly in a state of tension, resulting in micro tears, bleeding, and effusion, which cause pain. They are more vulnerable if the lumbar spine is affected by degenerative changes.

In addition, rupture of the interspinous ligament due to trauma is also a cause of chronic low back pain when scar tissues are formed.[21]

Chinese Medicine Etiology and Pathology

Qi and Blood Stagnation

According to Chinese medicine, blood is stagnant when Qi is stagnant; injury is imparted by external violence or chronic indulgence and long-term emotional imbalance, resulting in Qi stagnation. In general, Qi stagnation leads to blood stasis, because Qi is a "commander" of blood. Qi and blood stagnation cause local blockages, and pain is the result of the blockages.

Manifestations

- There is no notable history of injury, but there is a history of excessive anterior flexion of the waist. For example, the patient's work requires lumbar flexion for a long period of time each day.

- The lumbar range of motion is limited.
- Tearing pain in the posterior central lower back that is exacerbated by flexion.
- In some cases, lumbar hyperextension may also induce pain.
- A referred pain to the sacrum and buttock may occur in interspinous ligament injury.

Physical Examination

- Begin by obtaining the patient's medical history:
  - Look out for general health conditions, such as a job involving excessive anterior flexion of the waist.
- Inspection:
  - Usually, there are no visible signs of redness and swelling.
- Palpation:
  - The movable supraspinal ligaments may be palpable over the spinous processes.
  - There is tenderness on the spinous processes or at the intervertebral spaces.
- Range of motion:
  - The range of motion in the lower back is limited to flexion.
  - Pain may present on lumbar hyperextension.

Differential Diagnosis

1) Vertebral epiphysitis:
   - The patients are relatively young; the disease is seen most frequently in rapidly growing children.
   - It causes juvenile kyphosis.[22]
   - The patients may have a family history of similar spinal lesions.
   - Roentgenograms show irregularity of the upper and lower surfaces of the vertebral bodies as the most constant finding.
   - A later change is constant anterior wedging of the vertebral body.
2) Low back pain during menstruation:
   - During the menstrual cycle, some people also suffer from low back pain. The pain experienced is typically located along the center portion of the lower back. Back pain for most people will begin a few days prior to menstruation and usually subside afterward. The good news is that low back pain during menstruation is usually not serious and will subside for the most part.[23]
3) Acute lumbar sprain:
   - The patient has a history of injury.
   - The pain cannot be eliminated by rest.

Treatment

Treatment principle:

- Move blood, dissolve stasis, harmonize Ying Qi, and relieve pain.

Herbal Treatments

Herbal formula:

- Ba Wei Yao Tong Fang 八味腰痛方.[24]

Formula ingredients:

Dang Gui 當歸 9 g, Chi Shao 赤芍 9 g, Xu Duan 續斷 9 g, Hong Hua 紅花 9 g, Niu Xi 牛膝 12 g, Yan Hu Suo 延胡索 9 g, Chuan Xiong 川芎 12 g.

Ingredient explanation:

- Dang Gui 當歸 nourishes blood, reduces edema, and stops pain.
- Chi Shao 赤芍 moves blood, reduces swelling, stops pain.
- Xu Duan 續斷 nourishes liver and kidney, strengthens bones and tendons.
- Hong Hua 紅花 moves blood, releases bruising, opens meridians, stops pain.
- Niu Xi 牛膝 expels wind dampness, tonifies liver and kidneys, directs herbs.
- Yan Hu Suo 延胡索 moves blood, promotes Qi circulation, and stops pain.
- Chuan Xiong 川芎 moves blood, expels wind, and stops pain.

Formula actions:

- Moves blood, strengthens kidney, and stops pain.

Modifications:

- For blood stagnation, add Di Bie Cong 地鱉蟲, Tao Ren 桃仁, Ru Xiang 乳香, and Mo Yao 沒藥.
- For blood deficiency, add Shu Di 熟地 and Bai Shao 白芍.
- For cold and weak constitution, add Rou Gui 肉桂 and Zhi Fu Zi 制附子.
- For heat and weak constitution, add Zhi Mu 知母 and Huang Bai 黃柏.
- For deficiency of Qi, add Dang Shen 黨參 and Huang Qi 黃芪.
- For injury during or after pregnancy, add Rou Gui 肉桂 and Sang Ji Sheng 桑寄生.

Acupuncture Treatments

1) Insert needles onto DU12 身柱Shenzhu, DU4 命門Mingmen, DU3 腰陽關Yaoyangquan, UB40 委中Weizhong, DU26 人中Renzhong.
2) Add moxibustion to DU4 命門Mingmen for three cones.
3) Apply cupping onto the acupuncture points on DU meridian immediately after removing the needles.

Bleeding Treatments

- UB40 委中 Weizhong

Point Selected is UB40 委中 Weizhong

Procedures:

1) Prior to the procedure, the tools should be prepared and ready to be used in the working area.
2) Set up a clean field with paper towels.
3) Place these tools in the clean field, including dry sterile cotton balls, medical alcohol wipes or cotton swabs soaked in alcohol, sterile gloves, a biohazard sharps container, a biohazard trash container, goggles and facemask, and paper towels. Also, a medical lancet or a traditional three-edge needle or the seven-star needle is a bleeding tool; a medical lancet or a traditional three-edge needle is selected for bleeding UB40 委中 Weizhong in this case.
4) The patient lies prone with the legs extended and UB40 委中 Weizhong facing upward.
5) The room should be well lit.
6) The therapist washes their hands thoroughly and wipes their hands with a single-use napkin. Put on sterile gloves.
7) Wipe the acupuncture point UB40 委中 Weizhong with a medical alcohol wipe or a cotton swab soaked in alcohol. Move from the center to the periphery.
8) Place the medical lancet or the traditional three-edge needle at an angle of 90 degrees to the skin.
9) In one quick move, prick the needle into the acupuncture point about 0.1 cun deep, eliciting a few drops of blood.
10) If more drops of blood are needed, absorb the blood with a sterile cotton ball.
11) Lastly, the point is pressed with a sterile cotton until the bleeding ceases.
12) Dispose of the tools in the proper containers.
13) Wash the hands thoroughly.

Electro-Acupuncture Treatment

- UB23 腎俞 Shenshu ~ UB25 大腸俞 Dachangshu 2/100 Hz for 20 minutes daily.

Tui Na Treatments

1) The patient sits on a stool; the therapist sits behinds the patient.
2) The therapist places one hand on the patient's shoulder while kneading the affected lumbar spine with the thumb of the other hand for several minutes. The force is light if the pain is of high intensity.
3) The palm rubs and chafes the Du meridian and the lumbar Huo Tuo Jia Ji region for a few minutes or until it is hot.

Patient Advisory

- The patient should be advised to avoid lumbar rotation for 1 week after injury.
- Keep the lumbar region warm.
- During recovery, do core strengthening exercises to provide more spinal stability; the exercise that is recommended includes swimming or stationary bicycling.
- When lifting objects from the floor, one should squat to lift; do not bend the lumbar region to lift.

Learn how to lift and bend properly. Follow these tips:

- If an object is too heavy, too big, or awkward, ask for assistance.
- Stand as close to the object as possible.
- Spread the feet apart wider than and at both sides of the object.
- Bend the knees and grasp the object with both hands.
- Straighten the upper body. DO NOT lean forward or twist sideways so as not to lose gravity and sprain the lumbar soft tissues.
- Tighten the abdominal and hip muscles.
- Hold the object as close to the body as possible.
- Lift using the leg muscles.
- Carry the object and walk with the eyes. Slow down and be cautious when walking on an uneven surface.

SUPRASPINAL LIGAMENT INJURY 棘上韌帶傷害

Introduction

Supraspinous ligament injury refers to tearing or inflammatory degeneration and calcification of the supraspinous ligament due to trauma or strain. It is more common in young and middle-aged manual workers. When treatment is incomplete, it often develops into chronic strain, causing back pain.[25]

Most people have few supraspinous ligament fibers between L4 and L5, which forms a structurally weak area. The supraspinal ligament is almost absent at L5 and the sacrum.

L5 is fully movable, but S1 is fused; this joint is the most vulnerable to injury and is under the greatest tension with lumbar flexion.

Some types of jobs that involve poor posture, slumping the lumbar region forward while standing for long periods of time, are the worst for the low back and ligament strain or sprain. These include factory workers, auto mechanics, surgeons, dentists, construction workers, manual laborers, and office workers. [26]

While the body is in excessive anterior flexion, the supraspinous ligament's fibers resist separation; thus, this is the first to fail, leading to pain when the lumbar region is in flexion and lifting heavy objects or in flexion for a long period of time.[27]

Anatomy

Supraspinal Ligament[28]

The supraspinous ligament is the part of the interspinous ligament that moves backward.[29] The supraspinous ligament is a long, thick, and strong fibrous cord that connects together the tips of the spinous processes spanning from the seventh cervical vertebra to the 3rd or 4th lumbar vertebrae.

The supraspinous ligament is well developed only in the upper lumbar region and may terminate at L3, although the most common termination site appears to be at L4. The ligament is almost always absent at L5/S1.

It is thicker and broader in the lumbar than in the thoracic region, where it merges into the thoracolumbar fascia.

Function

- Supraspinous ligaments limit hyperflexion along with other ligaments of the vertebral column and help stabilize the lower spinal area.[30]
- It is stretched in flexion, and its fibers resist the separation of spinous processes during forward flexion. During hyperflexion, the interspinous ligament and the supraspinous ligament are the first to fail.

Western Medicine Etiology and Pathology

When the lumbar region is in a fully flexed position to lift heavy objects, the sacrospinalis muscle is slack. The muscles of the buttocks and the muscles of the back of the thigh are contracted, and the lumbar vertebrae are used as a lever to lift the object. The lumbosacral region is often the hinge, and the ligament is not protected by the sacrospinalis muscle. The tension falls onto ligaments around the lumbar region. The supraspinal ligament is at the outermost layer, and it receives the greatest traction, so it is vulnerable to damage.

Chinese Medicine Etiology and Pathology

Qi and Blood Stagnation

According to Chinese medicine, blood is stagnant when Qi is stagnant; injury is imparted by external violence or chronic indulgence and long-term emotional imbalance, resulting in Qi stagnation. In general, Qi stagnation leads to blood stasis, because Qi is a "commander" of blood. Qi and blood stagnation cause local blockages, and pain is the result of the blockages.

Manifestations

- Most patients have a history of injury during lumbar flexion.
- The sensation of pain manifests at the level of the spine where it is injured. It is severe on forward flexion, but extension may release the pain. Thus, some patients prefer lumbar extension.[31]

- Tender spots are easily found at the injury site, especially in underweight patients.[32]
  - In some cases, pain may present on lumbar extension and/or side-flexion if the interspinous ligaments are also involved the injury.
  - The lumbar range of motion is limited, especially in lumbar flexion and rotation.[33]
  - The pain is less with bed rest.[34]

Physical Examination

- Begin by obtaining the patient's medical history:
  - Look out for general health conditions, such as history of injury during lumbar flexion.
- Inspection:
  - Usually, there is no visible sign.
- Palpation:
  - Local pain occurs during palpation.[35] Usually, the area of pain only appears on one or two spinous processes.[36]
- Range of motion:
  - The range of motion of the lower back is limited to flexion and rotation.
  - Pain may present on lumbar extension or side flexion.
- Other tests:
  - X-ray is usually negative.
  - MRI is able to clearly reveal the torn ligament.
  - The injection of 5–10 ml 1% procaine into the painful sites may stop the symptoms, but the pain comes back after the local anesthesia wears off.

Differential Diagnosis

1) Vertebral epiphysitis:
   - The patients are relatively young; the disease is seen most frequently in rapidly growing children.
   - It causes juvenile kyphosis.[37]
   - The patient may have a family history of similar spinal lesions.
   - X-rays showed irregularity of the upper and lower surfaces of the vertebral bodies as the most constant finding.
   - A later change is constant anterior wedging of the vertebral body.
2) Low back pain during menstruation:
   - During the menstrual cycle, some people also suffer from low back pain. The pain experience is typically located along the center portion of the lower back. Back pain for most people will begin a few days prior to menstruation and usually subside afterward. The good news is that low back pain during menstruation is usually not serious and will subside for the most part.[38]

3) Acute lumbar sprain:
   - The patient has a history of injury.
   - The pain cannot be eliminated by rest.

Treatment

Treatment principle:

- Move blood, dissolves stasis, harmonize Ying Qi, and relieve pain.

Herbal Treatments

Herbal formula:

- Ba Wei Yao Tong Fang 八味腰痛方.

Formula ingredients:

Dang Gui 當歸 9 g, Chi Shao 赤芍 9 g, Xu Duan 續斷 9 g, Hong Hua 紅花 9 g, Niu Xi 牛膝 12 g, Yan Hu Suo 延胡索 9 g, Chuan Xiong 川芎 12 g.

Ingredient explanation:

- Dang Gui 當歸 nourishes blood, reduces edema, and stops pain.
- Chi Shao 赤芍 moves blood, reduces swelling, stops pain.
- Xu Duan 續斷 nourishes liver and kidney, strengthens bones and tendons.
- Hong Hua 紅花 moves blood, releases bruising, opens meridians, stops pain.
- Niu Xi 牛膝 expels wind dampness, tonifies liver and kidneys, directs herbs.
- Yan Hu Suo 延胡索 moves blood, promotes Qi circulation, and stops pain.
- Chuan Xiong 川芎 moves blood, expels wind, and stops pain.

Formula actions:

- Moves blood, strengthens kidney, and stops pain.

Modifications:

- For blood stagnation, add Di Bie Cong 地鱉蟲, Tao Ren 桃仁, Ru Xiang 乳香, Mo Yao 沒藥.
- For blood deficiency, add Shu Di 熟地 and Bai Shao 白芍.
- For cold and weak constitution, add Rou Gui 肉桂 and Zhi Fu Zi 制附子.
- For heat and weak constitution, add Zhi Mu 知母 and Huang Bai 黃柏.
- For deficiency of Qi, add Dang Shen 黨參 and Huang Qi 黃芪.

- For injury during or after pregnancy, add Rou Gui 肉桂 and Sang Ji Sheng 桑寄生.

Acupuncture Treatments

1. Insert needles onto DU12 身柱 Shenzhu, DU4 命門 Mingmen, DU3 腰陽關 Yaoyangquan, UB40 委中 Weizhong, DU26 人中 Renzhong.
2. Add moxibustion to DU4 命門 Mingmen for three cones.
3. Apply cupping onto the acupuncture points on DU meridian immediately after removing the needles.

Bleeding Treatments

- UB40 委中 Weizhong

Point Selected is UB40 委中 Weizhong

Procedures:

1) Prior to the procedure, the tools should be prepared and ready to be used in the working area.
2) Set up a clean field with paper towels.
3) Place these tools in the clean field, including dry sterile cotton balls, medical alcohol wipes or cotton swabs soaked in alcohol, sterile gloves, a biohazard sharps container, a biohazard trash container, goggles and facemask, and paper towels. Also, a medical lancet or a traditional three-edge needle or the seven-star needle is a bleeding tool; a medical lancet or a traditional three-edge needle is selected for bleeding UB40 委中 Weizhong in this case.
4) The patient lies prone with the legs extended, UB40 委中 Weizhong facing upward.
5) The room should be well lit.
6) The therapist washes their hands thoroughly and wipes the hands with a single-use napkin. Put on sterile gloves.
7) Wipe the acupuncture point UB40 委中 Weizhong with a medical alcohol wipe or a cotton swab soaked in alcohol. Move from the center of the acupuncture point to the periphery.
8) Place the medical lancet or the traditional three-edge needle at an angle of 90 degrees to the skin.
9) In one quick move, prick the needle into the acupuncture point about 0.1 cun deep, eliciting a few drops of blood.
10) If more drops of blood are needed, absorb the blood with a sterile cotton ball.
11) Lastly, the point is pressed with a sterile cotton ball until the bleeding ceases.
12) Dispose of the tools in the proper containers.
13) Wash the hands thoroughly.

Electro-Acupuncture Treatment

UB23 腎俞 Shenshu ~ UB25 大腸俞 Dachangshu 2/100 Hz for 20 minutes daily.

Tui Na Treatments

Tui Na techniques:

- Pressing, slippery, pushing

Tui Na procedures:

1) The patient lies in the prone position with a small cushion underneath the abdomen; the therapist stands at the left side near the lumbar region.
2) The therapist presses a thumb on the affected spinous process and applies the slippery technique with the other thumb on the supraspinous ligaments several times.
3) Apply the pressing technique with the thumbs several times.
4) Apply pushing techniques with the thumbs along the ligaments several times.
5) Perform the same techniques as in step 4 but pushing with the hypothenar of a hand for several minutes until the skin is hot.
6) Apply the thumb tips, pressing on DU12 身柱 Shenzhu, DU4 命門 Mingmen, DU2 腰俞 Yaoshu, and UB40 委中 Weizhong for 30 seconds on each point.

Patient Advisory

- The patient should be advised to avoid lumbar rotation for 1 week after injury.
- Keep the lumbar region warm.
- During recovery, do core strengthening exercises to provide more spinal stability; the exercise that is recommended includes swimming or stationary bicycling.
- When lifting objects from the floor, one should squat to lift; do not bend the lumbar region to lift.

Learn how to lift and bend properly. Follow these tips:

- If an object is too heavy, too big, or awkward, ask for assistance.
- Stand as close to the object as possible,
- Spread the feet apart wider than and at both sides of the object.
- Bend the knees and grasp the object with both hands.
- Straighten the upper body. DO NOT lean forward or twist sideways so as not to lose gravity and sprain the lumbar soft tissues.
- Tighten the abdominal and hip muscles.

- Hold the object as close to the body as possible.
- Lift using the leg muscles.
- Carry the object and walk with the eyes. Slow down and be cautious when walking on an uneven surface.

THE THIRD LUMBAR TRANSVERSE PROCESS SYNDROME 第三腰椎橫突綜合徵

Introduction

The third lumbar transverse process syndrome is a commonly seen low back pain. The lumbar transverse processes are quite long, and the transverse process of L3 is the longest. The vertebra of L3 is positioned in the middle of the lumbar spine, and its vertebra and the transverse process play an important role in lifting the weight of the torso and protecting the cauda equina.[39] With such anatomical and physiological characteristics of the third lumbar vertebra, the transverse processes of the third lumbar vertebra are particularly susceptible to wear and tear.[40] Additionally, it is the most vulnerable part of the lumbar spine when the muscles and fascia connected with the transverse processes are suddenly pulled, contracted, or catch cold, resulting in pain.

Anatomy

The third lumbar vertebra is in the middle of the five lumbar vertebrae in the lower back portion of the spinal column. It also serves as an axis of movement of the lumbar region. In addition, the transverse processes of the third lumbar vertebra are thin and the longest among all the transverse processes of the lumbar vertebrae, but they are lacking transverse foramina. Its spinous process is very large and nearly rectangular in shape.[41]

The L3 spinal nerve roots exit the cauda equina through the intervertebral foramina on the left and right sides of the spinal canal. The L3 dermatome includes the anterior aspect of the femur and the medial aspect of the leg.[42]

Western Medicine Etiology and Pathology

With the unique anatomical and physiological characteristics of the third lumbar vertebra, which is located at the middle of the physiological lordosis of the lumbar spine and serves as an axis of lumbar movements, the lever effect on the transverse processes of the third lumbar vertebra is the strongest, and the traction stress borne by both its transverse processes is the greatest among all the lumbar vertebrae. In addition, there are many different sizes of soft tissues attached to the transverse process, and the pulling forces borne by the muscles, ligaments, and fasciae adhering to it are also the strongest.

When the soft tissues are suddenly pulled, such as by a sudden lumbar rotation or an awkward fall with abnormal lumbar loading, local laceration, hemorrhage, and hematoma occur, leading to muscle spasm and compression of the lateral branch of the posterior ramus of the spinal nerve, resulting in pain in the region of the injury, which may radiate to the ipsilateral buttock and medial thigh.[43]

Chinese Medicine Etiology and Pathology

1) Congenital deficiency
2) Kidney and liver deficiency
3) Fixed Bi syndrome
4) Qi and blood stagnation

Explanation

- *Internal factors*: According to Chinese medicine, congenital deficiency is one of the sources of this low back pain. The lumbar region is congenitally weak, and the person may play sports or have a job that requires lumbar physical labor. The repetitive lumbar activity consumes kidney Qi and liver Yin and blood, resulting in chronic traumatic inflammation, edema, exudation, fibrosis, fascia thickening, adhesion, and atrophy of the soft tissues surrounding the spinous processes.
- *External factors*: Due to the internal factors, the local blood circulation is poor and the muscles are vulnerable; when the lumbar region is in a position or condition of overload of traction forces, the exterior six climatic factors invade the lumbar meridians, resulting in blockages and manifesting as pain. The six climatic factors are wind, damp, cold, dry, heat, and summer heat, and the first three are the common ones causing the blockage.

Manifestations

1) The syndrome often occurs in young and middle-aged physical laborers.
2) The patients may have a history of lumbar trauma or strain.
3) The pain is constant, sore, or tight.
4) The pain location is fixed; it is on one side of the hip and waist, possibly radiating to the medial side of the knee along the leg, and it is aggravated by lumbar flexion or rotation, particularly in the morning after waking up.

Physical Examination

- Begin by obtaining the patient's medical history:
  - General health conditions include a history of lumbar overexertion and working in a cold and windy environment.
  - Injury may occur at the moment of lumbar flexing and rotating to the side to lift up a bulky or heavy object.
  - Careers that may cause vulnerability to this injury are jobs requiring long periods of sitting and leaning forward, such as software engineers or office workers who sit with a forward-leaning posture.
- Inspection:
  - There may be mild edema.
  - The pain is not aggravated by coughing or sneezing.

    - There is no abnormality in lower extremity skin perception and muscle strength.
- Palpation:
    - There is an obvious tenderness and a fibrous or hard strip on the transverse process of the third lumbar vertebra at the margin of the sacrospinalis muscle.
- Range of motion:
    - Lumbar flexion and rotation may aggravate the pain.
    - Straight leg raise test is negative.

Differential Diagnosis

Prolapse of the lumbar intervertebral disc:

- The location of the pain is on the spinous process.
- The pain radiates to the leg along the dermatomal distributions.
- Cough and sneezing increase pain.
- Straight leg raise test is positive.[44]

Lumbosacral strain:

- The location of the pain is not only at a small spot; it is wide.
- The pain is on either one side or both sides.
- The pain is sore on and off.
- The pain is aggravated after work or change of weather, and relieved after rest and heat.

Interspinous ligament injury:

- The pain is acute in onset along the midline of the lower back.
- The pain is exacerbated during flexion of the lumbar spine.
- Tenderness is over the interspinous space.[45]

Treatment

Treatment principle:

- Move blood, relax tendons, release spasm, reduce swelling, and stop pain.

Herbal Treatments

Herbal formula:

- Fu Yuan Huo Xue Tang 復元活血湯

Ingredients:

- Chai Hu 柴胡 12 g, Tian Hua Fen 天花粉 12 g, Dang Gui 當歸 15 g, Hong Hua 紅花 9 g, Gan Cao 甘草 6 g, Chuan Shan Jia 穿山甲 9 g, Da Huang 大黃 6 g, Tao Ren 桃仁 9 g.

Formula explanations:

- Chai Hu 柴胡 spreads liver Qi and relieves stagnation.
- Tian Hua Fen 天花粉 relieves toxicity, expels pus, and reduces swelling.
- Dang Gui Wei 當歸 tonifies, moves, and harmonizes the blood, disperses cold, stops pain due to blood stasis, reduces swelling, expels pus, generates flesh, and alleviates pain.
- Hong Hua 紅花 moves the blood, dispels blood stasis, opens the channels, unblocks menstruation, and alleviates pain.
- Gan Cao 甘草 moderates spasms, alleviates pain, and moderates and harmonizes the harsh properties of other herbs.
- Shan Jia 穿山甲 moves the blood, dispels blood stasis, reduces swelling, promotes the discharge of pus, and expels wind-dampness from the channels.
- Da Huang 大黃 drains fire, clears heat, transforms dampness, promotes urination, moves the blood, and dispels blood stasis.
- Tao Ren 桃仁 breaks up blood stasis and moves blood circulation.

Formula actions:

- Moves blood stasis, alleviates pain and swelling.

Contraindications:

- Pregnancy.

Acupuncture Treatment

Acupuncture point formula:

UB17 膈俞 Geshu, UB23 腎俞 Shenshu, UB25 大腸俞 Dachangshu, UB40 委中 Weizhong, UB39 委陽 Weiyang, Hua Tuo Jia Ji 華佗夾脊 L2-L4, LI4 合谷 Hegu, SI3 後溪 Houxi and UB62 申脈 Shenmai, Ashi acupoints 阿是穴 (a tender point that is not included in the classic acupuncture points): 45 degrees angle toward the spine with warming needle for 15 minutes.

Cupping Treatment

- Apply cupping techniques after the treatments of acupuncture and Tui Na on UB23 腎俞 Shenshu and UB25 大腸俞 Dachangshu.

Moxibustion Treatment

- Warming needling techniques on UB23 腎俞 Shenshu, UB25 大腸 Dachangshu.

Bleeding Treatments

- Point selected is UB40 委中 Weizhong.

Procedures:

1) Prior to the procedure, the tools should be prepared and ready to be used in the working area.
2) Set up a clean field with paper towels.
3) Place these tools in the clean field, including dry sterile cotton balls, medical alcohol wipes or cotton swabs soaked in alcohol, sterile gloves, a biohazard sharps container, a biohazard trash container, goggles and facemask, and paper towels. Also, a medical lancet or a traditional three-edge needle or the seven-star needle is a bleeding tool; a medical lancet or a traditional three-edge needle is selected for bleeding UB40 委中Weizhong in this case.
4) The patient lies prone with the legs extended, UB40 委中Weizhong facing upward.
5) The room should be well lit.
6) The therapist washes their hands thoroughly and wipes their hands with a single-use napkin. Put on sterile gloves.
7) Wipe the acupuncture point UB40 委中 Weizhong with a medical alcohol wipe or a cotton swab soaked in alcohol. Move from the center of the acupuncture point to the periphery.
8) Place the medical lancet or the traditional three-edge needle at an angle of 90 degrees to the skin.
9) In one quick move, prick the needle into the acupuncture point about 0.1 cun deep, eliciting a few drops of blood.
10) If more drops of blood are needed, absorb the blood with a sterile cotton ball.
11) Lastly, the point is pressed with a sterile cotton ball until the bleeding ceases.
12) Dispose of the tools in the proper containers.
13) Wash the hands thoroughly.

Electro-Acupuncture Treatment

- UB23 腎俞 Shenshu ~ UB25 大腸 Dachangshu 2/100 Hz for 20 minutes daily.

Tui Na Treatments

Tui Na techniques:

- Rolling, stretching, slippery, pressing, kneading, chafing

Tui Na procedures:

1) The patient is in a prone position; the therapist stands next to the lumbar region.
2) Apply rolling techniques on sacrospinalis muscle six to eight times.

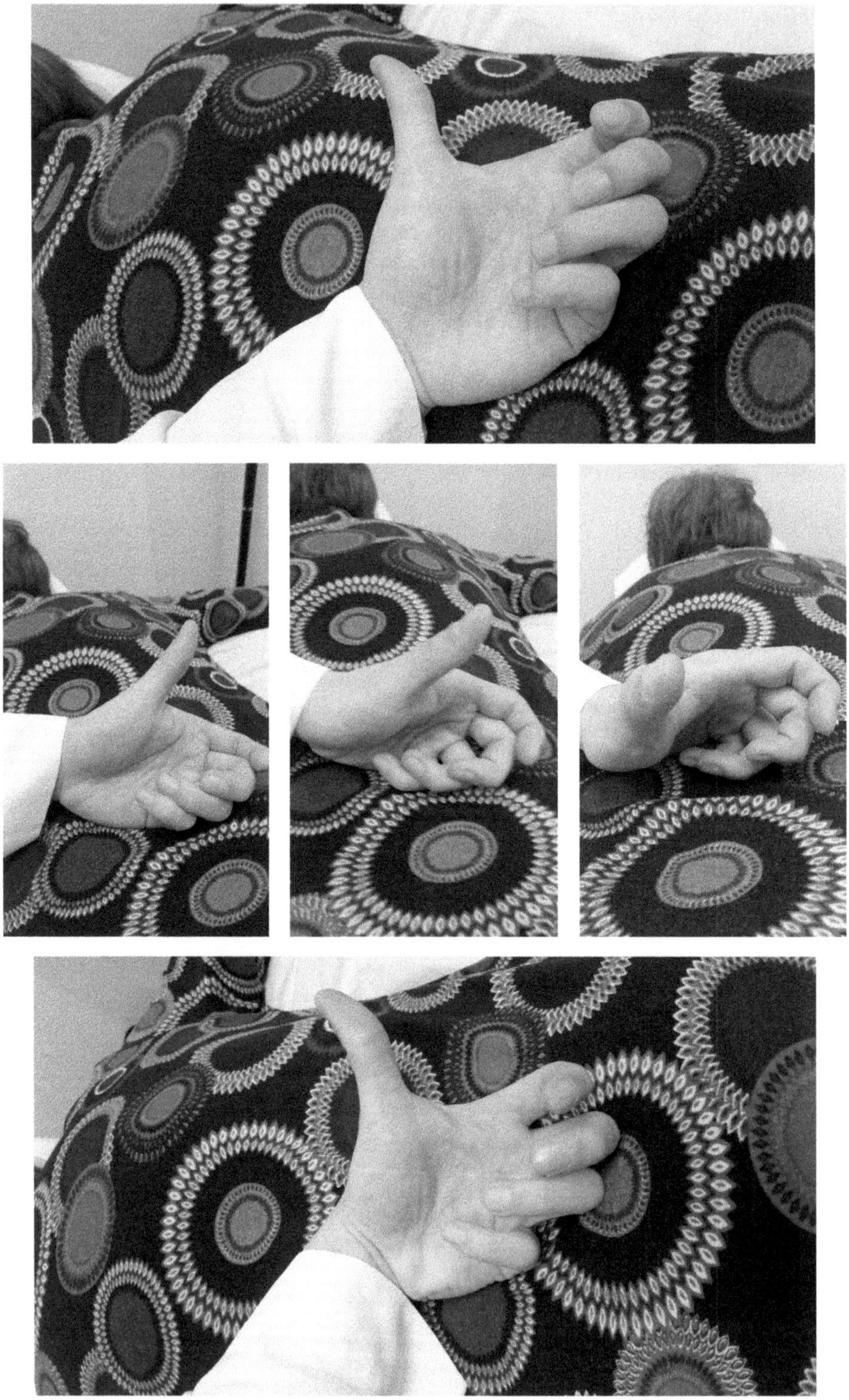

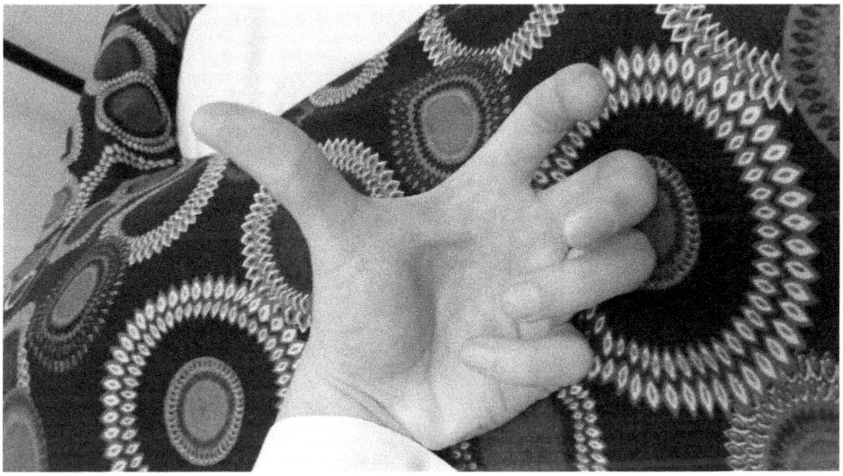

3) Apply stretching techniques to the lumbar region; the procedures are as below:
   a) The patient is prone; the therapist stands at the contralateral side.
   b) The therapist places one hand on the lumbar region, lifts up the non-affected leg with the other hand, and crosses the leg over to the affected side, with the hand on the affected lumbar region pressing perpendicularly for 1 minute.

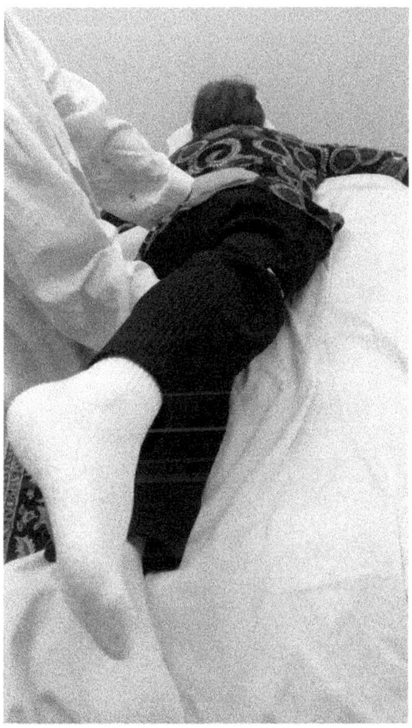

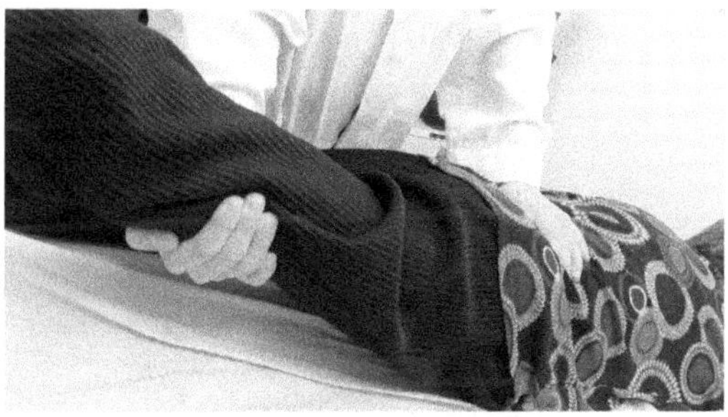

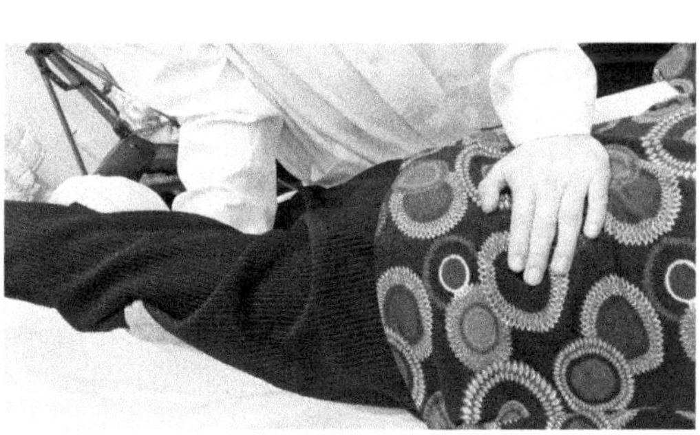

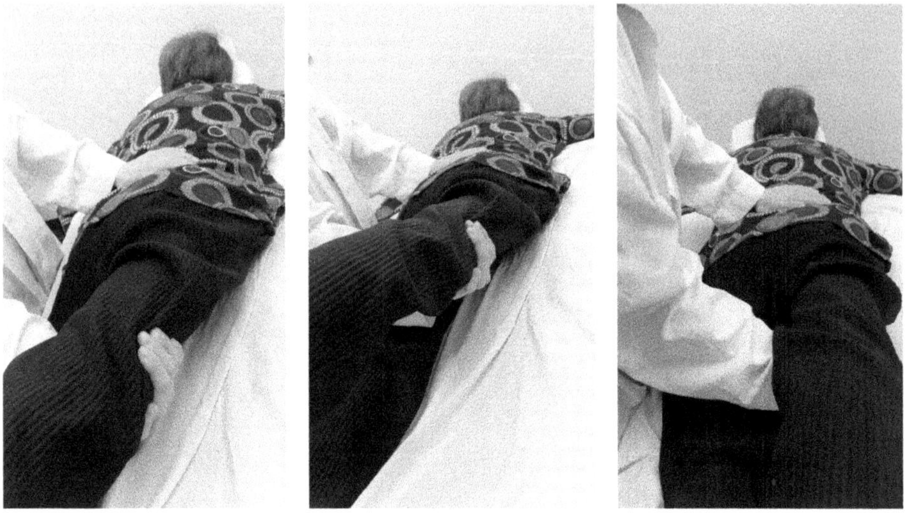

4) Apply finger pressing and kneading techniques on the affected spinous process, UB23 腎俞 Shenshu, UB25 大腸俞 Dachangshu, for 5 minutes.
5) Apply stretching techniques to the lumbar region; the procedures are as below:
   a) The patient is prone; the therapist stands at the contralateral side.
   b) The therapist places one hand on the affected lumbar region, lifts up the non-affected shoulder with the other hand, and lifts the shoulder toward the painful lumbar region, with the hand on the affected lumbar region pressing perpendicularly for 1 minute.

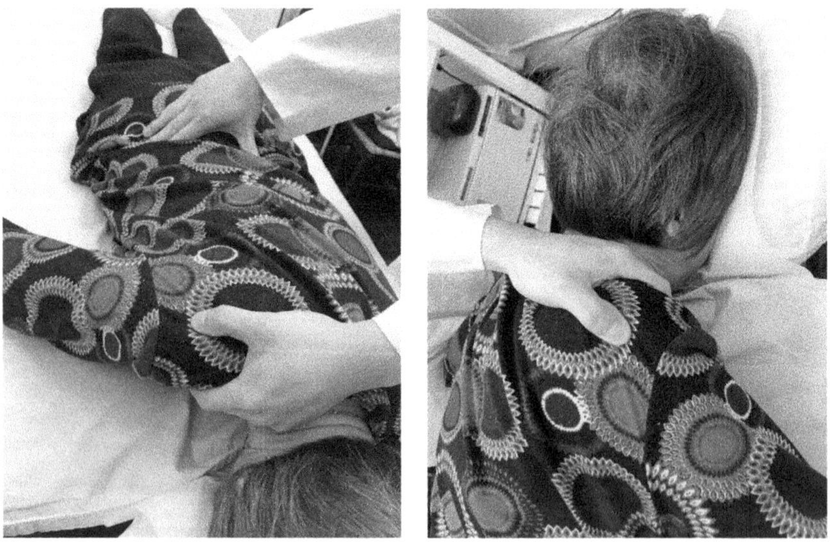

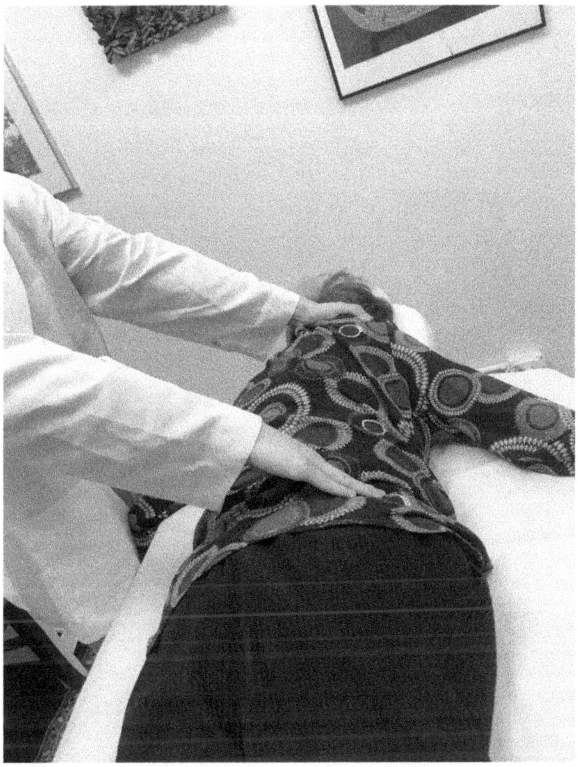

6) Apply slippery techniques to the hard strips for 2–3 minutes.
7) Apply pressing and kneading techniques to the ipsilateral buttock and the lateral sides of the thigh and leg for 3–5 minutes.

8) Apply pressing technique on GB30 環跳 Huantiao with elbow, UB54 秩邊 Zhibian, UB40 委中 Weizhong, UB57 承山 Chengshan with thumb tips for 1 minute on each point.

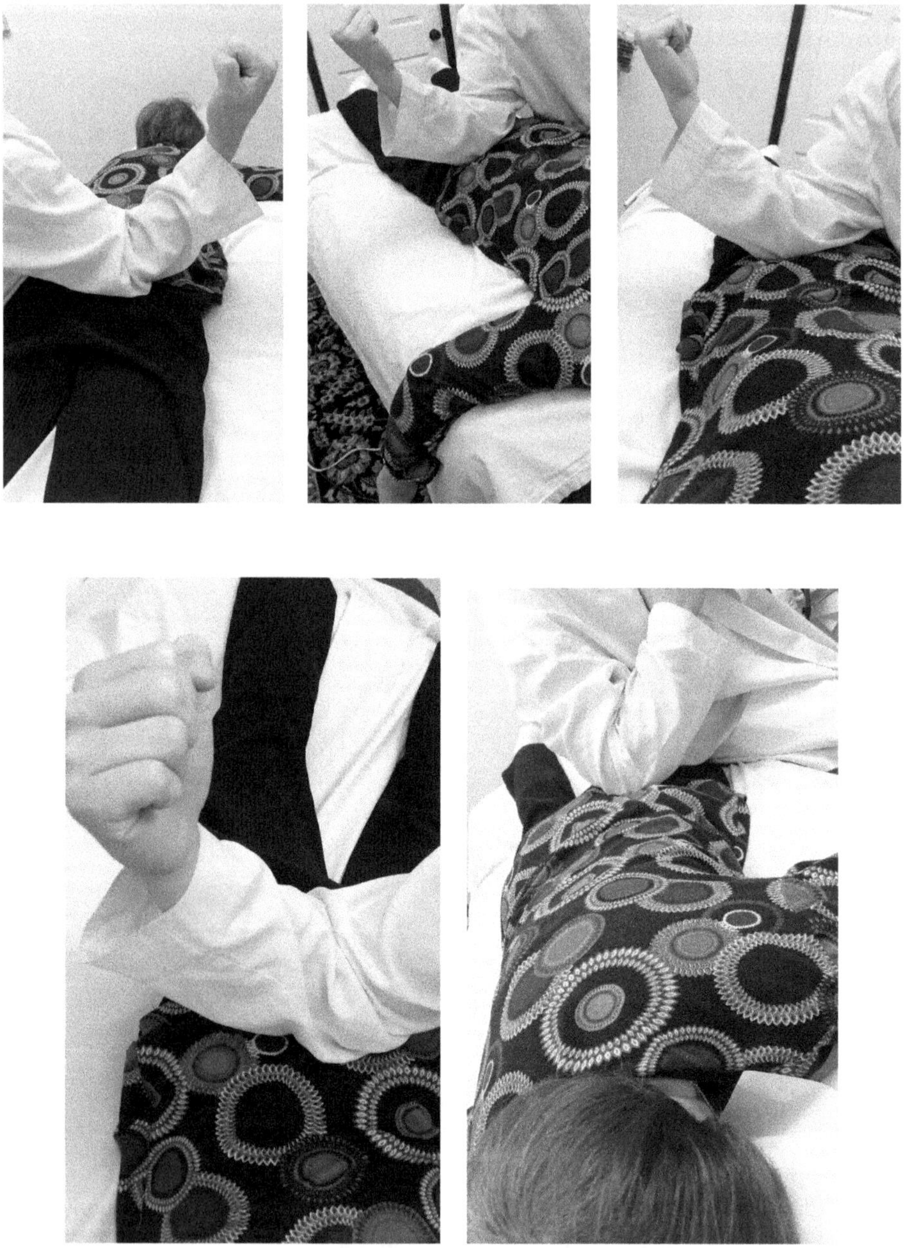

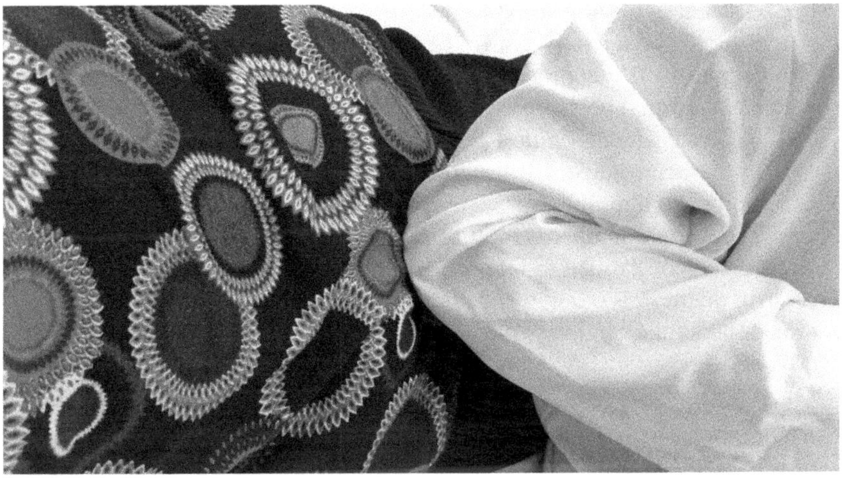

9) Apply chafing technique on both lumbar urinary bladder meridians until the skin is warm.

Patient Advisory

1) Patients should seek treatment as soon as possible when the problem is still in the acute state.
2) Patients should correct inaccurate posture.
3) Good sitting posture:
    - Be sure to straighten the back and avoid leaning forward. Place a cushion behind the lower back to support the normal lumbar curvature.
    - Flex the hips and knees at about 90 degrees.
    - Keep the legs shoulder width apart.
    - Keep the feet flat on the floor. If the feet cannot reach the floor, place the feet on a footrest.
    - Do not sit for longer than an hour. It is recommended to place a cup of tea within reach, with a few chrysanthemum flowers and goji berries with hot water, and to finish each cup within an hour. Goji berries help with eye tiredness, chrysanthemum flowers help to clear heat in the eye, and tea helps to clear the heat inside the body. With 250 cc of water, this is enough to make someone get up and walk to the bathroom. Someone sitting for 8 hours will finish drinking eight cups of water.
4) Good standing posture:
    - Be sure to stand with head and back erect. Keep the head upright on the midline between the shoulders.
    - Stand with legs slightly apart with both knees slightly flexed.
    - Keep the weight on the front third of the feet.
    - Be mindful not to lean your body weight to either side.

5) Good sleeping posture:
    - Sleeping on a relatively firm bed is good for sciatic pain recovery.
    - Supine is the appropriate sleeping posture; both sides of the body receive equal pressure.
    - Place a cushion or pillow below the knees to alleviate lumbar pressure.
    - In the acute stage, lift the lower legs higher by placing more pillows below the tibia.
    - The normal lumbar curvature can also be supported by placing a thin cushion or a rolled towel under the back.
6) A firm mattress is recommended for sleeping.
7) Keep the lumbar region warm.
8) Avoid overexertion.

SPINAL OSTEOARTHRITIS 腰椎退变性小关节損傷性關節炎

Introduction

Spinal osteoarthritis occurs more often among patients who are overweight. For patients under the age of 45, it is more common among men; After the age of 45, it is more common among women (National Institute on Aging 2017). Generally, osteoarthritis is more common among older patients; it is one of the most frequent causes of physical disability among older adults (National Institute on Aging 2017). However, it has also been seen in younger people as well in those who have jobs or do sports that put repetitive stress on certain lumbar joints.[46]

Anatomy

The spine is a column of 33 vertebrae. Five of the lower vertebrae are fused into the sacrum. The coccyx is formed from three to five (most often four) rudimentary vertebrae and does not contain a spinal canal, pedicles, laminae, or spinous processes.[47] The rest of the vertebrae are connected by facet joints. Tucked between the vertebrae are cushion-like pads called discs, whose function is to protect the spine and provide flexibility. Resting within the column of vertebrae lies the spinal cord, and the nerve roots are outside the special column.

Western Medicine Etiology and Pathology

Osteoarthritis is also known as degenerative joint disease. Spinal osteoarthritis is a degenerative change of the cartilage of the lumbar joints and discs. The intervertebral discs dehydrate, resulting in disc degeneration. Then, the degeneration leads to a narrowing between vertebrae, which causes pressure on the articular cartilage surface of the facet joint, causing bone to rub against bone. Bone spurs called osteophytes begin to form around the facet joints; the spur grows slowly over time, often without symptoms.

Then, the spur takes up space in the foramen, impinging on the surrounding soft tissues to cause swelling and pain and to restrict movement in a joint; the spur puts pressure on the nerves that leave the spinal column, resulting in weakness, tingling or numbness, and pain in the lumbar region and legs. In severe cases, this can also affect bladder and bowel function. As the spur grows bigger, it extends into the spinal canal, resulting in spinal stenosis.[48]

Excessive repetitive lumbar movements and injury from sports such as golfing or baseball cause traumatic change to the synovial capsule and can lead to facet joint degeneration, resulting in osteoarthritis.

Incomplete or wrong treatment of simple subarticular subluxation and incarcerated subluxation can cause traumatic inflammation of the facet joints, leading to osteoarthritis.

Chinese Medicine Etiology and Pathology

- *Liver and kidney deficiency*: Osteoarthritis originates from weakness of liver and kidney. The joint pain cannot be improved due to chronic joint diseases that consume the energy of liver and kidney, such as lumbar facet joint syndrome or prolapse of the lumbar intervertebral disc.
- *Bi syndrome*: The joints are already weak, and there is invasion of cold, damp, and wind to form blockages.
- *Qi and blood deficiency*: Blockages cause stagnation of Qi and blood flow, resulting in pain.

Manifestations

1) *Low back pain*: Intermittent stiffness in the lumbosacral region, dull or sore pain or discomfort. It is aggravated by exertion or holding a position for too long a period, and it is relieved by movement and bed rest, but there is pain when in bed for too long. Usually, the range of motion is still in the normal range. It is more commonly seen in older patients because there is degeneration of the facet joints.
2) The pain may be worse when the patient is exposed to cold, rain, and wind.
3) *Referred pain*: Sometimes there is weakness, tingling, or numbness and pain in the lumbar region and legs. In severe cases, this also can affect bladder and bowel function.

Physical Examination

- Begin by obtaining the patient's medical history:
    - General health conditions such as incomplete or wrong treatment of simple subarticular subluxation.
    - Sporting activities that involve repeated lumbar movements may be high risk; golfing is an example.

- Pain may be aggravated by exposure to wind, rain, and cold, by holding a position for too long, or by overexertion.
- Pain may be improved by a short period of bed rest or aggravated by a long period of bed rest.
- Inspection:
  - During an acute attack, the physiological curvature of the lumbar spine can disappear, the spinous processes of the lumbar spine are not aligned, and the distance between the spinous processes of the affected vertebrae increases.[49]
  - The lumbar muscles are hypertonic.
- Palpation:
  - There are tender spots at the affected joint; the pain is deep.
  - There is also percussion pain.
- Range of motion:
  - In most cases, the range of motion is still in the normal range, but some patients may complain of intermittent stiffness in the lumbosacral region.
- Range of motion:
  - Radiography may reveal a lumbar spur.

Differential Diagnosis

Prolapse of the lumbar intervertebral disc as well as lumbar vertebral fractures and abnormal growths such as tumors. The radiographic imaging is of obvious importance for making the differential diagnosis.

Treatment

Treatment principle:

- Tonify kidney and liver, move blood, drain damp, and stop pain.

Herbal Treatments

Treatment principle:

- Tonifiy liver and kidney, strengthen bones and tendons.

Herbal formula:

- Bu Shen Zhuang Jin Tang 補腎壯筋湯.

Formula ingredients:

Sheng Di Huang 生地黃 15 g, Shan Zhu Yu 山茱萸 15 g, Qing Pi 青皮 6 g, Bai Shao 白芍 10 g, Xu Duan 續斷 10 g, Du Zhong 杜仲 10 g, Dang Gui 當歸 10 g, Fu Ling 茯苓 10 g, Wu Jia Pi 五加皮 10 g, Niu Xi 牛膝 6 g.

Ingredient explanation:

- Sheng Di Huang 生地 cools blood and tonifies kidney Yin.
- Shan Zhu Yu 山茱萸 tonifies liver and kidney Yin and strengthens kidney Yang.
- Qing Pi 青皮 soothes liver Qi and breaks up stagnant Qi, dries dampness, and transforms phlegm.
- Bai Shao 白芍 nourishes blood, preserves Yin, calms liver Yang, softens tendons, and alleviates spasm and pain.
- Xu Duan 續斷 nourishes liver and kidney, strengthens bones and tendons, alleviates pain.
- Du Zhong 杜仲 tonifies the kidneys and liver, strengthens bones and tendons.
- Dang Gui 當歸 nourishes blood, reduces edema, and stops pain.
- Fu Ling 茯苓 drains damp, tonifies spleen.
- Wu Jia Pi 五加皮 expels wind and damp, nourishes and warms liver and kidney, strengthens bones and tendons.
- Niu Xi 牛膝 expels wind dampness, tonifies liver and kidneys, directs herbs downward.

Modifications:

- For strengthening tendon and bone, add Gui Ban 龜膠 and Gou Qi Zi 枸杞.
- For tonifying Qi, add Dang Shen 黨參, Huang Qi 黄芪, and Bai Zhu 白术.

Acupuncture Treatments

Acupuncture point formula:

- Local acupuncture points: UB17 膈俞 Geshu, UB18 肝俞 Ganshu, UB20 脾俞 Pishu, UB23 腎俞 Shenshu, UB25 大腸俞 Dachangshu.
- Distal acupuncture points: LI4 Hegu, SI3 後溪Houxi, UB62 申脈 Shenmai, UB40 委中 Weizhong, SP8 地機 Diji, SP9 陰陵泉 Yinlingquan, KD3 太溪 Taixi.

Cupping Treatments

UB23 腎俞 Shenshu, UB25 大腸俞 Dachangshu.

Moxibustion Treatments

Warming needling techniques on UB23 腎俞 Shenshu, UB25 大腸俞 Dachangshu.

Bleeding Treatments

Point selected is UB40 委中 Weizhong.

Procedures:

1) Prior to the procedure, the tools should be prepared and ready to be used in the working area.
2) Set up a clean field with paper towels.
3) Place these tools in the clean field, including dry sterile cotton balls, medical alcohol wipes or cotton swabs soaked in alcohol, sterile gloves, a biohazard sharps container, a biohazard trash container, goggles and facemask, and paper towels. Also, a medical lancet or a traditional three-edge needle or the seven-star needle is a bleeding tool; a medical lancet or a traditional three-edge needle is selected for bleeding UB40 委中 Weizhong in this case.
4) The patient lies prone with the legs extended, UB40 委中 Weizhong facing upward.
5) The room should be well lit.
6) The therapist washes their hands thoroughly and wipes the hands with a single-use napkin. Put on sterile gloves.
7) Wipe the acupuncture point UB40 委中 Weizhong with a medical alcohol wipe or a cotton swab soaked in alcohol. Move from the center of the acupuncture point to the periphery
8) Place the medical lancet or the traditional three-edge needle at an angle of 90 degrees to the skin.
9) In one quick move, prick the needle into the acupuncture point about 0.1 cun deep, eliciting a few drops of blood.
10) If more drops of blood are needed, absorb the blood with a sterile cotton ball.
11) Lastly, the point is pressed with a sterile cotton ball until the bleeding ceases.
12) Dispose of the tools in the proper containers.
13) Wash the hands thoroughly.

Electro-Acupuncture Treatments

UB23 腎俞 Shenshu ~UB25 大腸俞 Dachangshu 2/100 Hz for 20 minutes daily.

Tui Na Treatments

Tui Na techniques:

- Kneading, pushing, rotating

Tui Na procedures:

1) The patient is prone; the therapist stands at the left-hand side.
2) Applies thumb pads, kneading UB23 腎俞 Shenshu and UB25 大腸俞 Dachangshu to tonify Qi and warm meridians.
3) Applies pushing techniques with palms on the lumbar region until warm.

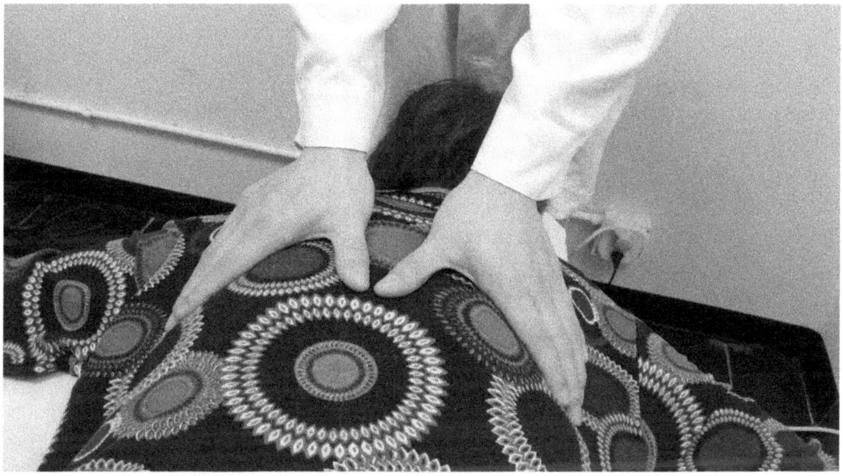

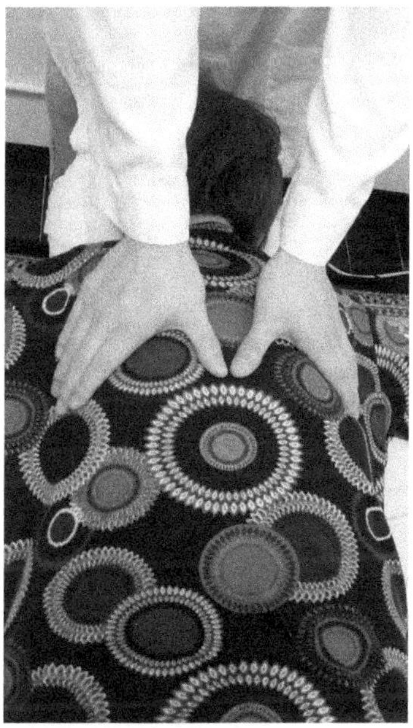

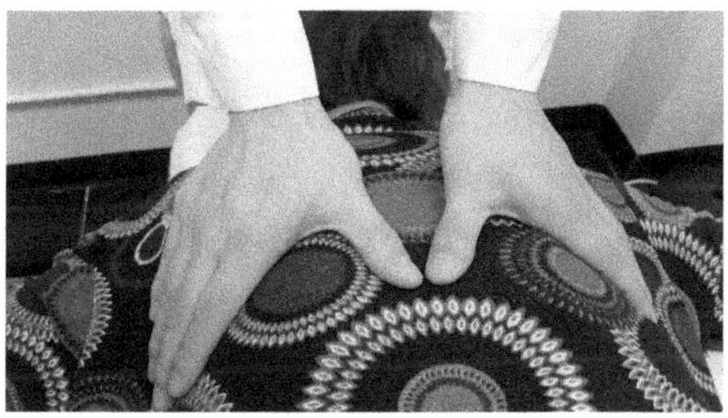

4) Applies rotating techniques on the hip joint.
   a) The patient and the therapist remain in the same position.
   b) The therapist places one hand on the lumbar region and places the other hand under the quadriceps tendon on the contralateral leg.
   c) Lifts up the contralateral leg, at the same time pressing the affected lumbar region with the hand.
   d) Rotates the contralateral leg in circular motions 5–10 times.
   e) Repeats the same procedures with the ipsilateral leg 5–10 times.

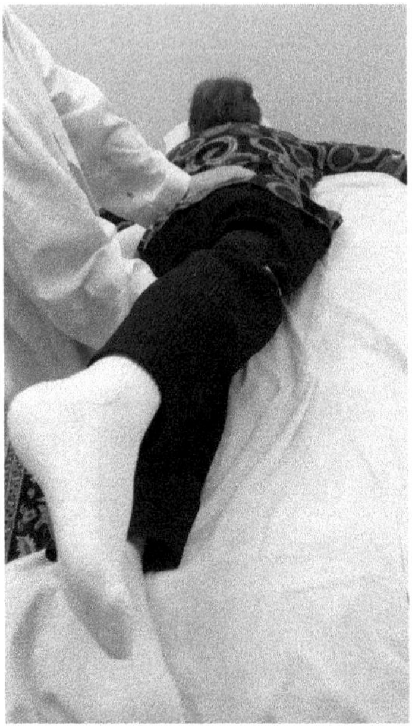
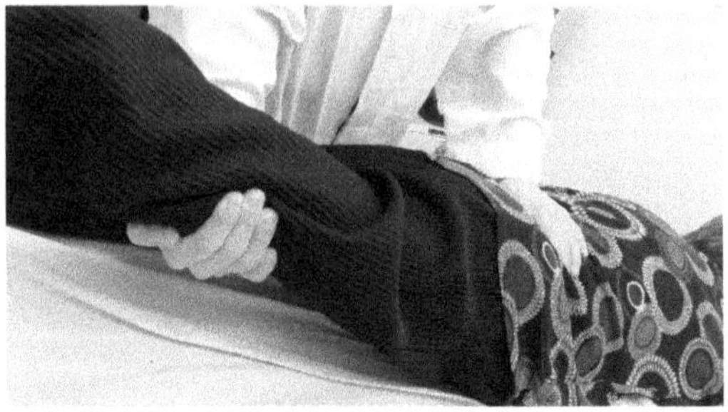

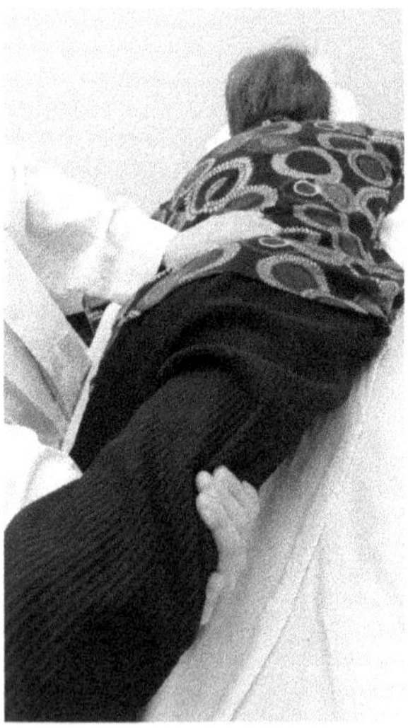
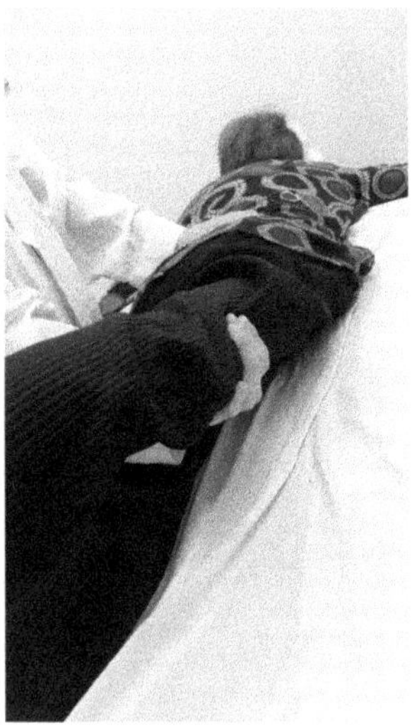
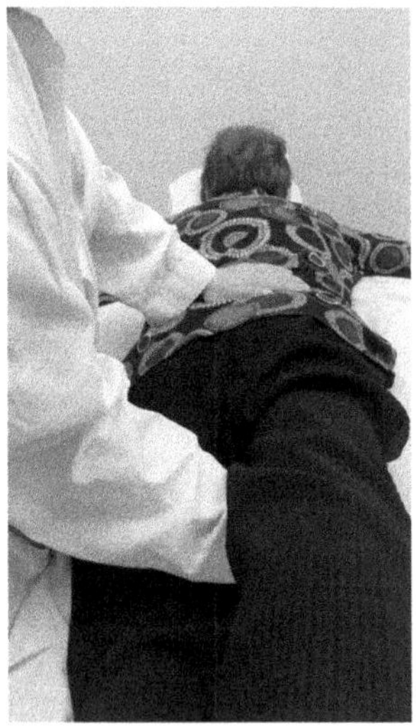

Patient Advisory

- Keep the lumbar region warm.
- Avoid physical indulgence.
- Keep correct postures in sitting and standing; avoid favoring one side.
- Start appropriate exercises to strengthen the lower back but avoid overexertion:

1) Lie prone on the floor with legs straight and slightly apart.
2) Place both hands on the floor in front of the shoulders.
3) Lift up the upper trunk with the arms slowly as high as possible without feeling low back pain.
4) Be sure that only the upper trunk is lifted up, and the hips and legs are still on the floor.
5) Hold the position for 5 to 30 seconds depending on the lumbar discomfort. The exercise is practiced breathing naturally and free from pain.
6) Be mindful that the lumbar muscles should remain fully relaxed.
7) To return to the beginning, slowly flex the elbows to lower the upper trunk until the chest reaches the floor.
8) Repeat the procedures three to five times.
9) The pace of the movements should be slow; the lumbar muscles work harder when moving slowly.

- High protein foods that are rich in vitamins and minerals are recommended.
- Avoid food that is deep fried, raw, or cold.

LUMBAR SPINAL STENOSIS 腰椎椎管狭窄症

Introduction

Lumbar spinal stenosis is the condition of a narrowing passageway within the spine whereby the nerve roots leaving the spinal column are compressed. It is another common cause of sciatica. Some people with spinal stenosis may not have symptoms.

Anatomy

The spinal canal is the cavity that contains the spinal cord within the vertebral column, and many columns form the vertebrae through which the spinal cord passes. The canal is enclosed within the intervertebral foramen of the vertebrae. In the intervertebral spaces, the canal is protected by the ligamentum flavum posteriorly and the posterior longitudinal ligament anteriorly. Lumbar spinal stenosis is a narrowing of the lumbar spinal canal50 resulting in symptoms.

Lumbar stenosis is most commonly a result of age-related osteoarthritis; it is degenerative arthritis and a degenerative disc disease. It typically affects people over the age of 50.

Western Medicine Etiology and Pathology

Each of the following factors can reduce the normal space available for the nerves in the spinal canal and can make the spinal nerves more vulnerable to irritation, inflammation, and resultant symptoms of lumbar spinal stenosis:

- Bulging or herniated discs.
- Age-related enlargement of ligamentous structures.

- Formation of bone spurs (osteophytes).
- Spondylolisthesis, which is the condition of slippage of one vertebra on another due to degeneration of the lumbar discs.
- Spondylosis, which is the condition of loss of the normal height of the discs between the vertebrae of the spine.
- Tumors growing in the spine or tumors that originated in another part of the body and spread to the spine.51

Chinese Medicine Etiology and Pathology

According to Chinese medicine, lumbar spinal stenosis is due to Kidney deficiency, indulgence consuming more Kidney essence, and multiple chronic lumbar injuries. Eventually, the lumbar meridians are invaded by Wind, Damp, and Cold, causing Qi and Blood blockages that result in pain.

Manifestations

The symptoms may include:

- Low back pain.
- Radiating pain, tingling, and weakness on the buttocks and legs, similar to the pattern of sciatica.
- Cramping in the calves when walking; patients often have to take short rests in order to walk any distance.
- In really severe cases, patients lose normal bowel or bladder function.
- Pain may improve on flexion; patients prefer pushing shopping carts or riding a bicycle.

Manifestations in Chinese medicine:

1) Kidney essence deficiency:
    - Symptoms: sore low back pain, weak knees and legs, aggravated by overexertion, and improved by bedrest. Loss of teeth, forgetfulness.
    - Tongue is pale with thin coat; pulse is deep, thread, and short.
2) Wind cold damp blockages:
    - Symptoms: heaviness and sensation of distention in the lumbar region. The sensation is intermittent, aggravated by rainy days and cold exposure and improved by warmth. Tongue is pale and wet with white coat; pulse is deep, tight, and slow.
3) Qi and blood stagnation:
    - Symptoms: low back pain that is aggravated by long periods of sitting but also worse on walking or standing. There is referred pain down the legs. Pale face, fatigue. Tongue is purple with thin coat; pulse is wiry and choppy.

Physical Examination

- Begin by obtaining the patient's medical history:
  - General health condition; it typically affects people over the age of 50.
  - A history of injury is not common, but fracture of the spine or dislocation from trauma may be a factor.[52]
- Inspection:
  - There are no typical signs of acute injury such as bruising and swelling, but the skin on the lumbar region should be inspected. Lumbar spinal stenosis caused by occult spinal dysraphisms or occult spina bifida may often have external signs, including patches of hair, nevi, hemangiomas, or dimples on the lower back in the midline (Alvarez and Russel 1998).
- Palpation:
  - It is not positive.
- Range of motion:
  - Pain may improve with flexion and be aggravated by hyperextension.
  - The straight leg raising test is usually negative in patients with lumbar stenosis.

Differential Diagnosis

- Prolapse of the lumbar intervertebral disc has very similar symptoms of low back pain and referring pain onto the dermatomal distribution on the legs. The straight leg raise test and Linder's Sign are typically positive in prolapse of the lumbar intervertebral disc.
- Cauda equina syndromes usually occur as a result of compression of the nerve roots in the lumbosacral spine distal to the conus medullaris, such as tumors of the lumbosacral spine. The symptoms associated with a lumbosacral spinal tumor typically include urinary incontinence. Pain worsens with recumbency and is relieved by walking; pain is usually more intense at night and may awaken the patient from sleep (Alvarez and Russell 1998). Most patients have severe pain in the lower back and buttocks, which may travel down the posterior thigh, passing the knee, to the calf and foot, as well as numbness and tingling in the "saddle area" (the rectal and genital areas and the inner thighs).

Treatment

Treatment principle:

1) Kidney essence deficiency: Tonifies kidney essence.
2) Wind cold damp blockages: Expels wind, drains damp, warms and opens meridians,
3) Qi and blood stagnation: Nourishes Qi and blood, moves blood, disperses bruising.

Herbal Treatments

Kidney Essence Deficiency

Herbal formula:

- Zuo Gui Wan 左歸丸.

Formula ingredients:

Shu Di Huang 熟地黃 24 g, Shan Yao 山藥 12 g, Gou Qi Zi 枸杞 12 g, Shan Zhu Yu 山茱萸 12 g, Chuan Niu Xi 牛膝 9 g, Tu Si Zi 菟絲子 12 g, Lu Jiao Jiao 鹿角膠 12 g, Gui Ban Jiao 龜板膠 12 g.

Ingredient explanation:

- Shu Di Huang 熟地黃 nourishes blood; nourishes liver and kidney Yin.
- Shan Yao 山藥 tonifies kidney Yin and the spleen Qi.
- Gou Qi Zi 枸杞子 nourishes blood, tonifies liver and kidney Yin.
- Shan Zhu Yu 山茱萸 nourishes liver and kidney, strengthens kidney Yang.
- Chuan Niu Xi 牛膝 moves blood, expels wind dampness, tonifies liver and kidneys, directs herbs downward.
- Tu Si Zi 菟絲子 tonifies kidney Yang, tonifies kidney and liver Yin.
- Lu Jiao Jiao 鹿角膠 nourishes and tonifies essence, Qi, and blood.
- Gui Ban Jiao 龜板膠 nourishes kidney Yin and Yang, strengthens bones, and nourishes blood.

Formula actions:

- Nourishes Yin.
- Tonifies the kidneys.
- Supplements Jing.
- Benefits marrow.

Contraindications:

- Contraindicated for those with signs of excess heat.

Cautions:

- This formula is not suited for long-term use.

Wind Cold Damp Blockages

Herbal formula:

- Du Huo Ji Sheng Tang 獨活寄生湯.

Formula ingredients:

Du Huo 獨活 9 g, Sang Ji Sheng 桑寄生 6 g, Du Zhong 杜仲 6 g, Niu Xi 牛膝 6 g, Xi Xing 细辛 3 g, Qin Jiao 秦艽 6 g, Fu Ling 茯苓 6 g, Rou Gui 肉桂

6 g, Fang Feng 防風 6 g, Chuan Xiong 川芎6 g, Dang Shen 党参 6 g, Zhi Gan Cao 甘草 3 g, Dang Gui 當歸 6 g, Bai Shao 白芍6 g, Shu Di Huang 熟地黄 6 g.

Ingredient explanation:

- Du Huo 獨活 expels wind, damp, and cold in lower Jiao and joints, removes chronic Bi stagnation in lower limb.
- Xi Xing 细辛 enters Shao Yin meridian to expel wind, damp, and cold.
- Qin Jiao 秦艽 expels wind and damp, opens meridians, benefits joints.
- Rou Gui 肉桂 warms meridians, expels cold, opens meridians to promote blood circulation
- Fang Feng 防風 expels wind and drains damp.
- Sang Ji Sheng 桑寄生, Du Zhong 杜仲, and Niu Xi 牛膝 are for deficiency of Qi and blood in liver and kidney, strengthen bone and tendons; Sang Ji Sheng 桑寄生 is able to expel wind, drains damp; Niu Xi 牛膝 is able to move blood, opens meridians to benefit joints and tendons.
- Dang Gui 當歸, Chuan Xiong 川芎, Shu Di Huang 熟地黄, and Bai Shao 白芍 are to nourish and harmonize blood.
- Dang Shen 党参, Fu Ling 茯苓, and Gan Cao 甘草 are for tonifying spleen Qi.

Formula actions:

1. Dispels wind, cold, and damp.
2. Relieves painful obstruction.
3. Supplements Qi and blood.
4. Tonifies the liver and kidney.

Modifications:

- For severe pain, add Zhi Chuan Wu 制川烏, Zhi Cao Wu 制草烏, Bai Hua She 白花蛇, Xu Duan 川斷, Gou Ji 狗脊, Tu Si Zi 菟絲子, Hong Hua 紅花, Di Long 地龍.
- For more cold symptoms, add Zhi Fu Zi 附子, Gan Jiang 乾薑.
- For more damp symptoms, omit Shu Di Huang 熟地黃 and add Fang Ji 防己, Yi Yi Ren 薏苡仁, and Cang Zhu 蒼朮.

Western uses:

- Chronic rheumatic arthritis
- Osteoarthritis
- Rheumatic sciatica
- Chronic low back pain

Contraindications:

- Damp heat excessive Bi syndrome.

Cautions:

- Avoid food that is raw, cold, or oily.
- Keep warm during windy and cold weather.

Qi and Blood Stagnation

Herbal formula:

- Bu Yang Huan Wu Tang 補陽還五湯.

Formula ingredients:

Huang Qi 黃耆 120 g, Dang Gui Wei 當歸尾 6 g, Chi Shao 赤芍 5 g, Di Long 地龍 3 g, Chuan Xiong 川芎 3 g, Tao Ren 桃仁 3 g, Hong Hua 紅花 3 g.

Ingredient explanation:

- Huang Qi 黃耆 tonifies Qi, strengthens the spleen, raises the Yang Qi of the spleen and stomach, tonifies Wei Qi, stabilizes the exterior, and tonifies the blood.
- Dang Gui Wei 當歸尾 invigorates, harmonizes, and nourishes the Blood and alleviates pain.
- Chi Shao 赤芍 clears heat, cools and invigorates the blood, disperses blood stasis, and eliminates pain.
- Di Long 地龍 unblocks and moves the channels, clears heat, releases spasms and convulsions.
- Chuan Xiong 川芎 moves the blood, promotes Qi movement, expels wind, and relieves pain.
- Hong Hua 紅花 moves the blood, dispels blood stasis, and alleviates pain.
- Tao Ren 桃仁 disperses blood stasis, moves blood circulation, descends Qi, and releases spasms.

Modifications:

- For moving blood, add Niu Xi 牛膝 6 g.
- For expelling wind, draining damp, and tonifying liver and kidney, add Sang Ji Sheng 桑寄生 6 g.
- For expelling wind, draining damp, and tonifying liver and kidney, add Wu Jia Pi 五加皮6 g.

Acupuncture Treatments

Acupuncture point formulas:

1) Kidney essence deficiency:

Rx: DU20 百會 Baihui, UB23 腎俞 Shenshu, UB25 大腸俞 Dachangshu, KD3 太溪 Taixi, KD6 照海 Zhaohai, KD7 復溜 Fuliu, SP6 三陰交 Sanyinjiao, KD1 涌泉 Yong Quan, SP10 血海 Xuehai.

2) Wind cold damp blockages:

Rx: DU20 百會 Baihui, GB20 風池 Fengchi, DU16 風府 Fengfu, UB20 脾俞 Pishu, UB23 腎俞 Shenshu, UB25 大腸俞 Dachangshu, SP9 陰陵泉 Yinlingquan, ST40 豐隆F englong, KD7 復溜 Fuliu, SP3 太白 Taibai; add moxa box to UB23 腎俞 Shenshu, UB25 大腸俞 Dachangshu; add warm needle for 3–5 cones to SP9 陰陵泉 Yinlingquan, KD7 復溜 Fuliu.

3) Qi and blood stagnation:

Rx: SI3 後溪 Houxi and UB62 申脈 Shenmai, UB17 膈俞 Geshu, UB23 腎俞 Shenshu, UB25 大腸俞 Dachangshu, UB39 委陽 Weiyang, UB40 委中 Weizhong, LI4 合谷 He Gu, SP10 血海 Xuehai, LIV8 曲泉 Ququan; apply bleeding techniques with seven-star needle to UB25 大腸俞 Dachangshu and add cupping onto UB25 大腸俞 Dachangshu immediately after removing the cups.

Cupping Treatment

UB23 腎俞S henshu, UB25 大腸俞 Dachangshu.

Moxibustion Treatment

Warming needling techniques on UB23 腎俞 Shenshu, UB25 大腸俞 Dachangshu.

Bleeding Treatments

Point selected is UB40 委中 Weizhong.

Procedures:

1) Prior to the procedure, tools should be prepared and ready to be used in the working area.
2) Set up a clean field with paper towels.
3) Place these tools in the clean field, including dry sterile cotton balls, medical alcohol wipes or cotton swabs soaked in alcohol, sterile gloves, a biohazard sharps container, a biohazard trash container, goggles and facemask, and paper towels. Also, a medical lancet or a traditional three-edge needle or the seven-star needle is a bleeding tool; a medical lancet or a traditional three-edge needle is selected for bleeding UB40 委中Weizhong in this case.
4) The patient lies prone with the legs extended, UB40 委中 Weizhong facing upward.
5) The room should be well lit.
6) The therapist washes their hands thoroughly and wipes the hands with a single-use napkin. Put on sterile gloves.
7) Wipe the acupuncture point UB40 委中 Weizhong with a medical alcohol wipe or a cotton swab soaked in alcohol. Move from the center of the acupuncture point to the periphery.

8) Place the medical lancet or the traditional three-edge needle at an angle of 90 degrees to the skin.
9) In one quick move, prick the needle into the acupuncture point about 0.1 cun deep, eliciting a few drops of blood.
10) If more drops of blood are needed, absorb the blood with a sterile cotton ball.
11) Lastly, the point is pressed with a sterile cotton ball until the bleeding ceases.
12) Dispose of the tools in the proper containers.
13) Wash the hands thoroughly.

Electro-Acupuncture Treatment

UB23 腎俞 Shenshu ~UB25 大腸俞 Dachangshu 2/100 Hz for 20 minutes daily.

Tui Na Treatments

Tui Na techniques:

- Rolling, kneading, grasping, pressing

Tui Na procedures:

1) The patient is in the prone position; the therapist stands beside the lumbar region.
2) Applies rolling techniques with a hand on the affected lumbar region for a minute or two.

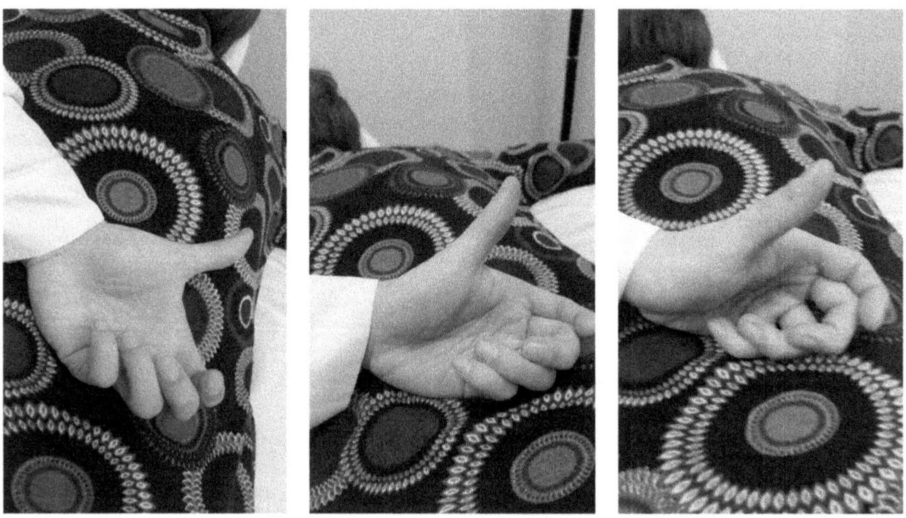

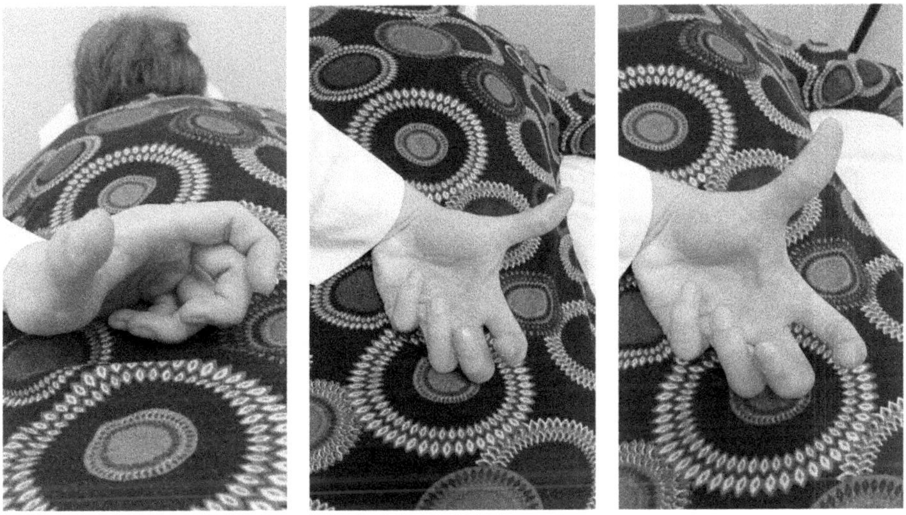

3) Repeat step 2 with a forearm for 2 minutes.
4) Then, change the technique to kneading with the heel of a hand on the lumbar region, and gradually move the kneading downward on the leg along the UB meridian to end on UB57 承山 Chengshan on the calf; repeat this step for about 10 minutes.

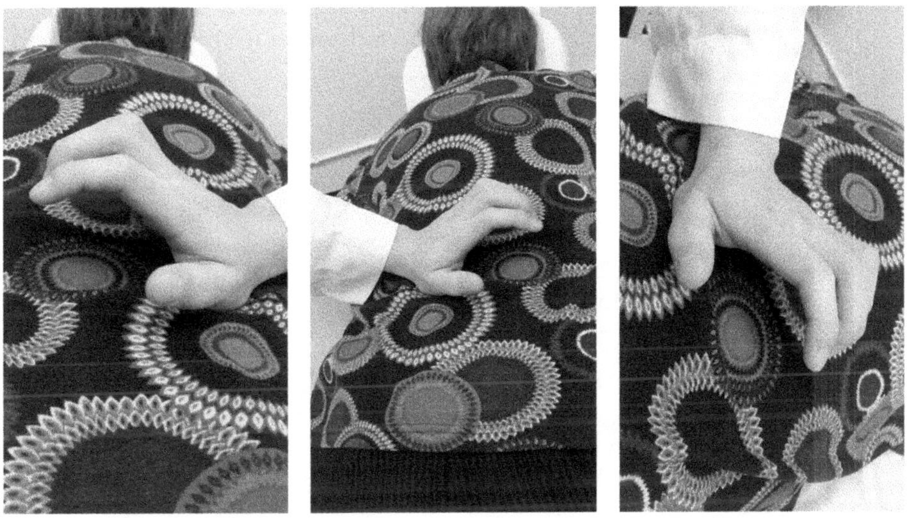

5) Applies grasping techniques on both sides of the lumbar region for about 5 minutes or until warm.

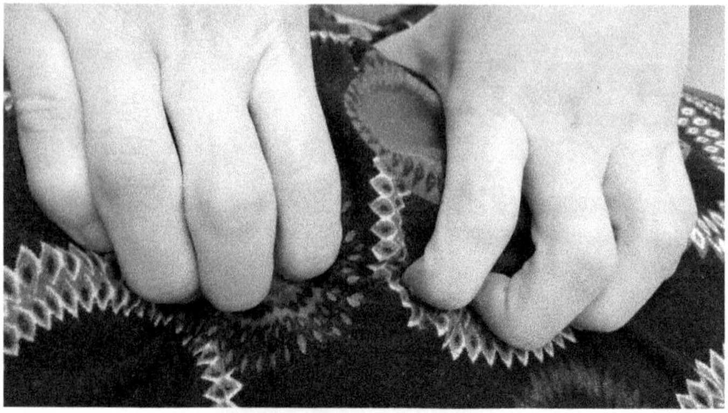

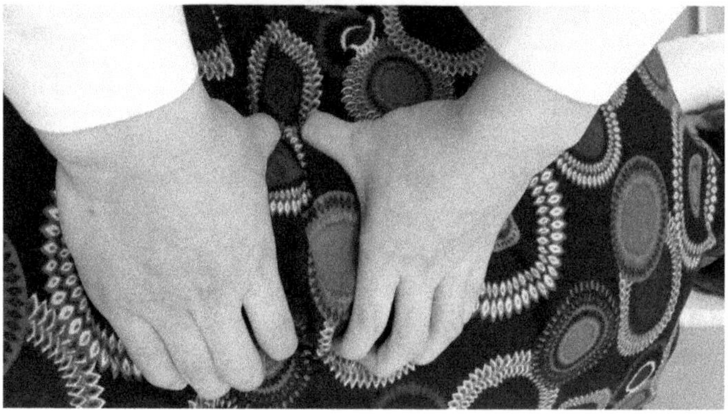

6) Applies pressing and kneading with thumb pads on DU2 腰俞 Yaoshu, UB23 腎俞 Shenshu, and UB25 大腸俞 Dachangshu for 1 minute each.

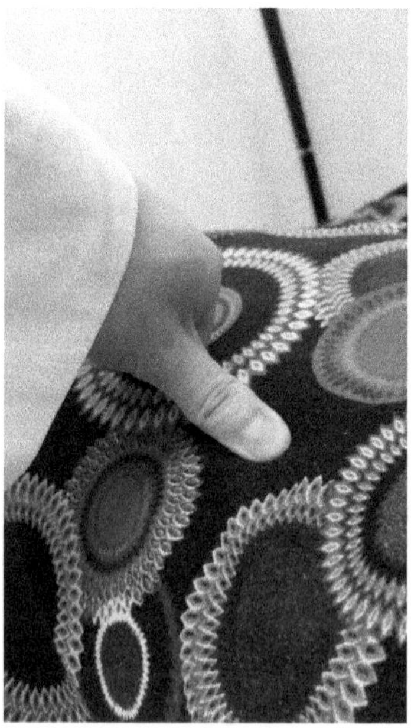

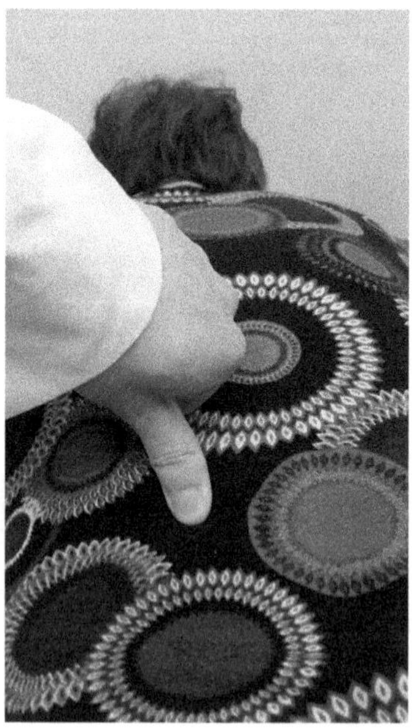

7) Repeat step 4 for 3–5 minutes.

Patient Advisory

1) The patient is recommended to avoid catching cold and wind.
2) Avoid holding a position for too long a period.
3) Lift heavy objects from the legs, not from the back.
4) Practice good posture.

Good sitting posture:

a) Be sure to straighten the back and avoid leaning forward. Place a cushion behind the lower back to support the normal lumbar curvature.
b) Flex the hips and knees at about 90 degrees.
c) Keep the legs shoulder width apart.
d) Keep the feet flat on the floor. If the feet cannot reach the floor, place the feet on a footrest.
e) Do not sit for longer than an hour. It is recommended to place a cup of tea within reach, with a few chrysanthemum flowers and goji berries with hot water, and to finish each cup within an hour. Goji berries help with eye tiredness, chrysanthemum flowers help to clear heat in the eye, and tea helps to clear the heat inside the body. With 250 cc of water, this is enough to make someone get up and walk to the bathroom. Someone sitting for 8 hours will finish drinking eight cups of water.

Good standing posture:

a) Be sure to stand with head and back erect. Keep the head upright on the midline between the shoulders.
b) Stand with legs slightly apart and both knees slightly flexed.
c) Keep the weight on the front third of the feet.
d) Be mindful not to lean your body weight to either side.

Good sleeping posture:

a) Sleeping on a relatively firm bed is good for sciatic pain recovery.
b) Supine is the appropriate sleeping posture; both sides of the body receive equal pressure.
c) Place a cushion or pillow below the knees to alleviate lumbar pressure.
d) In the acute stage, lift the lower legs higher by placing more pillows below the tibia.
e) The normal lumbar curvature also can be supported by placing a thin cushion or a rolled towel under the lower back.

Exercise the lumbar region regularly:

a) Lie prone on the floor with legs straight and slightly apart.
b) Place both hands on the floor in front of the shoulders.
c) Lift up the upper trunk with the arms slowly as high as possible without feeling low back pain.
d) Be sure that only the upper trunk is lifted up, and the hips and legs are still on the floor.

e) Hold the position for 5 to 30 seconds depending on the lumbar discomfort. The exercise is practiced breathing naturally and free from pain.
f) Be mindful that the lumbar muscles should remain fully relaxed.
g) To return to the beginning, slowly flex the elbows to lower the upper trunk until the chest reaches the floor
h) Repeat the procedure three to five times.
i) The pace of the movements should be slow; the lumbar muscles work harder when moving slowly.

LOW BACK PAIN IN CHINESE MEDICINE

INTRODUCTION

The treatment of low back pain in Chinese medicine is unique, effective, and safe. Chinese medicine looks at low back pain in an entirely different way than Western medicine. The model of treatment in Chinese medicine also is totally different in terms of differentiating the type of low back pain. The concepts of the lumbar region and the etiology, pathology, and diagnosis are expressed in the language of Chinese medicine; even the symptoms are described differently than in Western medicine. Because the treatment of lumbago in Chinese medicine is effective and unique, this chapter will only discuss the therapeutic modalities in traditional Chinese medicine.

Low back pain is very common; about 8 out of 10 adults experience low back pain during their lifetimes.[53] It has caused 2.63 million annual emergency department visits in the United States.

Depending on the type of back pain, or if there is a reason to suspect that a specific condition is causing the back pain, Western medicine providers may or may not go through a series of tests such as X-ray, MRI, CT scans, blood tests, bone scans, and nerve studies. They may recommend medication, from over-the-counter pain relievers such as nonsteroidal anti-inflammatory drugs (NSAIDs) to prescribed muscle relaxants and narcotics that contain opioids, such as oxycodone or hydrocodone.[54]

CHINESE MEDICINE ETIOLOGY AND PATHOLOGY

The concept of pain is described as follows in Chinese medicine:

> If there is free flow, there is no pain;
> If there is no free flow, there is pain.[55]

This can be explained as follows: if there is a stagnation in the meridians from Qi and blood, particularly through the lumbar region, there is pain. When the blockages are removed, there is free flow of Qi and blood, and the pain subsides.

1) Invasion of external pathogens 外邪侵襲

   An external invasion of wind, cold, dampness, or heat may invade the lower back region, causing blockages in the meridians, which impede the flow of Qi and blood, causing pain.

The theories behind this are:

1) The immune system, called Defensive Qi in Chinese medicine, is weak, and the person is susceptible to invasion by external pathogens.
2) The external pathogens are much stronger than the person's immune system. For example, when a healthy and strong young person gets stuck in an avalanches and buried in snow, the immune system starts to become weak over time, because the coldness impedes the person's Yang Qi.

2) Qi stagnation and blood stasis 氣滞血瘀

The Qi and blood can be stagnant due to physical or mental trauma.

1) *Mental trauma*: This is an internal cause of pain. The liver is associated with emotional stress and wind, and the liver governs the tendons. Emotional stress changes the quality of tendons, making them vulnerable.

   The liver is associated with wind, and wind is associated with movements; its attack is fast and is often recognizable by signs of pain.
2) *Physical trauma*: This is an external cause of pain. Bodily injury causes blood stasis in the meridians, and then, Qi flow is impeded. As a result of Qi and blood stagnation in meridians, pain occurs, according to "If there is no free flow, there is pain."

3) Kidney deficiency 肾虧體虚

According to traditional Chinese medicine, low back–related diseases are connected with the kidneys. The lumbar region is the "mansion of the kidneys," meaning that the lower back is the most closely related to lumbar medical conditions but not limited to the health of the Chinese kidney system (the Chinese kidney system is associated with many systems in Western medicine, such as the endocrine system, the respiratory system, the circulatory system, and the immune system). Accordingly, low back pain is closely associated with deficiency of the Chinese kidney system.[56]

The lumbar region needs to be nourished by kidney essence. This can be adversely affected by congenital deficiency, prolonged illness, indulgence in sexual activity, excessive bleeding during menstruation, excessive lumbar strain, multiple pregnancy, malnutrition with excessive mental exertion, or an irregular diet that does not provide enough supply to meet the demand of kidney.

When the essence is undernourished, "the mansion of the kidneys" is vulnerable, and low back pain occurs.

Manifestations

1) Cold damp wind invasion 寒濕腰痛

Low back pain occurs after exposure to cold and damp environment or weather. It is aggravated on rainy days with a heavy sensation and stillness in the lumbar region. The pain is less when using hot compression. There is limitation of extension and flexion of the lower back and a cold sensation in the lower back. Pain may also radiate to the buttock and lower limbs. The patient is usually averse to cold or has a slight fever. Tongue is pale with white and sticky coat. Pulse is deep and weak or deep and slow.

2) Wind damp heat invasion 濕热腰痛

Low back pain is acute in onset with burning, a hot sensation on touching, and heaviness. Low back movement is limited. The pain is aggravated by exposure to warmth or on warmer days and with warm compression; it is less painful with cold compression. There is local redness, swelling, and dislike of pressure, accompanied by fever, thirst but disinclination to drink, dizziness, light-headedness, yellow urine, and loose stools. Tongue is red with yellow sticky coat. Pulse is rapid, slippery or rapid, wiry.[57]

3) Blood stasis 瘀血腰痛

There is a history of injury. The pain is stabbing, and the location is fixed. The patient cannot tolerate pressure. The pain is greater at night and less during the day with limitation of movement. Tongue is dim with bruise marks. Pulse is deep and choppy.

4) Kidney essence deficiency 肾虚腰痛

Low back pain is sore, accompanied by lassitude and weakness of the loins and knees, poor hearing, dizziness, poor memory and concentration, loose teeth, or tiredness.

The pain is aggravated by fatigue and alleviated by bed rest and pressure.

a) *Yang deficiency*: Cold hands, feet, lower abdomen, and lower back accompanied by lassitude. Tongue is pale. Pulse is deep and thin.
b) *Yin deficiency*: Five-center heat, dry mouth, insomnia, forgetfulness, and tinnitus. Tongue is red. Pulse is wiry, thin, and deep.

5) Liver Qi stagnation 氣鬱腰痛

Low back pain is associated with emotional stress, depression, anxiety, and anger. Pain is distended at the genital region, wandering pain that comes and goes, hypochondriac pain, sighing. Tongue has distended sublingual veins. Pulse is wiry.

6) Stone Lin low back pain 石淋腰痛

Low back pain is unilateral or bilateral, the pain is cramping and colicky in the lower back and perineum. Difficulty in urination with sand and stones, sudden interruption of micturition; dark yellow urine or possible hematuria in severe cases. Tongue is red with thin, yellow coating. Pulse is wiry and rapid.

Treatment

Herbal Treatments

Cold Damp Wind Invasion 寒濕腰痛

Treatment principle:

- Disperses cold, drains damp, warms meridians.

Herbal formula:

- Shen Shi Tang 滲濕湯.

Formula ingredients:

Bai Zhu 白術9 g, Gan Jiang 乾薑 3 g, Bai Shao 白芍3 g, Zhi Fu Zi 附子 3 g, Fu Ling 白茯苓 3 g, Ren Shen 人蔘 3 g, Gui Zhi 桂枝 1.5 g, Zhi Gan Cao 炙甘草 1.5 g.

Ingredient explanation:

- Bai Zhu 白術 tonifies the spleen Qi, dries dampness.
- Gan Jiang 乾薑 warms middle Jiao, expels interior cold, dispels wind and dampness.
- Bai Shao 白芍 softens the liver and relieves pain.
- Zhi Fu Zi 附子 disperses cold and dampness, warms the meridians, and stops pain.
- 白茯苓Fu Ling 茯苓 promotes urination, drains dampness, and strengthens the spleen.
- Ren Shen 人蔘 tonifies Qi and Yang.
- Gui Zhi 桂枝 warms the meridians and relieves pain.
- Zhi Gan Cao 炙甘草 tonifies the spleen, harmonizes the other herbs, and guides the herbs to all 12 meridians.

Formula actions:

- Drains damp, releases joints.

Wind Damp Heat Invasion 濕热腰痛

Treatment principle:

- Clears heat, drains damp, relaxes tendons.

Herbal formula:

- Er Miao San 二妙散.

Formula ingredients:

- Huang Bai 黄柏15 g, Cang Zhu 蒼術15 g.

Ingredient explanation:

- Huang Bai 黃柏 15 g drains damp and heat, especially from the lower Jiao.
- Cang Zhu 蒼術 15 g strongly dries dampness, tonifies the spleen, and clears dampness from the lower Jiao.

Formula actions:

- Eliminates heat and dampness.

Modifications:

- For atrophy disorder, add Xi Xian Cao 豨簽草, Mu Gua 木瓜, Bi Xian 萆薢.
- For more sore pain in the lower limbs, add Yi Yi Ren 薏苡仁, Mu Gua 木瓜, Bing Lang 檳榔.
- For severe pain, add 牛膝Niu Xi, 木瓜Mu Gua, 五加皮Wu Jia Pi.

Blood Stasis 瘀血腰痛

Treatment principle:

- Move blood, regulate Qi, and stop pain.

Herbal formula:

- Shen Tong Zhu Yu Tang 身痛逐瘀湯.

Formula ingredients:

Dang Gui 當歸 9 g, Chuan Xiong 川芎 9 g, Tao Ren 桃仁 9 g, Hong Hua 紅花 9 g, Mo Yao 沒藥 6 g, Wu Ling Zhi 五靈脂 6 g, Qin Jiao 秦艽 9 g, Qiang Huo 羌活 9 g, Xiang Fu 香附 3 g, Chuan Niu Xi 川牛膝 9 g, Di Long 地龍 6 g, Gan Cao 甘草 6 g.

Ingredient explanation:

- Dang Gui 當歸, Chuan Xiong 川芎, Tao Ren 桃仁, and Hong Hua 紅花 move blood, disperse bruising, and stop pain due to blood stasis.
- Qin Jiao 秦艽 and Qiang Huo 羌活 expel wind, cold, and dampness, unblock painful obstruction, and alleviate pain.
- Wu Ling Zhi 五靈脂, Mo Yao 沒藥, and Xiang Fu 香附 move Qi and blood and stop pain due to Qi and blood stagnation.
- Chuan Niu Xi 川牛膝 and Di Long 地龍 unblock and promote movement in the channels and collaterals, stop spasms and convulsions.
- Gan Cao 甘草 harmonizes the herbs.

Formula actions:

- Moves Qi and blood.
- Unblocks painful obstruction in meridians.
- Stops pain.

Contraindications:

- Contraindicated during pregnancy.

Modifications:

- For heat symptoms, add Chai Hu 柴胡, Huang Bai 黃柏.
- For Qi deficiency, add Dang Shen 黨參, Huang Qi 黃芪.
- For pain in lumbar region and leg, add Xu Duan 續斷, Du Zhong 杜仲, Sang Ji Sheng 桑寄生.
- For severe pain, add Quan Xie 全蠍, Wu Gong 蜈蚣.
- For cold symptoms, take out Qin Jiao 秦艽and add Zhi Chuan Wu 制川烏.
- For cold symptoms and severe pain, add Gui Zhi 桂枝and Xi Sin 細辛.
- For wind traveling pain, add Fang Feng 防風.
- For radiating pain to the leg, omit Wu Ling Zhi 五靈脂and add Shen Jin Cao 伸筋草.
- For damp heat joint edema, add Cang Zhu 蒼朮, Huang Bai 黃柏, and Fang Ji 防己.
- For stiffness in joints, add Shen Jin Cao 伸筋草, Quan Xie 全蠍, and Wu Gong 蜈蚣.
- For mild joint edema and more damp Bi pain, add Fang Ji 防己, Yi Yi Ren 薏苡仁, and Cang Zhu 蒼朮.

Kidney Essence Deficiency 肾虚腰痛

Treatment principle:

- Tonify kidney Yang for Yang deficiency; tonify kidney Yin for Yin deficiency.

Herbal formula for kidney Yang deficiency:

- You Gui Wan 右歸丸.

Formula ingredients:

Shu Di Huang 熟地黃240 g, Shan Yao 山藥120 g, Gou Qi Zi 枸杞子120 g, Tu Si Zi 菟絲子120 g, Du Zhong 杜仲120 g, Lu Jiao Jiao 鹿角膠120 g, Shan Zhu Yu 山茱萸90 g, Dang Gui 當歸90 g, Zhi Fu Zi 附子60 g, Rou Gui 肉桂60 g.

Ingredient explanation:

- Zhi Fu Zi 附子 warms Ming Men Fire.
- Rou Gui 肉桂 warms the kidney Yang and Ming Men Fire and assists the generation of Qi and blood.
- Lu Jiao Jiao 鹿角膠 tonifies the kidneys, nourishes and tonifies essence, Qi, and blood.
- Shu Di Huang 熟地黃 nourishes blood; nourishes liver and kidney Yin.

- Shan Zhu Yu 山茱萸 nourishes liver and kidney Yin and strengthens kidney Yang.
- Shan Yao 山藥 tonifies kidneys; benefits the Yin and the spleen.
- Qi Gou Qi Zi 枸杞子 nourishes blood, tonifies the liver and kidney Yin.
- Tu Si Zi 菟絲子 tonifies kidney Yang, tonifies kidney and liver Yin.
- Du Zhong 杜仲 tonifies the kidneys and liver, strengthens the sinews and bones, and lowers blood pressure.
- Dang Gui 當歸 moves and tonifies blood, disperses cold, and stops pain due to blood stasis.

Herbal formula for kidney Yin deficiency:

- Rx. Zuo Gui Wan 左歸丸.

Formula ingredients:

Shu Di Huang 熟地黃24 g, Shan Yao 山藥12 g, Shan Zhu Yu 山茱萸12 g, Gou Qi Zi 枸杞子12 g, Tu Si Zi 菟絲子12 g, Lu Jiao Jiao 鹿角膠12 g, Gui Ban Jiao 龜板12 g, Niu Xi 牛膝9 g.

Ingredient explanation:

- Shu Di Huang 熟地黃 nourishes blood; nourishes liver and kidney Yin.
- Shan Yao 山藥 tonifies kidneys; benefits the Yin and the spleen.
- Gou Qi Zi 枸杞子 nourishes blood, tonifies the liver and kidney Yin.
- Shan Zhu Yu 山茱萸 nourishes liver and kidney Yin and strengthens kidney Yang.
- Niu Xi 牛膝 expels wind dampness, tonifies liver and kidneys, directs herbs downward.
- Tu Si Zi 菟絲子 tonifies kidney Yang, tonifies kidney and liver Yin.
- Lu Jiao Jiao 鹿角膠 tonifies the kidneys, nourishes and tonifies essence, Qi, and blood.
- Gui Ban Jiao 龜板 nourishes kidney Yin and Yang, strengthens bones, and nourishes blood.

Liver Qi Stagnation 氣鬱腰痛

Treatment principle:

- Moves liver Qi and stops pain.

Herbal formula:

- Xiao Yao San 逍遙散.

Formula ingredients:

Chai Hu 柴胡 10 g, Dang Gui 當歸 10 g, Bai Shao 白芍 10 g, Bai Zhu 白朮 10 g, Fu Ling 茯苓 10 g, Zhi Gan Cao 甘草 5 g, Sheng Jiang 生薑 10 g, Bo He 薄荷 5 g.

Ingredient explanations:

- Chai Hu 柴胡 relieves liver Qi stagnation, pacifies the liver, relieves the Shao Yang, and reduces fever.
- Dang Gui 當歸 nourishes the blood, regulates menses, and invigorates and harmonizes the blood.
- Bai Shao 白芍 calms liver Yang, alleviates pain, nourishes blood, and regulates menstruation.
- Bai Zhu 白朮 tonifies the spleen and tonifies Qi.
- Fu Ling 茯苓 strengthens the spleen and harmonizes the Middle Jiao.
- Zhi Gan Cao 甘草 tonifies spleen Qi, stops pain, and harmonizes the other herbs.
- Bo He 薄荷 relieves liver Qi Stagnation, disperses stagnant heat, and enhances Chai Hu's ability to relieve the liver.
- Sheng Jiang 生薑 harmonizes, prevents rebellious Qi, and normalizes the flow of Qi at the center.

Stone Lin Low Back Pain 石淋腰痛
Treatment principle:

- Drains damp, opens blockage.

Herbal Formula

- Shi Lin Tong Pian 石淋通片 6 pills, 3 times a day.

Acupuncture Treatments

Acupuncture point formula:

- *Cold damp wind invasion* 寒濕腰痛
  DU4 命門 Mingmen, UB25 大腸俞 Dachangshu, SP9 陰陵泉 Yinlingquan, UB40 委中 Weizhong.
- *Wind damp heat invasion* 濕热腰痛
  SP9 陰陵泉 Yinlingquan, SP6 三陰交 Sanyinjiao, UB40 委中 Weizhong.
- *Blood stasis* 瘀血腰痛
  DU4 命門 Mingmen, UB40 委中 Weizhong, UB17 膈俞 Geshu, SP10 血海 Xuehai
- *Kidney essence deficiency* 肾虚腰痛
  UB23 腎俞 Shenshu, KD3 太溪 Taixi, DU2 腰俞 Yaoshu, UB40 委中 Weizhong.
  Kidney Yang deficiency: add Ren4 關元 Guanyuan, Ren6 氣海 Qihai.
  Kidney Yin deficiency: add KD6 照海 Zhaohai, GB39 懸鐘 Xuanzhong.
- *Liver Qi stagnation* 氣鬱腰痛
  LI4 合谷 He Gu, LIV3 太沖 Taichong, SJ5 外關 Waiguan, GB34 陽陵泉 Yanglingquan, UB18 肝俞 Ganshu, UB40 委中 Weizhong, UB23 腎俞 Shenshu, GB39 懸鐘 Xuanzhong.

- *Stone Lin low back pain* 石淋腰痛
  - UB39 委陽 Weiyang w/ KD2 然谷 Rangu exerts downward pulling.
  - UB39 委陽 Weiyang is the lower He Sea of Sanjiao and also exerts downward pulling on the kidney.
  - KD2 然谷 Rangu; UB28 膀胱俞 Pangguangshu; UB22 三焦俞 Sanjiaoshu; Ren3 中極 Zhongji; SP9 陰陵泉 Yinlingquan; ST28 水道 Shuidao; UB63 金門 Jinmen; Ren6 氣海 Qihai; Ren5 石門 Shimen.

Cupping Treatments

UB23 腎俞 Shenshu with needle in, UB40 委中 Weizhong with bleeding.

Gua Sha Treatments

Lower back, GB meridian on the leg, and UB meridian on the leg.

Moxibustion Treatments

Warming needling techniques on UB23 腎俞 Shenshu, UB25 大腸俞Dachangshu.

Bleeding Treatments

UB40 委中 Weizhong.

Point selected is UB40 委中 Weizhong.

Procedures:

1) Prior to the procedure, tools should be prepared and ready to be used in the working area.
2) Set up a clean field with paper towels.
3) Place these tools in the clean field, including dry sterile cotton balls, medical alcohol wipes or cotton swabs soaked in alcohol, sterile gloves, a biohazard sharps container, a biohazard trash container, goggles and facemask, and paper towels. Also, a medical lancet or a traditional three-edge needle or the seven-star needle is a bleeding tool; a medical lancet or a traditional three-edge needle is selected for bleeding UB40 委中 Weizhong in this case.
4) The patient lies prone with the legs extended, UB40 委中 Weizhong facing upward.
5) The room should be well lit.
6) The therapist washes their hands thoroughly and wipes the hands with a single-use napkin. Put on sterile gloves.
7) Wipe the acupuncture point UB40 委中 Weizhong with a medical alcohol wipe or a cotton swab soaked in alcohol. Move from the center of the acupuncture point to the periphery.
8) Place the medical lancet or the traditional three-edge needle at an angle of 90 degrees to the skin.
9) In one quick move, prick the needle into the acupuncture point about 0.1 cun deep, eliciting a few drops of blood.

10) If more drops of blood are needed, absorb the blood with a sterile cotton ball.
11) Lastly, the point is pressed with a sterile cotton ball until the bleeding ceases.
12) Dispose of the tools in the proper containers.
13) Wash the hands thoroughly.

Seven-Star Needle Treatment

UB 23 腎俞 Shenshu: Cupping after bleeding.
Point selected is UB 23 腎俞 Shenshu.
Procedures:

- Prior to the procedure, tools should be prepared and ready be used in the working area.
- Set up a clean field with paper towels.
- Place these tools in the clean field, including dry sterile cotton balls, medical alcohol wipes or cotton swabs soaked in alcohol, sterile gloves, a biohazard sharps container, a biohazard trash container, goggles and facemask, and paper towels. Also, a medical lancet or a traditional three-edge needle or the seven-star needle is a bleeding tool; the seven-star needle is selected for bleeding UB 23 腎俞 Shenshu in this case.
- The patient lies prone with the legs extended, UB 23 腎俞 Shenshu facing upward.
- The room should be well lit.
- The therapist washes their hands thoroughly and wipes the hands with a single-use napkin. Put on sterile gloves.
- Wipe the acupuncture point UB 23 腎俞 Shenshu with a medical alcohol wipe or a cotton swab soaked in alcohol. Move from the center of the acupuncture point to the periphery.
- Place the seven-star needle at a perpendicular angle to the skin, and hold the end of the handle.
- Flick the flexible handle to tap the needles onto the acupuncture point lightly and superficially to allow the skin to develop a mild rash. The flexible handle is flicked at a fast speed, so it actually takes only a short time to get the rash result.
- Lastly, put a cup on to release more bleeding.
- Dispose of the tools in the proper containers.
- Wash the hands thoroughly.

Electro-Acupuncture Techniques

UB23 腎俞 Shenshu~UB40 委中 Weizhong 2/100 Hz for 20 minutes daily.

Tui Na Treatments

Tui Na techniques:

- Kneading, pressing, shaking, stretching, rolling, rubbing

Tui Na procedures:

- The patient lies in a prone position. The Tui Na therapist stands beside the patient.
- Apply kneading and pressing on the lower back with the heel of the hands to promote local Blood circulation and relax the spasming muscles.

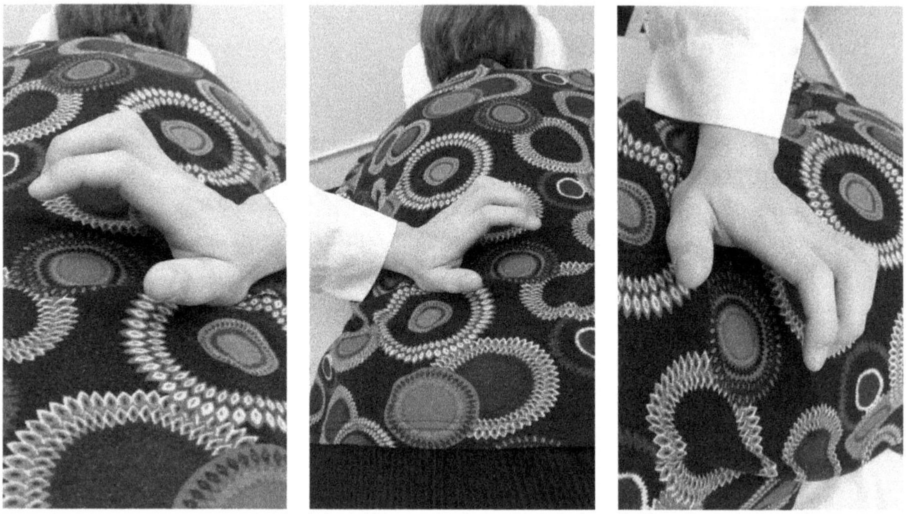

- Pull and shake the ankle for the lower back vertebrae.

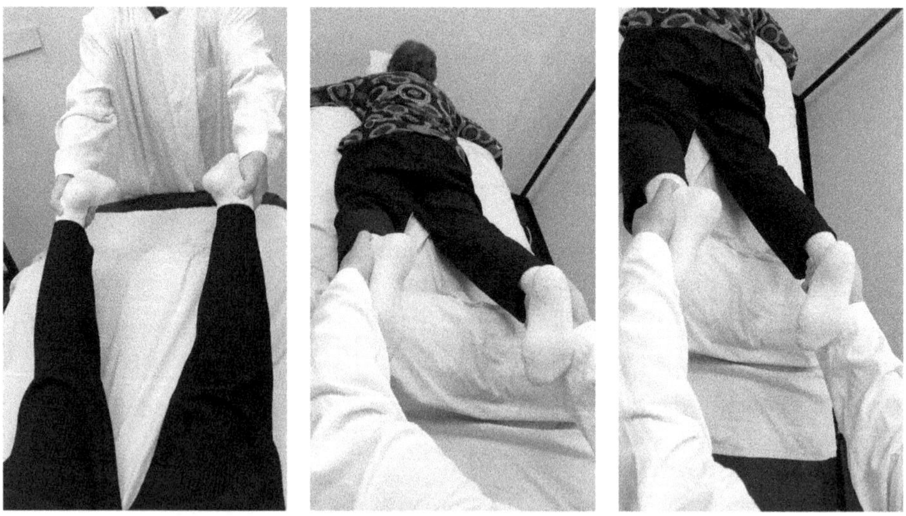

- Grasp the patient's lower back with your thumbs and your other four fingers for 10 to 20 minutes.

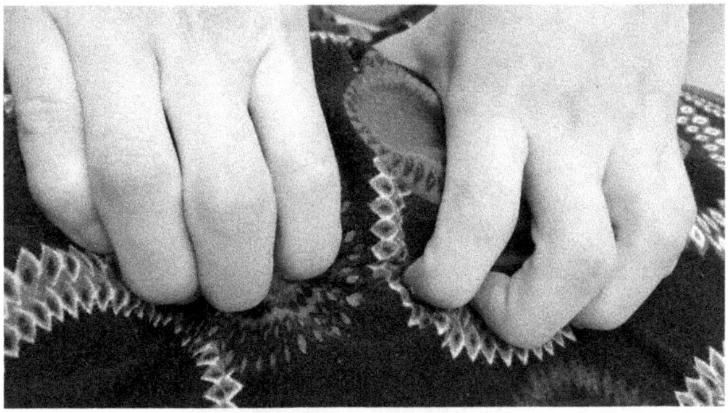

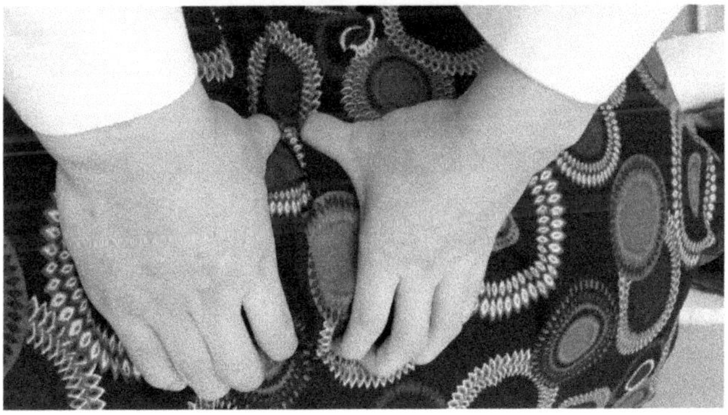

- Roll the muscles beside the spinal cord with both fists.

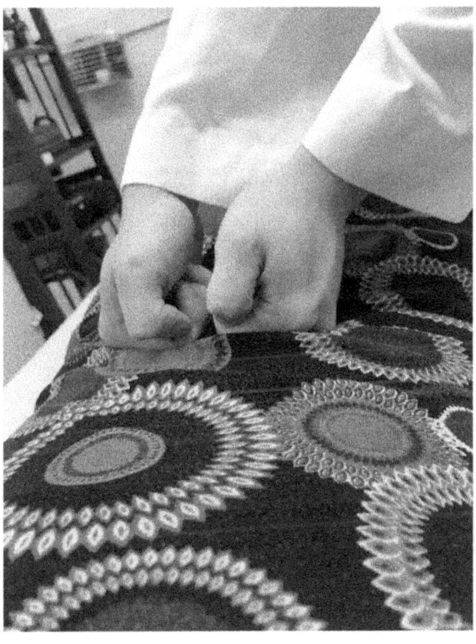

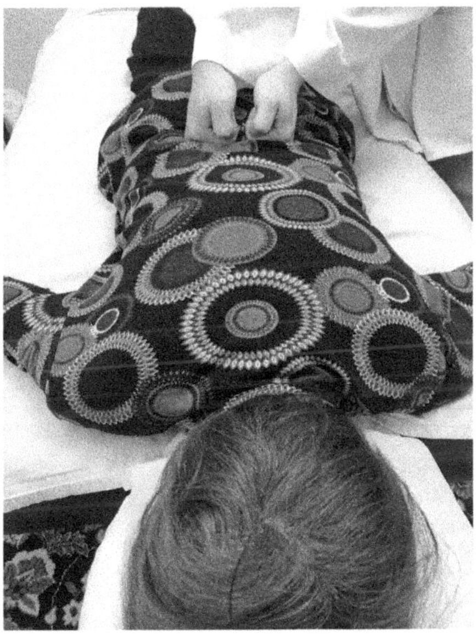

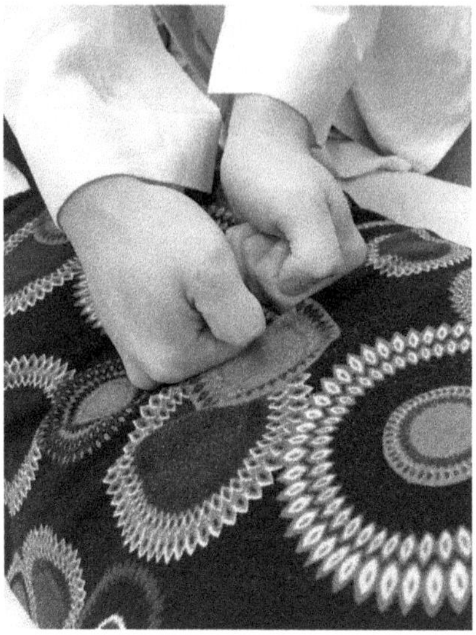

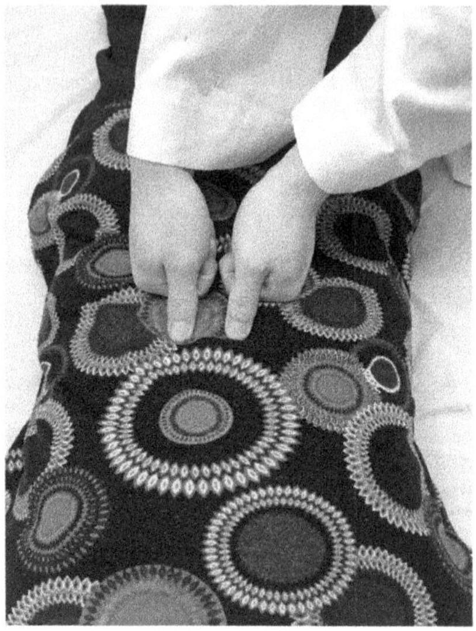

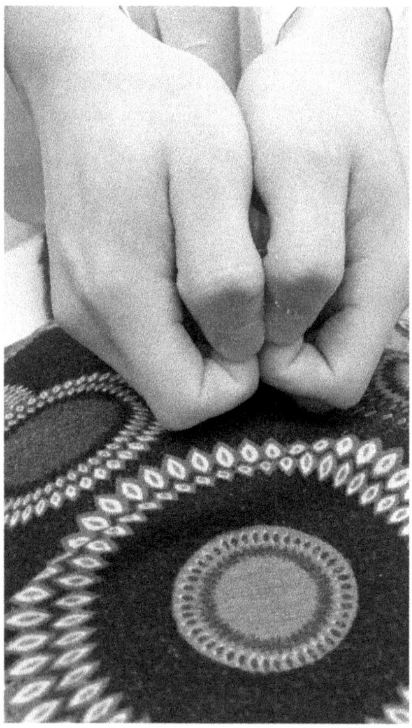

- Elbow press on the hips.

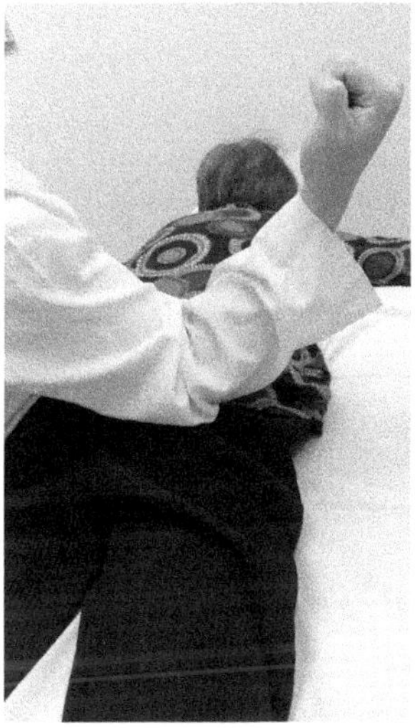
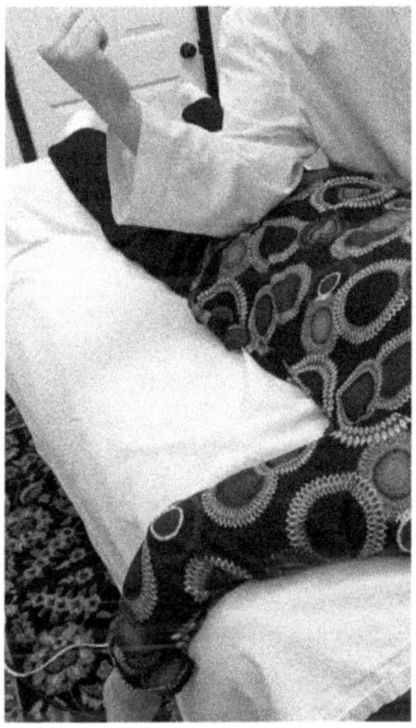
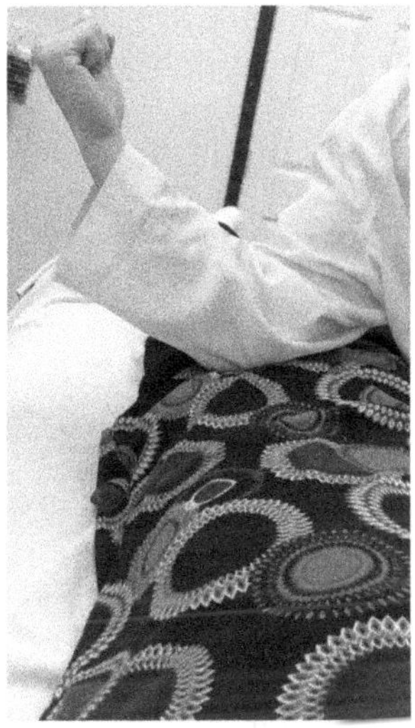

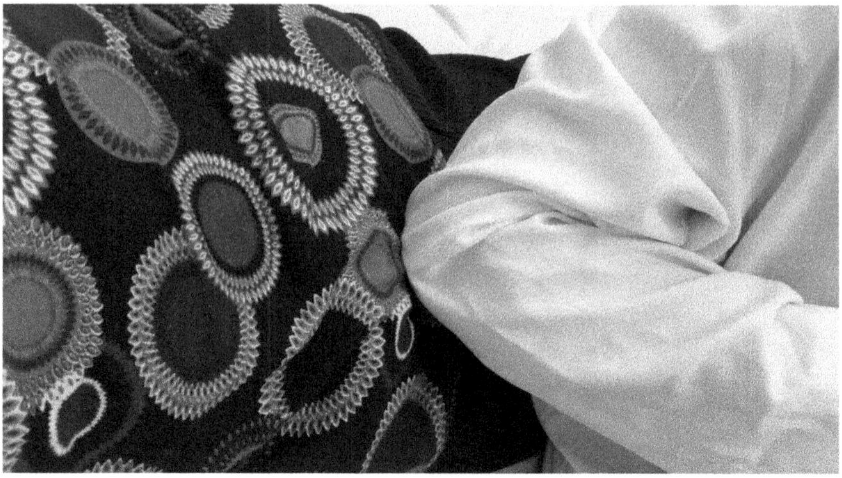

- Stretch the leg.

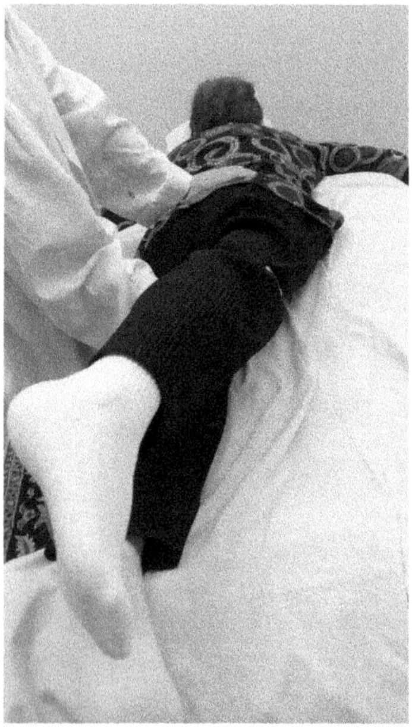

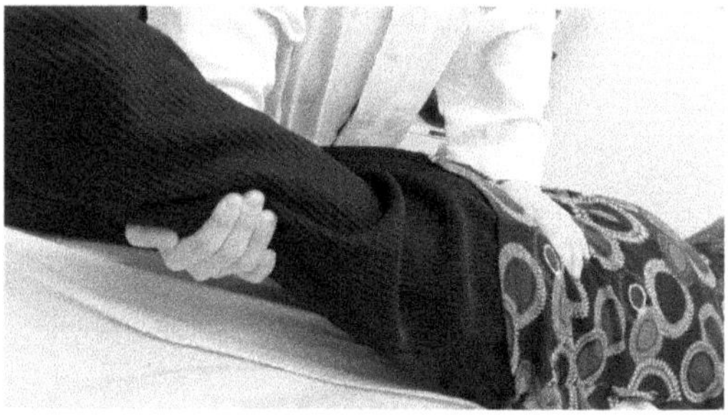

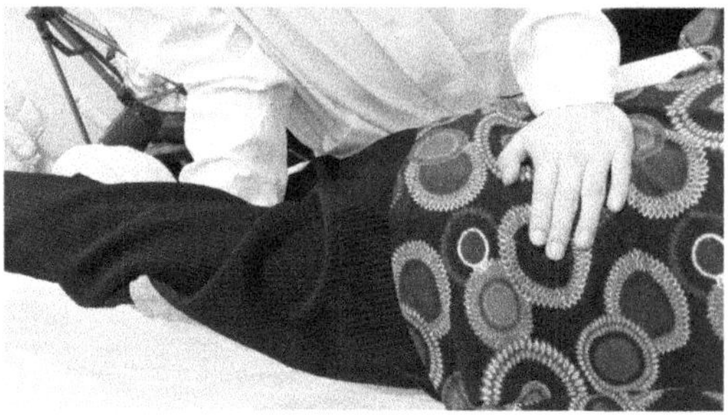

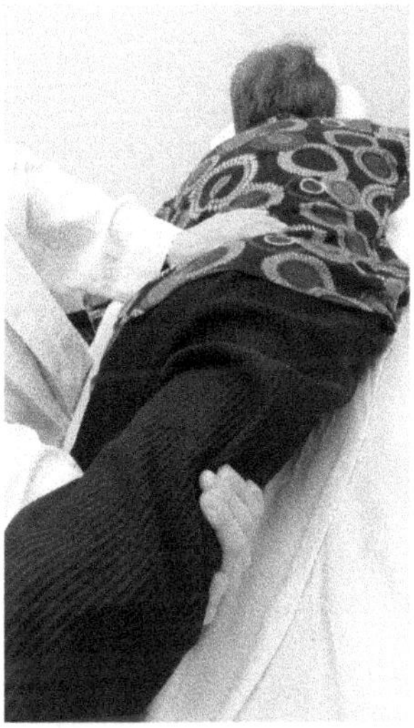

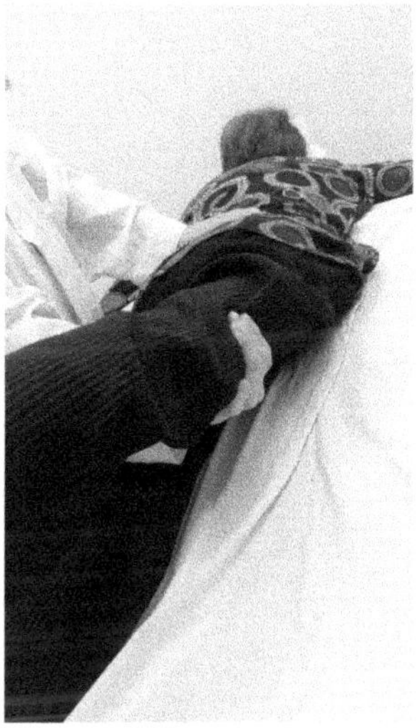

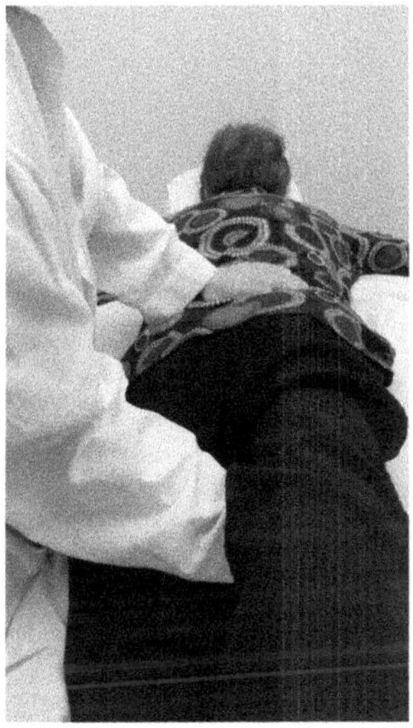

- Stretch the lower back.

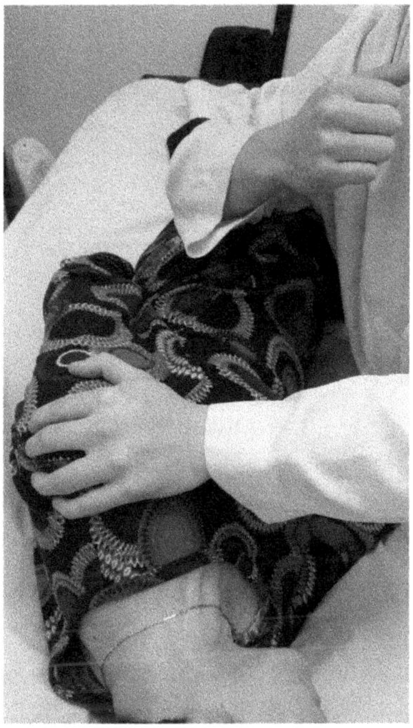

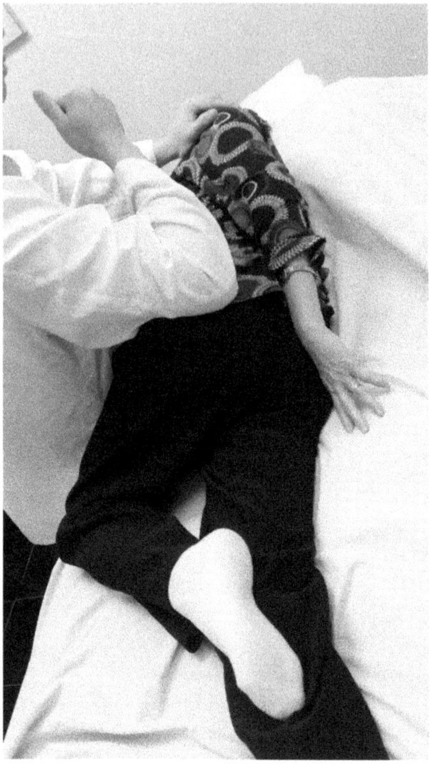

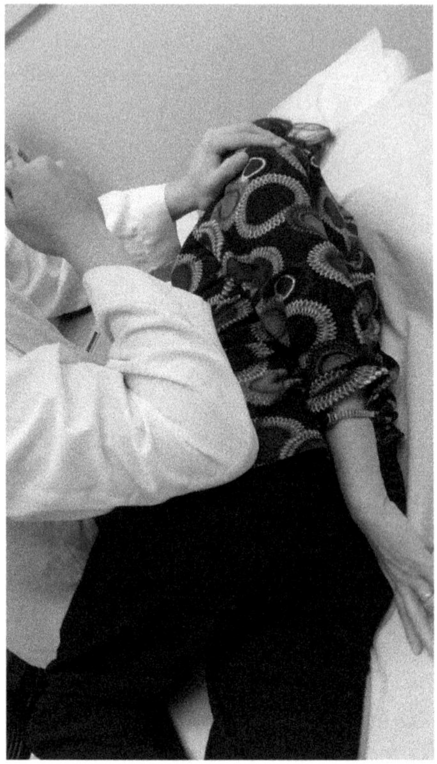

- Roll, knead, and rub on UB 23, GB 30, and urinary bladder meridians on the legs.

Patient Advisory

1) The patient is recommended to sleep on a firm bed and avoid cold in the lower back region.
2) Avoid caffeine, tobacco, and alcohol.
3) Avoid fatty food. Fatty food, caffeine, tobacco, and alcohol cause damp and heat that block meridians.
4) Avoid raw and cold food, as this causes damp cold, the factors blocking meridians.
5) Avoid strong-tasting food and hot spicy food, as these foods cause spleen Qi deficiency. spleen Qi deficiency produces damp, which blocks meridians, resulting in pain.
6) Avoid gaining weight, as obesity tends to provoke symptoms of low back pain.
7) Avoid staying up late at night; be in bed before 11 pm.
8) Avoid indulgence in sexual activity, as this drains kidney Qi.

9) Avoid overexertion in physical activity, as this drains kidney Qi.
10) Foods rich in calcium are recommended, such as milk, yogurt, eggs, cheese, and soy products.
11) Fresh vegetables and fruits are recommended.
12) A heating pad and massage are recommended for promoting the circulation of Qi and blood.
13) Appropriate exercise is recommended, 30 minutes daily. Routine stretching exercise is recommended after the acute stage; the lower body stretching exercise is helpful for sciatic pain recovery:
    a) Lie supine on a firm pad or carpet with both legs extended
    b) Flex the affected knee and hip, and place both hands on the patella
    c) Continue to flex the leg and move the knee gently toward the opposite shoulder
    d) Once the limit before feeling pain is reached, hold the position for 5 to 30 seconds. Be mindful that gentle stretching is better than overstretching. Repeat 5–10 times.
    e) Practice the same procedures on the other leg 5–10 times.
    f) Practice the same procedures on both legs 5–10 times.
    g) The frequency is twice daily, in the morning on waking and at night before sleep.

14) Keep good posture so that Qi and blood flow can be free.

Good sitting posture:

a) Be sure to straighten the back and avoid leaning forward. Place a cushion behind the lower back to support the normal lumbar curvature.
b) Flex the hips and knees at about 90 degrees.
c) Keep the legs shoulder width apart.
d) Keep the feet flat on the floor. If the feet cannot reach the floor, place the feet on a footrest.
e) Do not sit for longer than an hour. It is recommended to place a cup of tea within reach, with a few chrysanthemum flowers and goji berries with hot water, and to finish each cup within an hour. Goji berries help with eye tiredness, chrysanthemum flowers help to clear heat in the eye, and tea helps to clear the heat inside the body. With 250 cc of water, this is enough to make someone get up and walk to the bathroom. Someone sitting for 8 hours will finish drinking eight cups of water.

Good standing posture:

a) Be sure to stand with head and back erect. Keep the head upright on the midline between the shoulders.
b) Stand with legs slightly apart and with both knees slightly flexed.
c) Keep the weight on the front third of the feet.

d) Be mindful not to lean your body weight to either side.

Good sleeping posture:

a) Sleeping on a relatively firm bed is good for sciatic pain recovery.
b) Supine is the appropriate sleeping posture; both sides of the body receive equal pressure.
c) Place a cushion or pillow below the knees to alleviate lumbar pressure.
d) In the acute stage, lift the lower legs higher by placing more pillows below the tibia.
e) The normal lumbar curvature also can be supported placing a thin cushion or a rolled towel under the lower back.

NOTES

1. https://books.google.com/books?id=8DtFAgAAQBAJ&pg=PT327&lpg=PT327&dq=The+internal+pressure+on+a+disc+can+increase+from+,+Orthopedics+101&source=bl&ots=HrSF80XUPd&sig=ACfU3U3XI6l1L2YoPhfW5k5T5F18MYDUcQ&hl=en&sa=X&ved=2ahUKEwjAra6H5P7mAhUF7J4KHSIICW0Q6AEwAHoECAcQAQ#v=onepage&q=The%20internal%20pressure%20on%20a%20disc%20can%20increase%20from%20%2C%20Orthopedics%20101&f=false
2. www.ncbi.nlm.nih.gov/pmc/articles/PMC4160246/pdf/pone.0107436.pdf
3. https://books.google.com/books?id=yZc6DwAAQBAJ&pg=PA461&lpg=PA461&dq=intradiscal+pressure+seems+to+increase+while+sitting&source=bl&ots=New6J1kyO9&sig=ACfU3U2N_41mXOYHxbuzDfwyfu9d25UyCg&hl=en&sa=X&ved=2ahUKEwiH-K31if_mAhXIFjQIHRWIBrYQ6AEwEHoECAkQAQ
4. https://books.google.com/books?id=QrLACwAAQBAJ&pg=PA284&lpg=PA284&dq=how+to+release+synovial+membrane+incarceration&source=bl&ots=xeC06WyZRU&sig=ACfU3U2mq8PPHp2M888FdGurwlLQcqRioA&hl=en&sa=X&ved=2ahUKEwikkq7HwNTkAhXiFzQIHUwxD04Q6AEwDHoECAkQAQ#v=onepage&q=how%20to%20release%20synovial%20membrane%20incarceration&f=false
5. www.med66.com/new/40a188aa2009/20091112liuhon185912.shtml
6. https://books.google.com/books?id=iQIUFqiidYkC&pg=PA363&lpg=PA363&dq=that+were+necessary+to+make+repetitive+excessive+hyperextensions&source=bl&ots=J4j4dJfZEO&sig=ACfU3U0Ad5gjE0b64K7ediAIo4BbcqBl_Q&hl=en&sa=X&ved=2ahUKEwiZvaG64f_mAhUSJTQIHUVlC6kQ6AEwAHoECAYQAQ#v=onepage&q=that%20were%20necessary%20to%20make%20repetitive%20excessive%20hyperextensions&f=false

   Marc Safran, James E. Zachazewski, David A. Stone. *Instructions for Sports Medicine Patients*. Page 363.
7. https://journals.lww.com/acsm-csmr/FullText/2010/01000/Lumbar_Facet_Syndromes.13.aspx
8. Rebellion in Chinese medicine means the Qi flow goes to the opposite direction in the meridians. For example, lung Qi is supposed to be descending, but ascending of lung Qi causes coughing.
9. The symbol "~" is used to indicate that the electro machine should be hooked up to these acupuncture points.
10. When a person is upright, the body weight usually transmits to the sacrum and ilia. While sitting or supine, the person's weight transmits to the ischial tuberosity.

https://en.wikipedia.org/wiki/Ischial_tuberosity
11. www.healthline.com/health/si-joint-pain#sacroiliac-joints
12. www.healthline.com/health/si-joint-pain#sacroiliac-joints
13. 竖脊肌 (erector spinae) 又名骶棘肌(sacrospinalis)
14. www.sciencedirect.com/veterinary-science-and-veterinary-medicine
15. www.hindawi.com/journals/ecam/2016/7210705/
16. The patient is able to move but afraid to, because moving causes more pain.
17. Devastated blood is where heat hurts the blood.
18. https://pdfs.semanticscholar.org/c564/35a680378408fda23f0ef427701d08d9e4ac.pdf
19. https://en.wikipedia.org/wiki/Interspinous_ligament
20. www.physio-pedia.com/Interspinous_ligament
21. www.a-hospital.com/w/棘间韧带损伤
22. https://emedicine.medscape.com/article/311959-overview. Juvenile kyphosis is a deformity in the thoracic or thoracolumbar spine in which pediatric patients have increased kyphosis along with backache and localized changes in the vertebral bodies.
https://journals.lww.com/jbjsjournal/Abstract/1925/07040/VERTEBRAL_EPIPHYSITIS__A_CAUSE_OF_SPINAL_DEFORMITY.5.aspx
http://jb39.com/jibing/QingNianXingTuoBei263421.htm
23. www.spinemd.com/news-philanthropy/tips-to-reduce-lower-back-pain-during-menstruation
24. 陸銀華治傷經驗 p66
25. https://baike.baidu.com/item/棘上韧带棘上韧带损伤是指由于外伤或劳损而致棘上韧带的撕裂或炎性反应变性及钙化所致。多见于青壮年体力劳动者，为多发病。治疗不彻底时, 常会迁延成慢性劳损, 引起腰背部疼痛。
26. https://spineconnection.org/back-pain-conditions/ligament-injuries/
27. www.physio-pedia.com/Supraspinous_ligament
28. https://physio-pedia.com/Supraspinous_ligament
29. https://baike.baidu.com/item/棘上韧带
30. http://benbenjamin.com/webinars/LowBackWebinar5.pdf
31. www.a-hospital.com/w/棘上韧带损伤
32. www.a-hospital.com/w/棘上韧带损伤
33. www.a-hospital.com/w/棘上韧带损伤
34. http://drhuang.com/Chinese/health/medicine/book/中医伤科按摩学/54.htm
35. http://benbenjamin.com/webinars/LowBackWebinar5.pdf
36. http://drhuang.com/Chinese/health/medicine/book/中医伤科按摩学/54.htm
37. https://emedicine.medscape.com/article/311959-overview. Juvenile kyphosis is a deformity in the thoracic or thoracolumbar spine in which pediatric patients have increased kyphosis along with backache and localized changes in the vertebral bodies.
https://journals.lww.com/jbjsjournal/Abstract/1925/07040/VERTEBRAL_EPIPHYSITIS__A_CAUSE_OF_SPINAL_DEFORMITY.5.aspx
http://jb39.com/jibing/QingNianXingTuoBei263421.htm
38. www.spinemd.com/news-philanthropy/tips-to-reduce-lower-back-pain-during-menstruation
39. www.spine-health.com/conditions/spine-anatomy/all-about-l3-l4-spinal-segment
40. www.healthline.com/human-body-maps/l3-third-lumbar-spine-vertebrae#1
41. www.innerbody.com/anatomy/skeletal/l3-3rd-lumbar-vertebra#continued
42. www.spine-health.com/conditions/spine-anatomy/all-about-l3-l4-spinal-segment
43. www.ontcm.com/show-31-62-1.html
44. https://baike.baidu.com/item/%E8%85%B0%E4%B8%89%E6%A8%AA%E7%AA%81%E7%BB%BC%E5%90%88%E7%97%87
45. www.ncbi.nlm.nih.gov/pmc/articles/PMC2989641/

46. www.webmd.com/osteoarthritis/spinal-osteoarthritis-degenerative-arthritis-of-the-spine#2
47. https://radiopaedia.org/articles/coccyx?lang=us
48. www.depuysynthes.com/patients/aabp/understandingconditions/facetjointsyndrome
49. https://baike.baidu.com/item/腰椎退变性小关节损伤性骨关节炎
2.体征　小关节处有固定性压痛，压痛点深在，有叩击痛及传导痛。急性发作时，腰椎生理弯曲可消失，腰椎棘突排列不齐，患椎棘突间距离增大，在此处加压，可出现一凹陷。
50. https://en.wikipedia.org/wiki/Spinal_cavity
51. www.medicinenet.com/lumbar_stenosis/article.htm#what_causes_lumbar_spinal_stenosis
52. https://zhuanlan.zhihu.com/p/32249431 (5) 脊柱外伤：脊柱受外伤时，特别是外伤较重引起脊柱骨折或脱位时常引起椎管狭窄。
53. www.ninds.nih.gov/Disorders/Patient-Caregiver-Education/Fact-Sheets/Low-Back-Pain-Fact-Sheet
54. www.mayoclinic.org/diseases-conditions/back-pain/diagnosis-treatment/drc-20369911
55. www.spineuniverse.com/treatments/alternative/traditional-chinese-medicine-tcm-back-pain-1 Traditional Chinese Medicine (TCM) and Back Pain
56. www.spineuniverse.com/treatments/alternative/traditional-chinese-medicine-tcm-back-pain-1
57. www.zysj.com.cn/lilunshuji/neikexue/76-9-6.html http://what-when-how.com/treatment-of-pain-with-chinese-herbs-and-acupuncture/lower-back-pain-treatment-of-pain-with-chinese-herbs-and-acupuncture-part-1/

4 Hip Pain

LUMBOSACRAL STRAIN 腰肌勞損

INTRODUCTION

Lumbosacral strain is a chronic soft tissue injury in the lumbar region that causes low back pain. Strain means the tearing of the lumbosacral muscles, tendons, and fascia. Among the chronic back pain diseases, lumbosacral strain is a common pain. It is often found in middle-aged patients, especially patients with assigned female reproductive systems.

ANATOMY

The lumbosacral region of the spine is the part of the low back between the bottom of the ribcage and the top of the buttocks, and consists of five lumbar vertebrae (L1–L5) and five sacrum bones (S1–S5) in between the hips which are five separate bones at birth but calcify and become a single bone soon after we are able to walk and stand.

WESTERN MEDICINE ETIOLOGY AND PATHOLOGY

- Habitual improper posture or long-term improper positional labor.
- Improper treatment after an acute injury of the soft tissues in the lumbar region or reoccurrence of the strain, causing bleeding, fibrosis, and compression on nerves.
- Congenital malformation.
- Constitutional weakness or improper nursing after pregnancy.

CHINESE MEDICINE ETIOLOGY AND PATHOLOGY

- According to Chinese medicine, the lumbosacral region is the mansion of the kidney.

Kidney deficiency is the constitution; the patients then suffer from physical trauma or exposure to external pathogens such as damp, cold, heat, and wind, leading to Qi and blood stagnation from blockages in the meridians, which is called Bi syndrome. There are two causative factors:

a) This is an acute onset from chronic illness. When the lumbar is exposed to external pathogens, the symptoms are more excessive, superficial, and rapid onset.

b) When the pain is under kidney deficiency, the cause of the symptoms is from the interior weakness; it is more deficient and deep so the pain is chronically reoccurring.

Manifestations

1) *Lumbago*: Chronic reoccurrence of lower back pain. On and off soreness on one or both sides of the lumbosacral region, it is aggravated after work or change of weather, and relieved after rest, appropriate exercise, and improved posture and heat packs.
2) *Pain during flexion*: When pain occurs from leaning forward, it could be relieved after lumbar extension or striking the lumbosacral region. Patients often prefer hitting the lumbar with both fists.
3) *Worsening pain during weather changes*: Under exposure to cold, damp, and wind.
4) *Radiating pain*: Pain in the lower limb, pulling pain at the acute stage.
5) *Numbness of the leg at the sciatic nerve area*: It is negative in the chronic state, but may cause pain at the buttocks and the back of the thighs.

Differentiation

1) *Blood stasis type:* The pain location is fixed; it is constant stabbing pain, and it is more painful at night but less painful in the daytime.
2) *Damp heat type*: There is a warm sensation at the painful area, and it is aggravated during hot days or rainy days. The pain could be less after movements. Urine is dark. The tongue has a yellow greasy coating. The pulse is slippery and rapid.
3) *Damp cold type*: The pain is aggravated by cold exposure and rainy days. Limited range of motion and bed rest doesn't relieve the pain. The tongue is pale with a white greasy coating; the pulse is deep.
4) *Kidney deficiency type*: The lumbar is sore, and the patient prefers pressure on the affected region. The knees are weak. The pain is insidious when the patient is over-exerted and is less pronounced after bed rest. The face is pale. The patient has a dry mouth and is irritated. The patient prefers warm and dislikes cold. Often the extremities are cold. The tongue is pale; the pulse is deep and thin or thin and rapid.

Physical Examination

- Begin by obtaining the patient's medical history:
  - General health conditions such as leg length discrepancy, pregnancy.
  - Injury history to the sacroiliac joint such as car accident, falling from ladders.
  - Careers that may make the patient vulnerable to the injury, such as nurses and nursing aides who lift patients, or office workers who sit with poor postures such as crossed knees.

    - Sports injury such as falling and directly attacking the sacroiliac joint.
- Inspection:
    - If low back pain is aggravated by lumbar flexion, it could be relieved after lumbar extension.
    - Lightly tapping on the lumbosacral region may alleviate the pain.
- Palpation:
    - Wide tender area, muscle spasm.
    - Deviation of the spinal may be found in acute onset.
    - The lumbar region spasm.
- Range of motion:
    - Decreased lumbar lordosis.

Differential Diagnosis

Traumatic lumbosacral dislocation is an acute traumatic condition which is a displacement of the fifth lumbar vertebra in relation to the sacrum from a violent physical trauma such as a motor vehicle collision or a fall from a height.[1]

Treatment

Treatment principle:

- Relaxes tendons, opens and warms meridians.
- Releasees the local soft tissue spasm.
- Promotes local blood circulation.

Herbal Treatments

Kidney Deficiency Type

Herbal formula:

- Bu Shen Huo Xue Tang 補腎活血湯.

Formula ingredients:

Shu Di Huang 熟地 9 g, Du Zhong 杜仲 3 g, Gou Qi Zi 枸杞子 3 g, Bu Gu Zhi 補骨脂 9 g, Tu Si Zi 菟絲子 10 g, Dang Gui Wei 當歸尾 3 g, Mo Yao 沒藥 3 g, Shan Zhu Yu 山茱萸 3 g, 紅花1.5 g, Du Huo 獨活 3 g, Rou Cong Rong 肉蓯蓉 3 g.

Ingredient explanations:

- Shu Di Huang 熟地 nourishes blood, nourishes liver and kidney Yin.
- Dang Gui Wei 當歸尾 moves and harmonizes the blood.
- Tu Si Zi 菟絲子 tonifies kidney Yang, tonifies kidney and liver Yin.
- Gou Qi Zi 枸杞子 nourishes and tonifies liver and kidney blood and Yin.
- Bu Gu Zhi 補骨脂 tonifies kidney Yang.
- Shan Zhu Yu 山茱萸 nourishes liver and kidney, strengthens kidney Yang.
- Rou Cong Rong 肉蓯蓉 tonifies kidney Yang, benefits essence and marrow.

- Mo Yao 沒藥 moves blood, reduces swelling, and stops pain.
- Hong Hua 紅花 moves blood, relieves bruising, opens meridians, stops pain.
- Du Huo 獨活 expels wind and damp, stops pain.
- Du Zhong 杜仲 tonifies the kidneys and liver, strengthens bones and tendons, and lowers blood pressure.

Blood Stasis Type

Herbal formula:

- Di Long San 地龍湯.

Formula ingredients:

Di Long 地龍6 g, Rou Gui 肉桂3 g, Su Mu 蘇木9 g, Ma Huang 麻黄3 g, Dang Gui Wei 當歸尾12 g, Tao Ren 桃仁9 g, Huang Bai 黄柏6 g, Gan Cao 甘草6 g.

Ingredient explanations:

- Di Long 地龍 clears heat, unblocks and promotes movement in the meridians, extinguishes wind, and stops spasms.
- Rou Gui 肉桂 warms kidney Yang and Ming Men Fire and assists the generation of Qi and blood.
- Su Mu 蘇木 moves blood, reduces swelling, stops bleeding.
- Ma Huang 麻 induces sweating, releases the exterior, and warms and disperses cold pathogens.
- Huang Bai 黄柏 clears deficient kidney Yin heat.
- Dang Gui Wei 當歸尾 moves blood, disperses cold, and alleviates pain.
- Tao Ren 桃仁 breaks up blood stasis and invigorates blood circulation.
- Gan Cao 甘草 moderates spasms, alleviates pain, harmonizes the other herbs, and guides the herbs to all 12 meridians.

Formula action:

- Moves blood, disperses bruises, opens meridians.

Wind, Cold, and Damp Type

Herbal formula:

- Du Huo Ji Sheng Tang 獨活寄生湯.

Formula ingredients:

Du Huo 獨活9 g, Sang Ji Sheng 桑寄生6 g, Du Zhong 杜仲6 g, Niu Xi 牛膝6 g, Xi Xing 细辛3 g, Qin Jiao 秦艽6 g, Fu Ling 茯苓6 g, Rou Gui 肉桂6 g, Fang Feng 防風6 g, Chuan Xiong 川芎6 g, Dang Shen 党参6 g, Zhi Gan Cao 甘草3 g, Dang Gui 當歸6 g, Bai Shao 白芍6 g, Shu Di Huang 熟地黄6 g.

Ingredient explanations:

- Du Huo 獨活 expels wind, damp, and cold in lower Jiao and joints, removes chronic Bi stagnation in lower limbs.
- Xi Xing 细辛 enters Shao Yin meridian to expel wind, damp, and cold.
- Qin Jiao 秦艽 expels wind and damp, opens meridians, benefits joints.
- Rou Gui 肉桂 warms meridians, expels cold, opens meridians to promote blood circulation.
- Fang Feng 防風 expels wind and drains damp.
- Sang Ji Sheng 桑寄生, Du Zhong 杜仲, and Niu Xi 牛膝 are for deficiency of Qi and blood in the liver and kidneys, and strengthening bone and tendons; Sang Ji Sheng 桑寄生expels wind and drains damp; Niu Xi 牛膝moves blood and opens meridians to benefit joints and tendons.
- Dang Gui 當歸, Chuan Xiong 川芎, Shu Di Huang 熟地黄, and Bai Shao 白芍 nourish and harmonize blood.
- Dang Shen 党参, Fu Ling 茯苓, and Gan Cao 甘草 tonify spleen Qi.

Formula actions:

- Dispels wind, cold, and damp.
- Relieves painful obstruction.
- Supplements Qi and blood.
- Tonifies the liver and kidneys.

Modifications:

- For severe pain, add Zhi Chuan Wu 制川烏, Zhi Cao Wu 制草烏, Bai Hua She 白花蛇, Xu Duan 川斷, Gou Ji 狗脊, Tu Si Zi 菟絲子, Hong Hua 紅花, and Di Long 地龍.
- For more cold symptoms, add Zhi Fu Zi 附子 and Gan Jiang 乾薑.
- For more damp symptoms, omit Shu Di Huang 熟地黃 and add Fang Ji 防己, Yi Yi Ren 薏苡仁, and Cang Zhu 蒼朮.

Contraindications:

- Damp heat excessive Bi syndrome.

Damp Heat Type

Herbal formula:

- Si Miao San 四妙散.

Formula ingredients:

Cang Zhu 蒼朮 12 g, Huang Bai 黃柏 12 g, Yi Yi Ren 薏苡仁 30 g, Ren Dong Teng 忍冬藤 20 g, Bei Xie 萆薢 20 g, Mu Gua 木瓜 15 g, Fang J i 防己 15 g, Hai Tong Pi 海桐皮 15 g, Niu Xi 牛膝 15 g, Gan Cao 甘草 6 g.

Ingredient explanations:

- Cang Zhu 苍术 strongly dries dampness, tonifies the spleen, and clears dampness from the lower Jiao.
- Huang Bai 黄柏 drains damp-heat, especially from the lower Jiao, and drains kidney fire.
- Yi Yi Ren 薏苡仁 promotes urination, leaches out dampness, strengthens the spleen, expels wind-dampness, and clears damp-heat.
- Ren Dong Teng 忍冬藤 dispels wind-dampness, soothes the sinews, unblocks the channels, cools the blood, and clears damp-heat from the lower Jiao.
- Bei Xie 萆薢 expels wind-dampness and relaxes the sinews.
- Mu Gua 木瓜 promotes Qi and blood circulation and unblocks the channels.
- Fang Ji 防己 promotes urination and reduces edema especially in the lower body, expels wind-dampness, and alleviates pain.
- Hai Tong Pi 海桐皮 dispels wind-dampness, unblocks the channels and collaterals, promotes urination, and reduces edema.
- Niu Xi 牛膝 moves the blood, disperses blood stasis, clears damp-heat in the lower Jiao, and regulates water metabolism.
- Gan Cao 甘草 releases spasms and alleviates pain.

Acupuncture Treatments

Acupuncture point formula:

UB23 腎俞 Shenshu, UB25 大腸俞 Dachangshu, UB31 上髎 Shangliao, UB32 次髎 Ciliao, UB33 中髎 Zhongliao, UB34 下髎 Xialiao, GB30 環跳 Huantiao, UB40 委中 Weizhong, GB39 懸鐘 Xuanzhong, UB57 承山 Chengshan, LI4 合谷 He Gu.

Electroacupuncture Treatments

UB23 腎俞Shenshu ~ UB25 大腸俞Dachangshu 2/100 Hz for 20 minutes daily.

Gua Sha Treatments

Lower back.

Seven-Star Needle Treatments

UB 23 腎俞 Shenshu and UB25 大腸俞 Dachangshu; cupping on these two points after the seven-star treatment.

Procedures:

1) Prior to the procedure, tools should be prepared and ready to be used at the work area.

2) Set up a clean field with paper towels.
3) Place these tools in the clean field, along with dry sterile cotton balls, medical alcohol wipes or cotton swabs soaked in alcohol, sterile gloves, a biohazard sharp container, a biohazard trash container, goggles and facemask, and paper towels. A medical lancet, traditional three-edge needle, or the seven-star needle are bleeding tools; the seven-star needle is selected for bleeding UB 23 腎俞 Shenshu and UB25 大腸俞 Dachangshu in this case.
4) The patient lies prone with the legs extended, UB 23 腎俞 Shenshu and UB25 大腸俞 Dachangshu are facing upward.
5) The room should be well lit.
6) After the therapist washes their hands thoroughly and wipes their hands with a single-use napkin, the therapist should put on gloves.
7) Wipe the acupuncture points UB 23 腎俞 Shenshu and UB25 大腸俞 Dachangshu with a medical alcohol wipe or a cotton swab that soaked in alcohol. Move from the center of each acupuncture point to the periphery.
8) Place the seven-star needle at a perpendicular angle to the skin, hold the end of the handle.
9) Flick the flexible handle to tap the needles onto the acupuncture point lightly and superficially to allow the skin to have a mild rash. The flexible handle is flicked at a fast speed so it actually takes only a short time to get the rash result.
10) Lastly, put a cup on to release more blood.
11) Dispose of the tools in the proper containers.
12) Wash the hands thoroughly.

Auricular Therapy

Lumbosacral, Shen Men, Kidney.

Tui Na Treatments

Tui Na techniques:

- Rolling, pressing, kneading, striking, rubbing, rotating, stretching

Tui Na procedures:

- The patient lies in a prone position. The Tui Na therapist stands beside the patient near the lumbar.
- Apply rolling techniques on both lumbar and urinary bladder meridians six times.

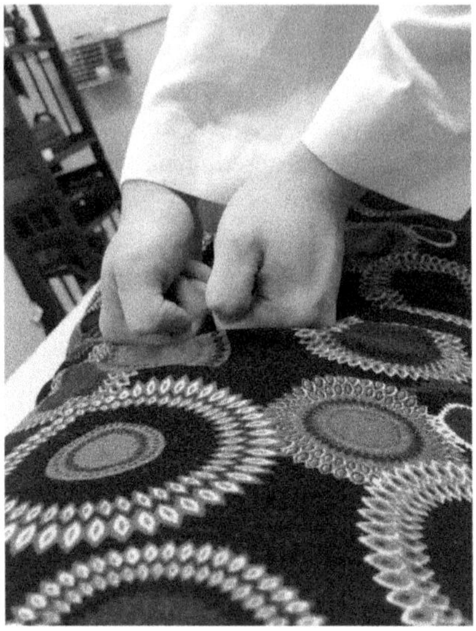

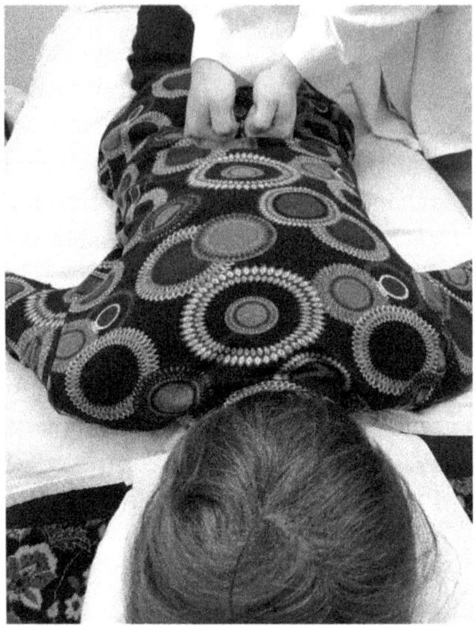

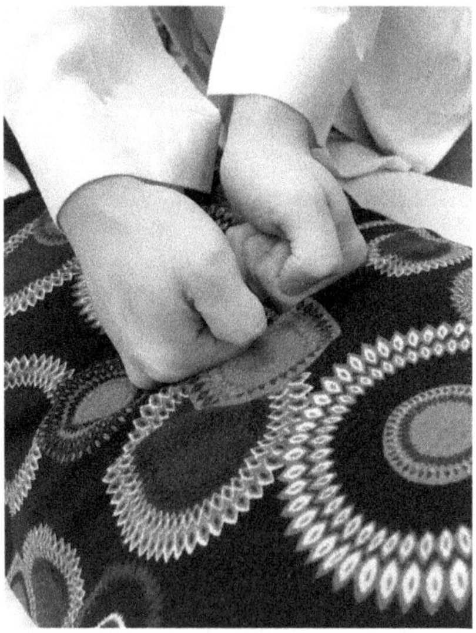

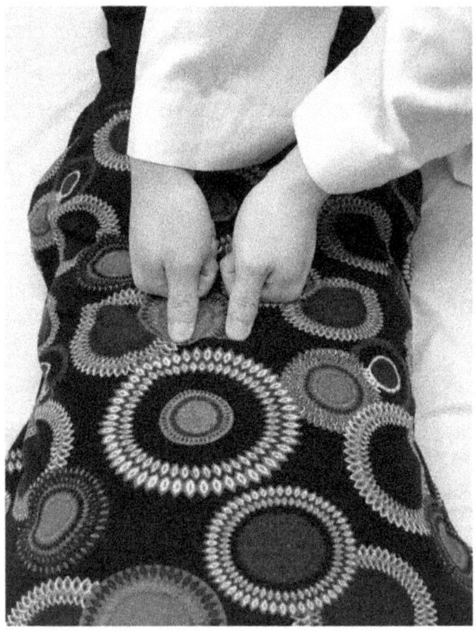

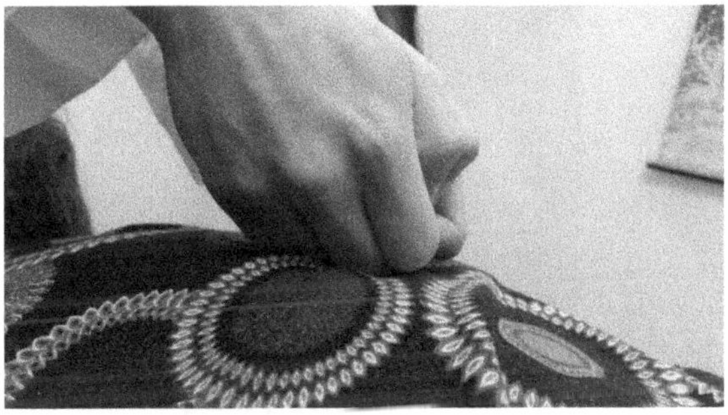

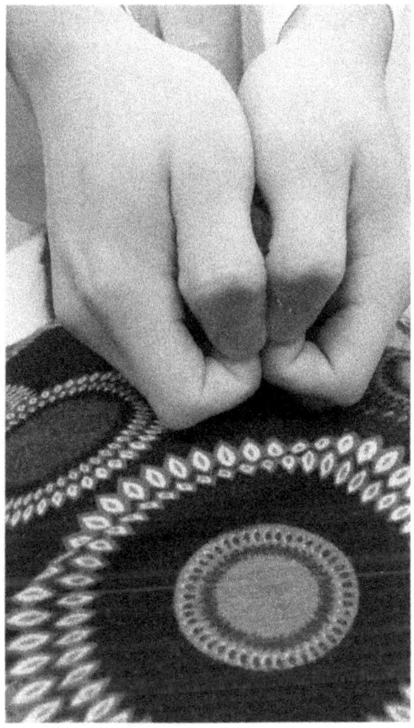

- Apply pressing and kneading techniques on UB25 大腸俞 Dachangshu, UB31 上髎 Shangliao, UB32 次髎 Ciliao, UB33 中髎 Zhongliao, UB34 下髎 Xialiao.
- Apply rubbing techniques on both lumbar and urinary bladder meridians.
- Apply rubbing techniques on the lumbosacral region.

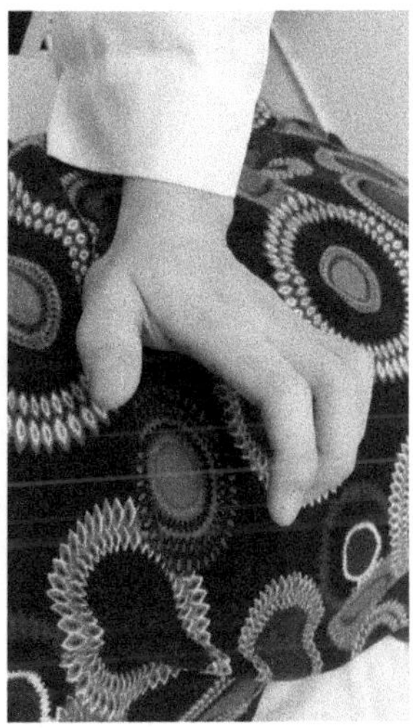

- Apply rotating techniques on the lower back.

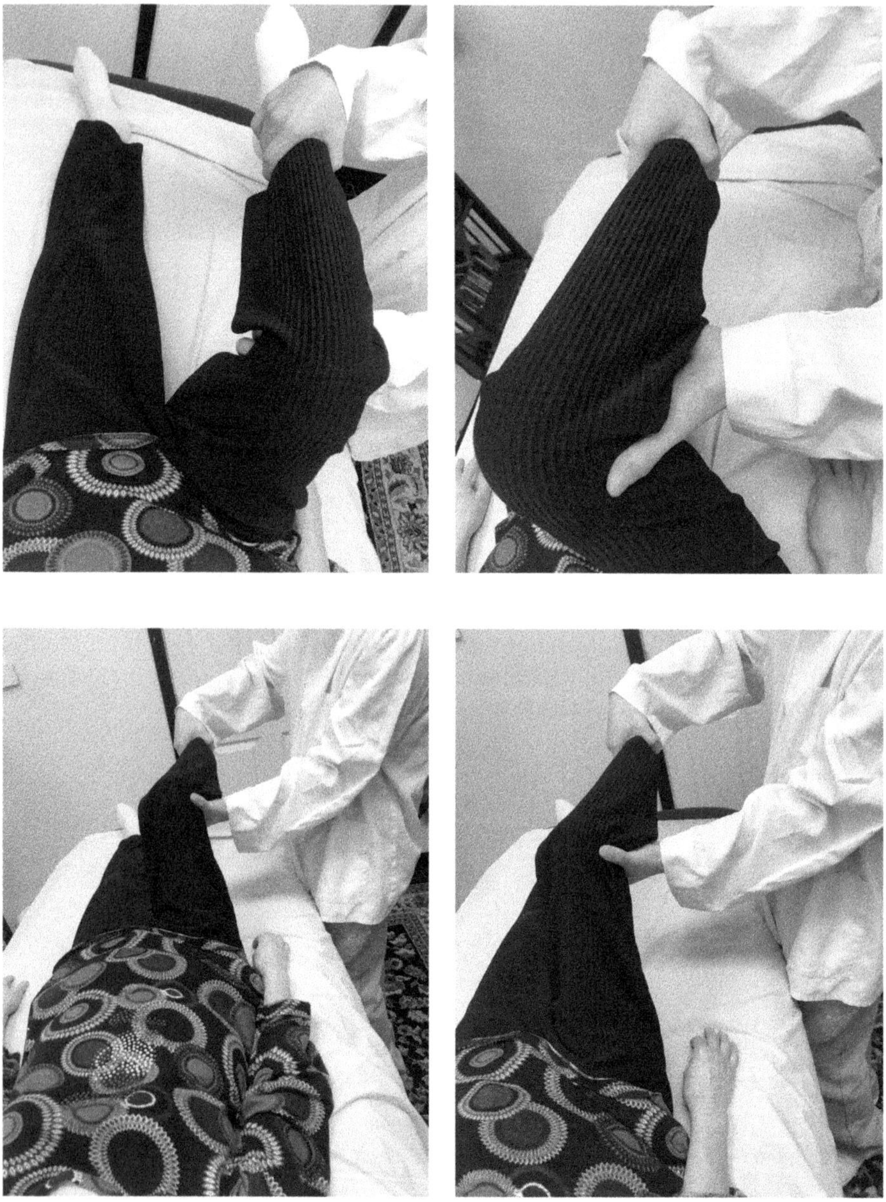

- Apply stretching techniques on the lower back.

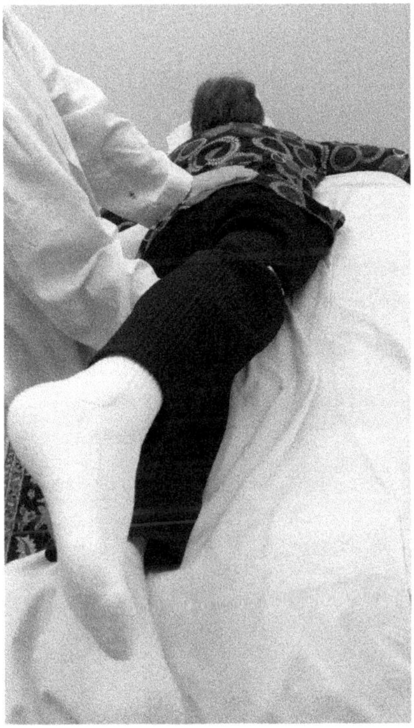

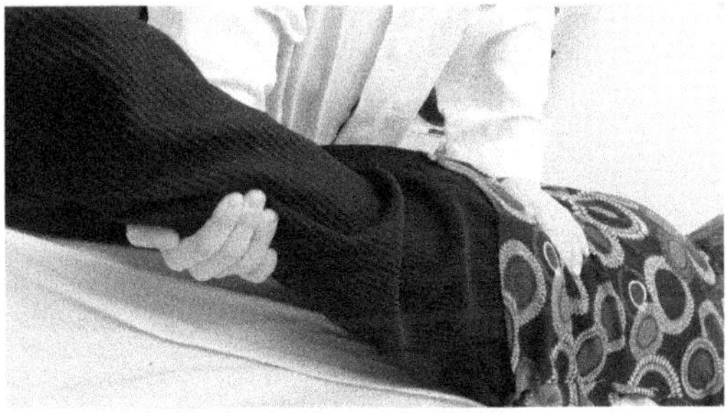

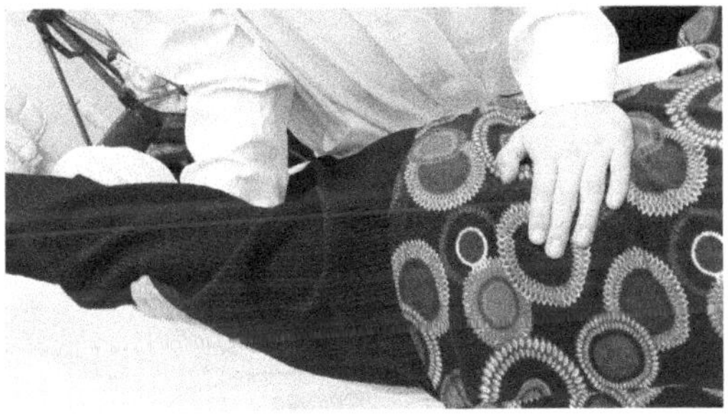

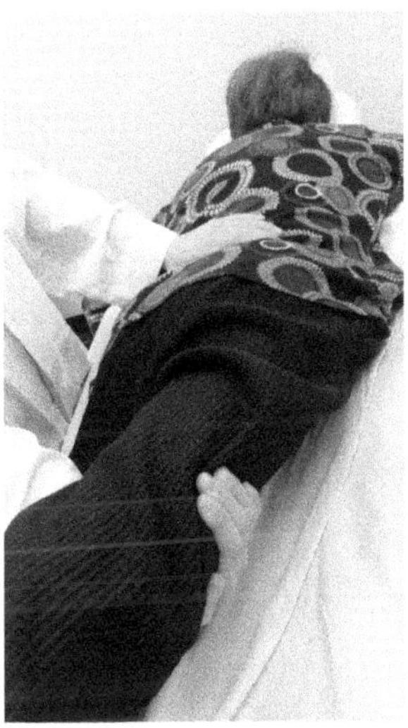

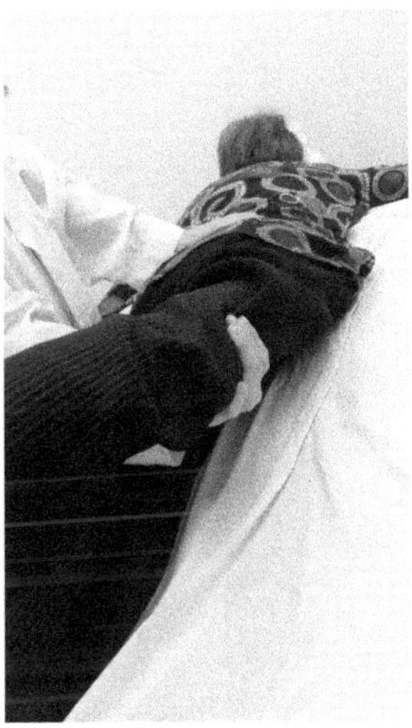

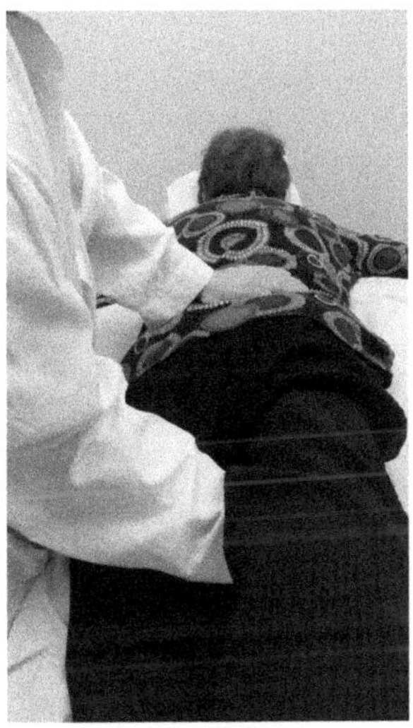

- Apply striking techniques on the erector spinae muscles until the skin is slightly red.

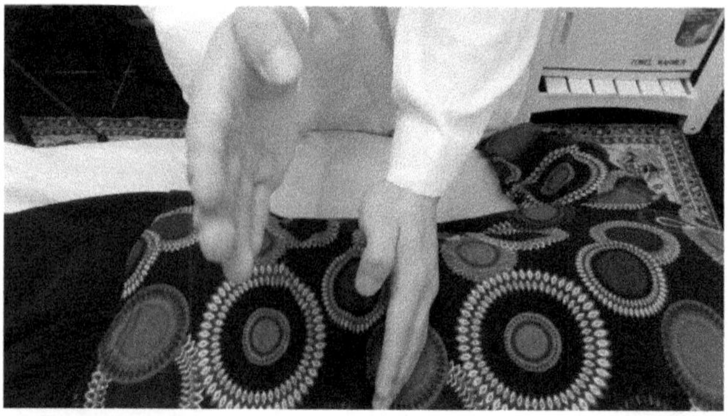

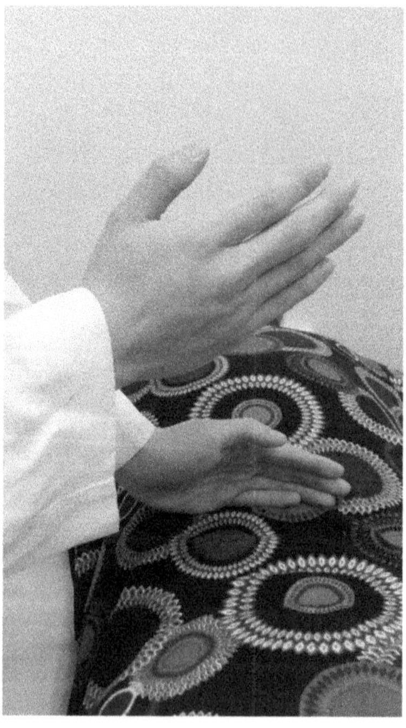

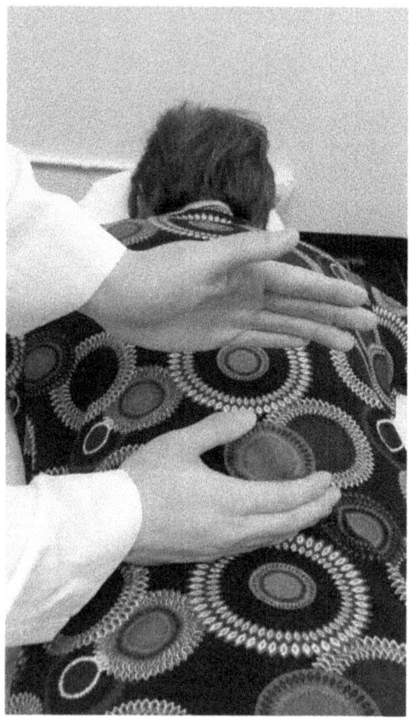

Patient Advisory

- Get plenty of bed rest whenever the pain is acute, and seek for therapy in the early state.
- Foods to warm the kidneys are recommended; they include walnut, goji berry, and lamb.
- Foods that are high in calcium such as milk, small fish, shrimp, and soy products are recommended.
- Avoid cold food, deep-fried food, and strong spicy food.
- Patients are advised to correct physical postures.
- Keep good postures so that Qi and blood flow can be free.

  Good sitting posture:

  a) Be sure to straighten the back, and avoid leaning forward. Place a cushion behind the low back to support the normal curvature of the lumbar.
  b) Flex the hips and knees at about 90 degrees.
  c) Keep the legs shoulder-width apart.
  d) Keep the feet flat on the floor. If the feet cannot reach the floor, place the feet on a footrest.
  e) Do not sit for longer than an hour. While sitting, a cup of tea with a few chrysanthemum flowers and goji berries with hot water, finished within an hour, is recommended. Goji berries help eye tiredness, chrysanthemum flowers help to clear Heat in the eye, and tea helps to clear the Heat inside the body. After 250cc water, someone will get up and walk to the bathroom. Sitting for eight hours will allow someone to finish drinking eight cups of water.

  Good standing posture:

  a) Be sure to stand with an erect head and back. Keep the head upright on the midline between the shoulders.
  b) Stand with both legs slightly apart and both knees slightly flexed.
  c) Keep the weight on the front third of the feet.
  d) Be mindful, and do not lean your body weight on either side.

  Good sleeping posture:

  a) Sleeping on a relatively firm bed is good for sciatic pain recovery.
  b) Supine is the appropriate sleeping posture, because both sides of the body receive equal pressure.
  c) Place a cushion or pillow below the knees to alleviate lumbar pressure.
  d) In the acute stage, lift the lower legs higher by placing more pillows below the tibia.
  e) The normal lumbar curvature also can be supported placing a thin cushion or a rolled towel at the lower back.

  Patients need to exercise the lumbar soft tissues to strengthen the back:

  a) Lie prone on the floor with legs straight and slight apart.
  b) Place both hands on the floor in front of the shoulders.
  c) Lift the upper trunk with the arms slowly to the point before there is any low back pain.

d) Be sure only the upper trunk is lifted, and the hip and legs are still on the floor.
e) Hold it for five to thirty seconds depending on the lumbar discomfort. The exercise is practiced with natural breath and no pain.
f) Be mindful that the lumbar muscles remain fully relaxed.
g) To return to the beginning, slowly flex the elbows to lower the upper trunk until the chest reaches the floor
h) Repeat the procedures three to five times.
i) The pace of the movements should be slow; the lumbar muscles work harder when moving slowly.

TRAUMATIC SYNDROME OF THE PIRIFORM MUSCLE 梨状肌综合徵

INTRODUCTION

Generally, patients with piriformis syndrome complain of sciatic nerve pain because piriformis syndrome is also one of the sciatic nerve irritated syndromes. The sharp and severe pain radiates to the leg including the thigh, calf, foot, and maybe the toes. Typically, the radiating pain with piriformis sciatica starts at the buttock instead of the lumbar.

ANATOMY

The piriformis muscle is a flat, pyramid-shaped[2] muscle located deep to the gluteus maximus in the buttocks near the top of the hip joint. It originates from the anterior aspect of the sacrum at the level of about S2 to S4. It passes laterally to exit the bony pelvis through the greater sciatic foramen and often attaches to the gluteal surface of the ilium, close to the posterior inferior iliac spine.[3] It inserts onto the superior and medial aspects of the greater trochanter. The nerves that innervate the muscle are from the anterior rami of S1 and S2 spinal nerves.[4] The arterial supply for this muscle is from the inferior gluteal, superior gluteal, and internal pudendal arteries, all branches of the internal iliac artery.[5]

This muscle is important in lower body movement because it stabilizes the hip joint by steadying the head of the femur inside the acetabulum of the hip bone[6] and lifts and rotates the thigh away from the body by contributing to the *retroversion* and *external rotation* of the hip joint, in addition to *abduction* of the thigh while in the flexed position.[7] This means that as the muscle works, it helps to turn the foot and leg outward.[8]

WESTERN MEDICINE ETIOLOGY AND PATHOLOGY

Spasm of the piriformis muscle causes compression or irritation on the sciatic nerve that passes through the muscle, resulting in sciatic pain.

There is variation of the location of the sciatic nerve and the piriformis muscle. In the vast majority of individuals, the sciatic nerve leaves the pelvis along with its surrounding nerves and blood vessels through the greater sciatic foramen below the piriformis muscle and divides into tibial nerve and common peroneal nerve distally to the piriformis.

In a small minority of individuals, the sciatic nerve divides proximally. There is a variety of different anatomic relations between the sciatic nerve and the piriformis muscle.

These variations include:

- The sciatic nerve may be undivided and may emerge above or through the piriformis muscle.
- The sciatic nerve may divide above the piriformis muscle; one branch leaves the pelvis below the muscle, the other passes through the piriformis.
- The sciatic nerve may divide above the piriformis muscle; one branch leaves the pelvic area above the muscle, the other passes through the piriformis.
- The sciatic nerve may divide above the piriformis; one branch travels in front of it, and the other travels behind it.
- Clinically, either the common tibial branch or the common peroneal branch of the sciatic nerve may pass through the piriformis muscle and have different clinical presentations.

Normally, a healthy piriformis muscle will not irritate the sciatic nerve; spasm of the muscle compresses the branch that passes through causing irritation or entrapment and resulting sciatic pain.

Spasm and hypertrophy of the piriformis muscle can result from a hip injury that applies firm pressure directly over the piriformis muscle or over-stretching or overuse of the piriformis muscle such as long-distance running or excessive crouching.

Chinese Medicine Etiology and Pathology

According to Chinese medicine, cold damp invasions are the first factors to cause this pain. It is more prevalent in the winter or after a patient has been outdoors in the cold, such as prolonged sitting in a cold seat or kayaking in cold weather with the environment of dampness, causing overexposure to extremes of weather involving cold, damp, and wind. These factors can lodge in the susceptible hardened area, blocking the flow of Qi and blood in meridians and resulting in pain.

Injury or overuse of the piriformis muscle causes Qi and blood to be stagnant in the acute state, which manifests as swelling and spasm of the piriformis muscle; the soft tissues are trying to heal themselves, a procedure which consumes the Yin and blood of the liver and kidneys, resulting in kidney and liver deficiency and manifesting as leg muscle and piriformis muscle atrophy.

Manifestations

Patients have an obvious injury or have been exposed to a cold environment. Patients are usually able to describe the history of injury such as lifting heavy objects from a squatting position. Pain may increase after sitting for longer than 15–20 minutes. Patients may have difficulty walking, and the pain may occur or increase with internal rotation of the ipsilateral leg such as cross-legged sitting or ambulation.[9]

Piriformis syndrome usually starts with tenderness, shooting, electrical sensation, tingling, or numbness in the buttocks, usually on one side only. Pain can be severe and extend down the length of the sciatic nerve.[10]

Piriformis syndrome can come on suddenly, usually following an injury, or it can appear gradually over several months.[11]

Physical Examination

- Begin by obtaining the patient's medical history:
  - Injury history to the sacroiliac joint such as lifting heavy objects from squatting position.
  - Careers that require long periods of sitting, sitting with legs crossed, or walking while bearing weight such as house-mover jobs.
  - Sports activities that require squatting.
- Inspection:
  - Ipsilateral foot is externally rotated when a patient who suffers from piriformis syndrome is relaxed in the supine position.
- Palpation:
  - Tenderness with palpation over the piriformis muscle, especially over the muscle's attachment superior to the greater trochanter.
  - Patients may also experience tenderness with palpation in the region of the sacroiliac joint, greater sciatic notch, and piriformis muscle.
  - Some patients have a palpable "sausage-shaped" mass in the buttock when lying in a prone position, which is a spasmed piriformis muscle.
- Range of motion:
  - Beatty's maneuver is a diagnostic test for piriformis syndrome. In this test, the patient lies on the unaffected side with the contralateral leg extended and lifts and holds the ipsilateral knee approximately four inches off the examination table. If sciatic symptoms are reproduced, the test result is positive.
  - The FAIR test is performed with the patient lying on the unaffected side in a lateral recumbent position with the contralateral leg extended, the ipsilateral side up, the hip flexed to an angle of 60 degrees, and the knee flexed to an angle of 60 degrees to 90 degrees. While stabilizing the hip, the examiner internally rotates and adducts the hip by applying downward pressure to the knee.

Differential Diagnosis

Prolapse of the Lumbar Intervertebral Disc versus Piriformis Syndrome

	Piriformis Syndrome	Prolapse of the Lumbar Intervertebral Disc
Location of Pain	Buttock and radiating along the dermatomal distribution	Lumbar and radiating along the dermatomal distribution
Tender Spots and	Deep in buttock	Beside the spinous process
Radiculopathy	Radiculopathy not obvious	Ipsilateral radiculopathy
Intra-Abdominal Pressure	Not obvious	Obvious
Piriformis Test	Positive	Negative
Piriformis Injection	Effective	Not effective
Straight Leg Raise Test (Physiopedia 2020)	Positive even greater than 70 degrees	Positive between 30–70 degrees
FAIR Test	Positive	Negative

TCM Diagnosis

- Cold damp Bi syndrome.
- Qi and blood stagnation.
- Liver and kidney deficiency.
- Wei syndrome.

Treatment

Treatment principle:

- Warm meridians, move Qi and blood, tonify liver and kidneys.

Herbal Treatments

Herbal formula:

- Formula for cold damp Bi syndrome: San Bi Tang 三痺湯.

Formula ingredients:

Huang Qi 黃耆 6 g, Xu Duan 續斷 6 g, Ren Shen 人參 6 g, Fu Ling 茯苓 6 g, Gan Cao 甘草 6 g, Dang Gui 當歸 6 g, Chuan Xiong 川芎 6 g, Bai Shao 白芍 6 g, Sheng Di Huang 生地 6 g, Du Zhong 杜仲 6 g, Niu Xi 川牛膝 6 g, Fang Feng 防風 6 g, Qin Jiao 秦艽 6 g, Rou Gui 桂心 6 g, Xi Xin 細辛 3 g, Du Huo 獨活 6 g, add Sheng Jiang 生薑 three slices, Da Zao 大棗 four pieces when cooking.

Ingredient explanations:

- Huang Qi 黃耆 tonifies Qi and blood and relieves numbness and pain.

- Xu Duan 續斷 tonifies the liver and kidneys, strengthens tendons and bones, promotes blood circulation, stops pain.
- Ren Shen 人參 tonifies Qi. When Ren Shen works together with Huang Qi 黄耆, they are good for Qi deficiency with general debility, anorexia, and spontaneous sweating.
- Fu Ling 茯苓 promotes urination, drains dampness, tonifies the spleen, and regulates the middle Jiao.
- Gan Cao 甘草 releases spasms, alleviates pain, harmonizes the other herbs, and guides the herbs to all 12 meridians.
- Dang Gui 當歸 moves and tonifies blood, disperses cold, and stops pain due to blood stasis.
- Chuan Xiong 川芎 moves blood and Qi, expels wind, and stops pain. When Chuan Xiong 川芎works together with Dang Gui 當歸, they are good for wind-damp invasion blocking the blood vessels causing pain, numbness, and paralysis.
- Bai Shao 白芍 nourishes the blood, calms liver Yang and liver wind, softens the liver, and alleviates pain. When Bai Shao 白芍 works together with Gan Cao 甘草, they are good for muscle spasms, pain, and numbness, especially in the calf, due to blood deficiency.
- Sheng Di Huang 生地 clears heat and cools the blood.
- Du Zhong 杜仲 tonifies the kidneys and liver, strengthens the tendons and bones, and aids the smooth flow of Qi and blood. When Xu Duan 續斷 works together with Du Zhong 杜仲, they are good partners to strengthen the lumbar by tonifying the liver and kidneys and securing the Chong and Ren meridians to release low back pain and weakness of knee joints, resulting in easier walking.
- Chuan Niu Xi 川牛膝 moves the blood, expels wind, drains dampness, tonifies the kidneys and liver, and strengthens the tendons and bones.
- Rou Gui 肉桂 warms kidney Yang and Ming Men Fire 命門火 and assists the generation of Qi and blood.
- Xi Xin 细辛 expels wind and internal cold and stops pain.
- Qin Jiao 秦艽 expels wind-dampness, opens the channels.
- Du Huo 獨活 dispels wind-dampness, stops pain.
- Fang Feng 防風 releases the exterior, expels external wind, expels wind-dampness.
- Da Zao 大棗 tonifies the spleen and stomach, nourishes the blood. With Sheng Jiang 生姜, enhances the effects of other Qi and blood, tonifying herbs by facilitating their digestion and absorption.
- Sheng Jiang 生姜 releases the exterior, induces perspiration, disperses cold, warms the middle Jiao, and reduces the toxicity of the other herbs.

Formula actions:

- Expels wind, drains damp, tonifies and moves blood, tonifies kidney Yang.

Modifications:

- For Bi syndrome with more wind, add Qiang Huo 羌活 and Gui Zhi 桂枝.
- For Bi syndrome with more cold, add Ma Huang 麻黃 and Zhi Fu Zi 附子.
- For Bi syndrome with more dampness, add Cang Zhu 蒼朮, Fang Ji 防己, and Bai Zhu 白朮.
- For Bi syndrome with more heat, add Sang Zhi 桑枝.
- For Bi syndrome with more blood stasis, add Tao Ren 桃仁 and Hong Hua 紅花.

Chronic Stage
Herbal formula: Du Huo Ji Sheng Tang 獨活寄生湯.

Formula ingredients:

Du Huo 獨活 9 g, Sang Ji Sheng 桑寄生 6 g, Du Zhong 杜仲 6 g, Niu Xi 牛膝 6 g, Xi Xing 细辛 3 g, Qin Jiao 秦艽 6 g, Fu Ling 茯苓 6 g, Rou Gui 肉桂 6 g, Fang Feng 防風 6 g, Chuan Xiong 川芎 6 g, Dang Shen 党参 6 g, Zhi Gan Cao 甘草 3 g, Dang Gui 當歸 6 g, Bai Shao 白芍 6 g, Shu Di Huang 熟地黃 6 g.

Ingredient explanations:

- Du Huo 獨活 expels wind, damp, and cold in lower Jiao and joints, removes chronic Bi stagnation in lower limbs.
- Xi Xing 细辛 enters Shao Yin meridian to expel wind and damp cold.
- Qin Jiao 秦艽 expels wind and damp, opens meridians, and benefits joints.
- Rou Gui 肉桂 warms meridians, expels cold, and opens meridians to promote blood circulation.
- Fang Feng 防風 expels wind and drains damp.
- Sang Ji Sheng 桑寄生, Du Zhong 杜仲, and Niu Xi 牛膝 treat deficiency of Qi and blood in the liver and kidneys and strengthen bone and tendons. Sang Ji Sheng 桑寄生 expels wind and drains damp. Niu Xi 牛膝moves blood and opens meridians to benefit joints and tendons.
- Dang Gui 當歸, Chuan Xiong 川芎, Shu Di Huang 熟地黃, and Bai Shao 白芍 nourish and harmonize blood.
- Dang Shen 党参, Fu Ling 茯苓, and Gan Cao 甘草 tonify spleen Qi.

Formula actions:

- Dispels wind, cold, and damp.
- Relieves painful obstruction.
- Supplements Qi and blood.
- Tonifies the liver and kidneys.

Contraindications:

- Damp heat excessive Bi syndrome.

Cautions:

- Avoid food that is raw, cold, or oily.
- Keep warm during windy and cold weather.

Modifications:

- With severe pain add Di Long 地龍, Xu Duan 續斷, Zhi Chuan Wu 制川烏, Zhi Cao Wu 制草烏, Bai Hua She 白花蛇, Xu Duan 川斷, Gou Ji 狗脊, Tu Si Zi 菟絲子, and Hong Hua 紅花.
- With more cold symptoms, add Zhi Fu Zi 附子 and Gan Jiang 乾薑.
- With more damp symptoms omit Shu Di Huang 熟地黃 and add Fang Ji 防己, Yi Yi Ren 薏苡仁, and Cang Zhu 蒼朮.

Formula for Qi and blood stagnation:
Herbal formula:

- Shu Jing Huo Xue Tang. 疏經活血湯.

Formula ingredients:

Dang Gui 當歸 2 g, Gan Cao 甘草 1 g, Bai Shao 白芍 2.5 g, Sheng Di Huang 生地黃 2 g, Cang Zhu 蒼朮 2 g, Niu Xi 牛膝 2 g, Chen Pi 陳皮 2 g, Tao Ren 桃仁 2 g, Wei Ling Xian 威靈仙 2 g, Chuan Xiong 川芎 1 g, Han Fang Ji 防己 1 g, Qiang Huo 羌活 1 g, Fang Feng 防風 1 g, Bai Zhi 白芷 1 g, Long Dan Cao 龍膽草 1 g, Fu Ling 茯苓 1 g, Sheng Jiang 生薑 3 g.

Ingredient explanations:

- Dang Gui 當歸 moves and tonifies blood, disperses cold, and stops pain due to blood stasis.
- Gan Cao 甘草 releases spasms, alleviates pain, harmonizes the other herbs, and guides the herbs to all 12 meridians.
- Bai Shao 白芍 nourishes blood, preserves Yin, calms liver Yang and liver Wind, softens tendons, and alleviates spasm and pain.
- Sheng Di Huang 生地黃 clears heat for heat symptoms due to Yin deficiency, cools the blood, nourishes Yin, generates fluids, and tonifies kidney Yin.
- Cang Zhu 蒼朮 strongly dries dampness and tonifies spleen.
- Niu Xi 牛膝 expels wind-damp, tonifies liver and kidneys, and directs herbs downward.
- Chen Pi 陳皮 regulates middle Jiao Qi, dries dampness, and transforms phlegm.
- Tao Ren 桃仁 breaks up stubborn blood stasis.
- Wei Ling Xian 威靈仙 expels wind-damp, stops pain.
- Chuan Xiong 川芎 moves blood and Qi, expels wind, and stops pain.
- Han Fang Ji 防己 expels wind-damp, stops pain, promotes urination, and reduces edema.

- Qiang Huo 羌活 expels wind, cold, and damp, unblocks painful obstructions, alleviates pain, and guides Qi to the Tai Yang and Du channels.
- Fang Feng 防風 releases the exterior, expels external wind, expels wind-damp.
- Bai Zhi 白芷 expels wind, drains dampness, dispels cold, and stops pain.
- Long Dan Cao 龍膽草 drains damp-heat from the liver and gall bladder channels.
- Fu Ling 茯苓 promotes urination and drains dampness.
- Sheng Jiang 生薑 releases the exterior, induces perspiration, disperses cold, warms the middle Jiao, and reduces the toxicity of the other herbs.

Formula actions:

- Unblocks meridians.
- Moves blood.
- Breaks up stubborn stasis.
- Expels wind and damp.

Contraindications:

- Avoid foods that are raw, cold, or damp in nature while taking this formula.

Modifications:

- For phlegm, add Zhi Tian Nan Xing 制天南星and Zhi Ban Xia 半夏.
- For Qi deficiency, add Ren Shen 人參, Bai Zhu 白朮and Gui Ban 龜板.
- For blood deficiency, add Dang Gui 黨參, Bai Shao 白芍, Shu Di Huang 熟地黃, and Chuan Xiong 川芎, E Jiao 阿膠, Dan Shen 丹參.

Acupuncture Treatments

Acupuncture point formula:

- *Patient position*: Lying laterally on the healthy side, the affected leg bent at both the hip and knee.
- *Acupuncture points*: UB31 上髎 Shangliao, UB32 次髎 Ciliao, UB33 中髎 Zhongliao, UB34 下髎 Xialiao, UB54 秩邊 Zhibian, GB30 環跳 Huantiao, GB39 懸鐘 Xuanzhong, UB40 委中 Weizhong, UB60 崑崙 Kunlun.
- For Bi syndrome with more wind and damp, add DU4 命門 Mingmen, UB12 大杼 Dazhu.
- For conditions showing more Qi and blood stagnation, add SP10 血海 Xuehai and UB17 膈俞 Geshu ipsilateral.
- For liver and kidney deficiency, add UB18 肝俞 Ganshu and UB23 腎俞 Shenshu; LIV8 曲泉 Ququan, LIV5 蠡溝 Ligou, SP6 三陰交 Sanyinjiao, and KD3 太溪 Taixi on the contralateral side.

Needling methods:

1) On UB31 上髎 Shangliao, UB32 次髎 Ciliao, UB33 中髎 Zhongliao, UB34 下髎 Xialiao, insert acupuncture needles into the sacral foramen.
2) Manipulate the needles for about two minutes to De Qi strongly.
3) On UB54 秩邊 Zhibian and GB30 環跳 Huantiao, insert acupuncture needles perpendicularly to De Qi as tight sensation. Apply even techniques to achieve the sensation of Qi as electrical current and the sensation radiating to the leg.
4) The rest of the acupuncture points are inserted with needles to only barely De Qi; keep the needle retention at 30 minutes and manipulate the needles once every 15 minutes.
5) Acupuncture treatment frequency is once a day.

Cupping Treatments

GB30 環跳 Huantiao, UB40 委中 Weizhong.

Moxibustion Treatments

Warming needling techniques on GB30 環跳 Huantiao, GB29 居髎 Juliao.

Bleeding Treatments

UB39 懸鐘 Xuanzhong, UB40 委中 Weizhong.

Electro-Acupuncture Treatments

GB30 環跳 Huantiao ~ UB4 0 委中 Weizhong 2/100 Hz for 20 minutes daily.

Tui Na Treatments

Tui Na techniques:

- Rubbing, pressing, plucking, kneading, pushing, rolling, grasping, and chafing

Tui Na procedures:

1) Acute piriformis injury:
   a) Patient is in a prone position with legs extended and the feet rotated inward to relax the piriformis muscles.
   b) The therapist stands near the affected side.
   c) Apply rubbing and pressing with hands on the sacrum and buttock regions for a few minutes.
   d) Palpate the affected buttock to locate the spasmed piriformis muscle and apply plucking techniques on piriformis muscle two to three times.
   e) Applies a pressing technique with thumb tip on UB40 委中Weizhong for one minute.

2) Chronic piriformis injury:

Step 1:

a) Patient is in a prone position; the therapist stands at the affected side.
b) Apply pressing and kneading techniques with the heels of hands on the affected buttock for a few minutes.
c) Overlap both thumbs and utilize them for pressing and pushing techniques along the muscle fiber five to seven times.
d) Apply pressing techniques with olecranon process of an elbow on the tender point on the affected buttock for five minutes.

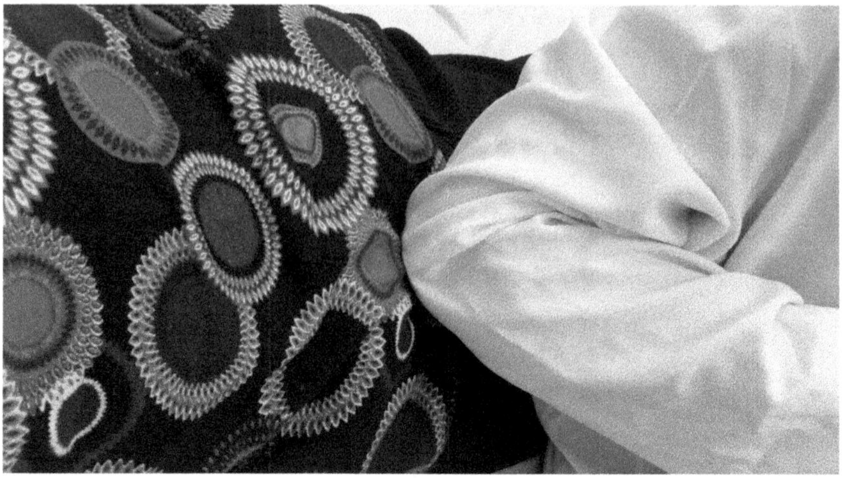

e) Apply rolling techniques on the leg along the dermatomal region for five minutes. The patient may need to lie on the lateral side if the leg pain is on the gall bladder meridian.

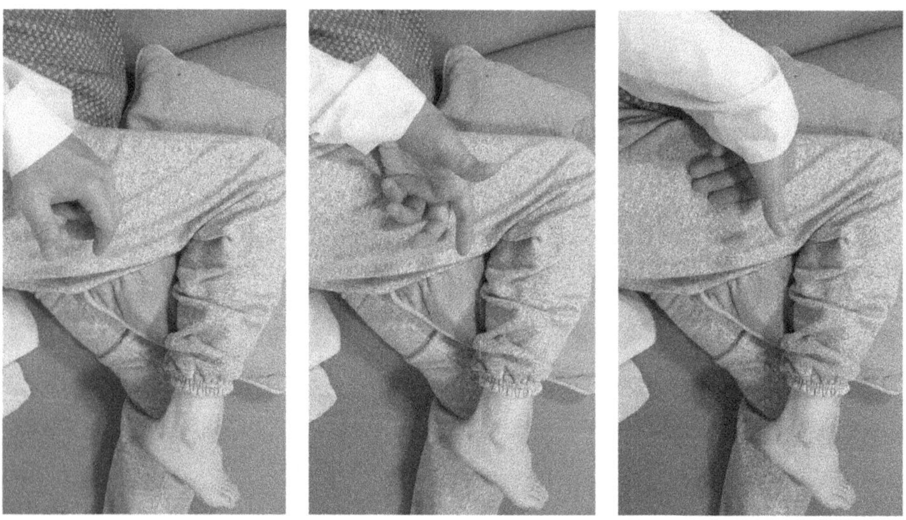

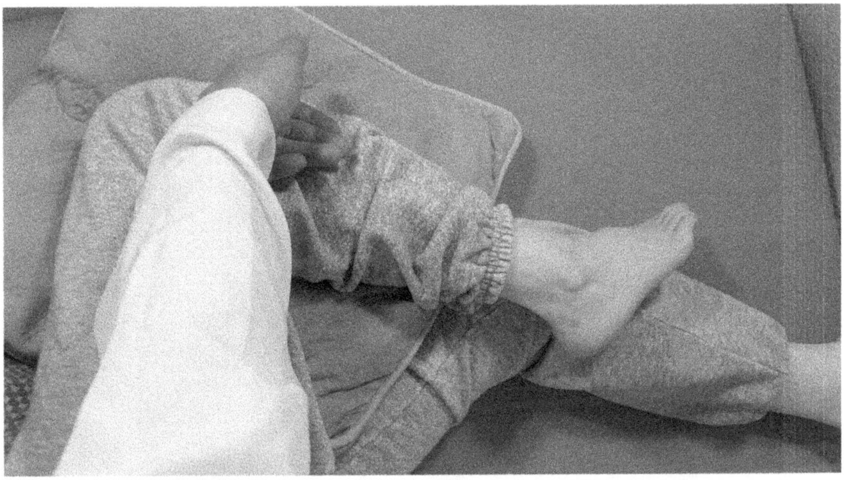

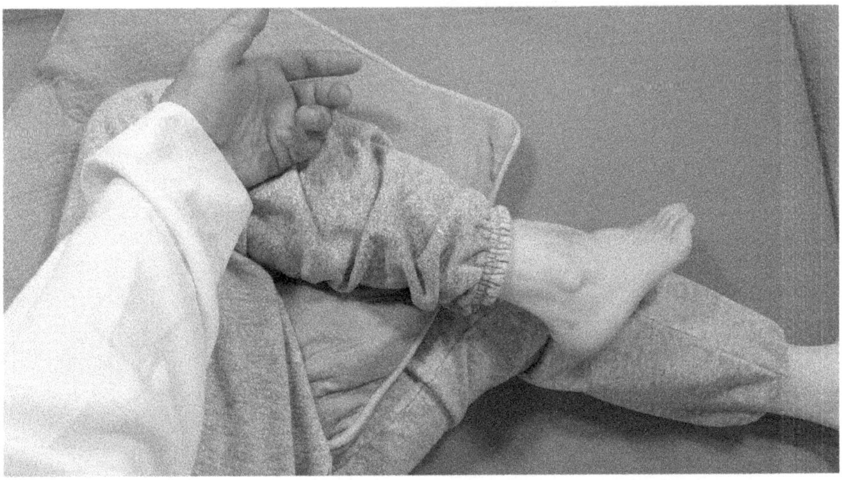

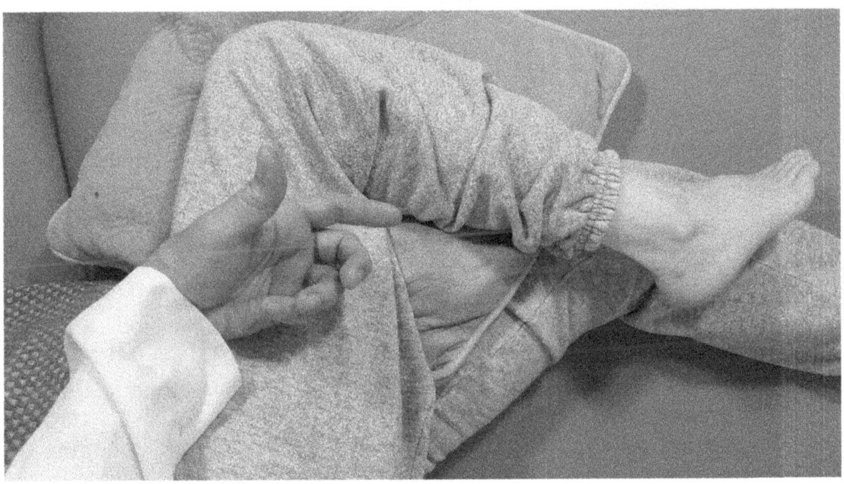

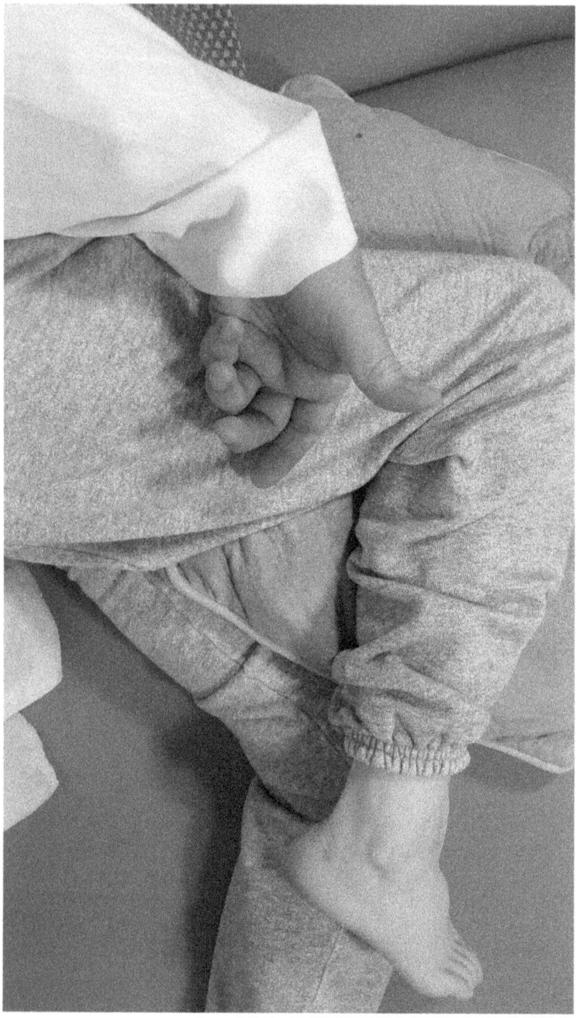

f) Apply pressing techniques on GB30 環跳Huantiao, UB40 委中Weizhong, UB57 承山Chengshan, UB60 崑崙Kunlun for one minute on each point.

Step Two:

a) Patient is in the supine position; the assistant secures the unaffected leg to prevent its movements.
b) The therapist stands at the affected side and grasps the ankle with one hand and places another hand on the knee, flexes the knee toward the ipsilateral chest, crosses the affected leg over to the non-affected leg (adduction), and pushes the knee over to the contralateral side (inward

rotation), crosses the foot to the non-affected leg, then extends the knee to extend the leg. The therapist repeats the procedures several times.

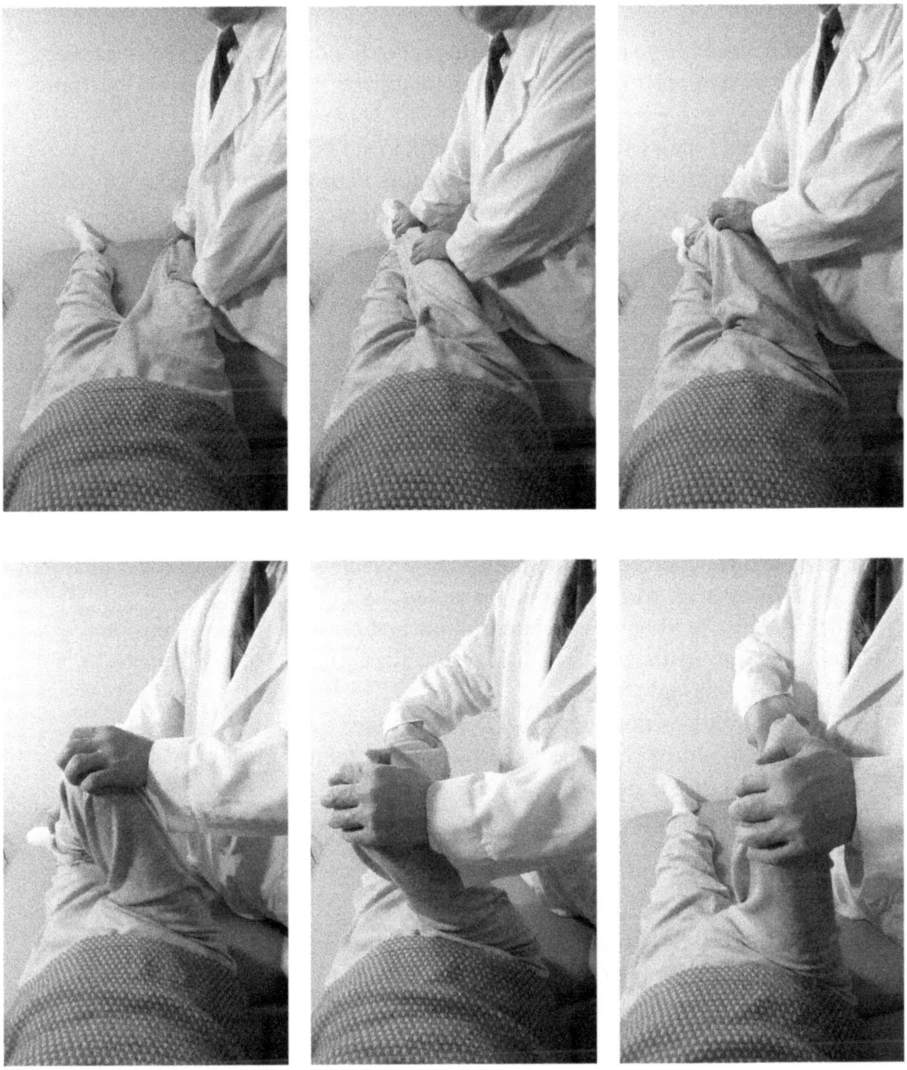

c) Apply the thumb pressing technique to the following acupuncture points: GB29 居髎 Juliao, GB31 風市 Fengshi, GB34 陽陵泉 Yanglingquan, GB39 懸鐘 Xuanzhong, and ST30 several times.

For cold sensation in the lumbar, buttock, and leg, Tui Na procedures may add chafing techniques on both sides of the lumbar, buttocks, and legs. The closing in Tui Na treatments is the phase of procedures to end the treatment. The main purpose is to promote blood and Qi flow. Palm striking is a common technique to promote the flow and thereby end the treatment.

Patient Advisory

- In an acute injury state, patients are recommended to avoid physical labor in the first three to five days, and treatment frequency should be more intensive.
- In a chronic injury state, patients should avoid heavy lifting.
- Patients are recommended to apply heating pads at home in a chronic state; 20 minutes daily is the recommended duration.
- Patients are recommended to avoid carrying a wallet in the back pocket, prolonged sitting, running on hills, or uneven surfaces.[12]
- Recommended foods that are those that tonify Yin and clear heat, including lotus seed, tomato, yam, lotus root, watermelon, pear, and chestnut.
- Avoid foods that are warm or hot and pungent include coffee, deep-fried food, lamb, hot pepper, onion, green onion, ginger, and tobacco.
- Recommended hip exercise:
  a) Lie supine on the bed or on a carpet floor with both legs extended. The head and back are flat on the floor.
  b) Cross the affected ankle over the contralateral quadricep tendon.
  c) Flex the contralateral leg until there is tension on the affected piriformis muscle. Keep the foot flat on the floor.
  d) Place a hand on the ipsilateral knee and press to increase tension.
  e) Lift the ipsilateral foot up off the floor if the buttock tension is not strong enough.
  f) Hold the position for five to thirty seconds.
  g) Gently return your legs to the original extended position.
  h) Repeat on the other side.
  i) Repeat the entire sequence two to three times.[13]

INJURY OF THE SUPERIOR CLUNEAL NERVE (ENTRAPMENT) 臀上皮神經損傷

Introduction

Injury of the superior cluneal nerve is not a rare clinical entity; it is another common cause of low back pain, and it is also called "pseudo-sciatica" (Konno et al. 2017). Approximately 50% of superior cluneal nerve disorder patients had leg symptoms.[14]

This pain causes hypersensitivity of the skin and subcutaneous tissues. Patients often refuse to sit on the ipsilateral side or prefer standing while being interviewed.

Clinically superior cluneal nerve entrapment is seen in older patients. The misdiagnosis rate is high.

Anatomy

The superior cluneal nerve originates from the lateral branches of the first to the third lumbar spine nerves.[15] These nerves are the terminal ends of the posterior primary[16] rami of lumbar spinal nerves and provide cutaneous innervation of the superior two-thirds of the buttock.[17]

The nerves pierce the back muscles on both the sacrospinalis muscle and the aponeurosis of the latissimus dorsi muscle emerges from the deep fascia just[18] over the iliac crest, comes out towards to surface, and innervates the skin over the superior two-thirds of the buttock.[19]

The nerves often go together with the arteries and veins when they pass through the iliac crest and the thoracolumbar fascia over the iliac crest. The lateral branch nerve of the L2 passes over the iliac crest together with vessels in the same fibrous sheath like in carpal tunnel syndrome (from outside to inside: N. → V. → A.) which is 1 cm long and covers the nerve and vessels closely inside.

The iliac crest is the place where the posterior layer of the thoracolumbar fascia[20] connects with the top of the iliac crest.[21] The direction of the fibrous tissues of the thoracolumbar fascia and iliac fascia is different.

The superior cluneal nerve lies on top of the iliac crest, just lateral to the quadratus lumborum muscle,[22] or it is 7 cm laterally to the midline, 1 cm inferiorly to the iliac crest.[23]

Western Medicine Etiology and Pathology

During physical labor such as leaning forward to pick up an object from the floor, the body leans forward and turns suddenly to either side. This is the main leading cause of an entrapment of the superior cluneal nerve. The entrapment increases pressure in the fibrous sheath and affects the cluneal nerve, resulting in pain.

Chinese Medicine Etiology and Pathology

According to Chinese medicine, this is due to Qi stagnation and over-exertion causing deficiency, and meridians are blocked by the external injury or the blockage from catching wind, damp, and cold. The diagnosis is muscle Bi syndrome or tendon Bi syndrome resulting in Qi and blood stagnation.

Manifestations

Superior cluneal nerve entrapment will present with pain over the posterior ilium and upper buttocks region. The therapist may find a tenderness along the rim of the posterior iliac crest, and a thickening tissue may be palpated.

Besides the pain over the buttock region, about half of the patients who suffer from injury of the superior cluneal nerve may also experience leg pain or tingling including all areas of the leg. Patients may complain of leg symptoms in a variety of areas from the groin to the sole of the foot. This often creates the potential for making a wrong diagnosis of other diseases that have similar radiculopathy such as sciatica, imitating radiculopathy from lumbar disc herniation, or lumbar spinal stenosis and piriformis syndrome.

1) *Lumbago*: Patients may have stabbing pain, sore pain, or bursting pain on one side of the lumbo-gluteal region. The patient has difficulty bending the lumbar, standing up after sitting for a period of time, or sitting down after standing for a moment.
2) *Radiating pain in the lower limb*: The pain may radiate to the lower limb in the acute stage but usually not beyond the knee.
3) *Palpation*: A movable, cord-like strip may be palpated directly below the midpoint of the Crista Iliaca, and the patient may experience distension, numbness, or unbearable pain under pressure on a strip.
4) *Deviation of the spinal cord*: Most patients may not have this deviation.
5) Numbness of the leg at the sciatic nerve area: Usually negative.
6) *Skin temperature*: Remains unchanged,
7) Straight Leg Raise Test: Negative.

Physical Examination

- Begin by obtaining the patient's medical history:
  - General health conditions such as leg length discrepancy, pregnancy.
  - Injury history to the sacroiliac joint such as a car accident or falling from a ladder.
  - Careers that may be vulnerable to injury, such as nurses and nursing aides who lift patients or office workers who sit with poor postures such as crossed knees.
  - Sports injury such as falling and directly attacking the sacroiliac joint.
- Inspection:
  - Usually there is an abnormal sign. Skin temperature remains unchanged.
- Palpation:
  - Tapping on the tender spot or thickened tissue may evoke a shock-like sensation. If this tapping produces an electrical shock-like sensation, it is considered a positive Tinel's sign but on the lower limb. This is strongly indicative of cluneal nerve involvement in the pain on the low back and leg.
  - A movable, cord-like strip may be palpated directly below the midpoint of the rim of the posterior Crista Iliaca.
- Range of motion:
  - Superior cluneal nerve entrapment patients often have a limitation in range of motion of the lower back; symptoms tend to worsen with flexion, and coupling this bending forward with rotation to the contralateral side further aggravates the pain symptoms.
  - The patient has difficulty bending the lumbar, standing up after sitting for a period of time, or sitting down after standing for a moment.

Differential Diagnosis

- *Sacral Iliac Joint (SI Joint) pain*:
  SI joint pain presents with localized pain that may be similar to the symptoms of superior cluneal nerve entrapment. Tenderness over the iliac crest is atypical.
- *Sciatica:*
  Patients who suffer from the other sciatic nerve pain due to radiculopathy from lumbar disc herniation, lumbar spinal stenosis, or piriformis syndrome usually also have tenderness in the gluteal regions and may also have tenderness on the iliac crest. The leg radiculopathy is on the L4-S1 dermatomal regions.
- *The third lumbar transverse process syndrome*:
  There is also referred pain to the buttock and leg; usually the referred pain doesn't pass below the knee. The trigger point is located at the L3 transverse process.
- *Piriformis syndrome*:
  Piriformis muscle is palpated at the affected buttock and is a trigger point. The piriformis test is positive.

Treatment

Treatment principle:

- Relaxes tendons, open the meridians.
- Releases the local soft tissue spasm.
- Promotes local blood circulation.

Herbal Treatments

Herbal formula:

- Du Huo Ji Sheng Tang 獨活寄生湯.

Formula ingredients:

- Du Huo 獨活 9 g, Sang Ji Sheng 桑寄生 6 g, Du Zhong 杜仲 6 g, Niu Xi 牛膝 6 g, Xi Xing 细辛 3 g, Qin Jiao 秦艽 6 g, Fu Ling 茯苓 6 g, Rou Gui 肉桂 6 g, Fang Feng 防風 6 g, Chuan Xiong 川芎 6 g, Dang Shen 党参 6 g, Zhi Gan Cao 甘草 3 g, Dang Gui 當歸 6 g, Bai Shao 白芍 6 g, Shu Di Huang 熟地黄 6 g.

Ingredient explanations:

- Du Huo 獨活 expels wind, damp, and cold in lower Jiao and joints, removes chronic Bi stagnation in lower limb.

- Xi Xing 细辛 enters Shao Yin meridian to expel wind, damp, and cold.
- Qin Jiao 秦艽 expels wind and damp, opens meridians, and benefits joints,
- Rou Gui 肉桂 warms meridians, expels cold, opens meridians to promote blood circulation.
- Fang Feng 防風 expels wind and drains damp.
- Sang Ji Sheng 桑寄生, Du Zhong 杜仲, and Niu Xi 牛膝 treat deficiency of Qi and blood in the liver and kidneys and strengthen bone and tendons. Sang Ji Sheng 桑寄生 expels wind and drains damp. Niu Xi 牛膝 moves blood and opens meridians to benefit joints and tendons.
- Dang Gui 當歸, Chuan Xiong 川芎, Shu Di Huang 熟地黄, and Bai Shao 白芍 nourish and harmonize blood.
- Dang Shen 党参, Fu Ling 茯苓, and Gan Cao 甘草 tonify spleen Qi.

Formula actions:

- Dispels wind, cold, and damp.
- Relieves painful obstruction.
- Supplements Qi and blood.
- Tonifies the liver and kidneys.

Contraindications:

- Damp heat excessive Bi syndrome.

Cautions:

- Avoid food that is raw, cold, or oily.
- Keep warm during windy and cold weather.

Modifications:

- With severe pain, add Di Long 地龍, Xu Duan 續斷, Zhi Chuan Wu 制川烏, Zhi Cao Wu 制草烏, Bai Hua She 白花蛇, Xu Duan 川斷, Gou Ji 狗脊, Tu Si Zi 菟絲子, and Hong Hua 紅花.
- With more cold symptoms, add Zhi Fu Zi 附子 and Gan Jiang 乾薑.
- With more damp symptoms, omit Shu Di Huang 熟地黃, and add Fang Ji 防己, Yi Yi Ren 薏苡仁, and Cang Zhu 蒼朮.

Acupuncture Treatments

Acupuncture point formula:

UB23 腎俞 Shenshu, UB54 秩邊 Zhibian, UB40 委中 Weizhong, UB57 承山 Chengshan, GB30 環跳 Huantiao and Ashi point (approximately 3 cm inferior to the highest iliac crest).

Ashi point needling techniques:

1) Insert a needle 0.35 mm × 75 mm perpendicularly to the Ashi point to the depth that the needle tip is able to touch the ilium. Apply even techniques to De Qi.
2) Insert four needles around the first needle as each needle approximately 2 cm distal superiorly, inferiorly, medially, and laterally to the first needle. The angle of these four needles is about 45 degrees toward the first needle. Needle retaining time is 30 minutes, manipulate the needles every ten minutes.

Gua Sha Treatments

Lower back, GB meridian on the hip, and UB meridian on the thigh

Cupping Treatments

UB 23 腎俞 Shenshu with a needle in inside, UB 40 委中 Weizhong with bleeding.

Seven-Star Needle Bleeding Treatments

- Ashi acupoint: Cupping after bleeding.
- Ashi acupoint (approximately 3 cm inferior to the highest iliac crest) with cupping treatment after bleeding.
- Point selected is Ashi point (approximately 3 cm inferior to the highest iliac crest).

Procedures:

- Prior to the procedure, tools should be prepared and ready to be used at the work area.
- Set up a clean field with paper towels.
- Place these tools in the clean field including dry sterile cotton balls, medical alcohol wipes or cotton swabs soaked in alcohol, sterile gloves, a biohazard sharp container, a biohazard trash container, goggles and facemask, paper towels, and a medical lancet or a traditional three-edged needle or the seven-star needle as a bleeding tool. The seven-star needle is selected for bleeding Ashi point (approximately 3 cm inferior to the highest iliac crest) in this case.
- The patient lies prone with the legs extended, Ashi point (approximately 3 cm inferior to the highest iliac crest) is facing upward.

- The room should be well lit.
- After the therapist washes hands thoroughly and wipes the hands with a single-use napkin, the therapist should put on sterile gloves.
- Wipe the acupuncture point Ashi point (approximately 3 cm inferior to the highest iliac crest) with a medical alcohol wipe or a cotton swab soaked in alcohol. Move from the center of the Ashi acupoint to the periphery.
- Place the seven-star needle at a perpendicular angle to the skin, hold the end of the handle.
- Flick the flexible handle to tap the needles onto the acupuncture point lightly and superficially to allow the skin to develop a mild rash. The flexible handle is flicked at a fast speed so it actually takes only a short time to get the rash result.
- Lastly, put a cup on to release more blood.
- Dispose of the tools in the proper containers.
- Wash the hands thoroughly.

Tui Na Treatments

Tui Na techniques:

- Plucking, rolling, pressing, kneading, stretching, rubbing, and pushing

Tui Na procedures:

1) The patient is in a prone position. The Tui Na therapist stands beside the patient.
2) Apply plucking techniques on the fibrous sheath perpendicularly several times.
3) Apply rolling, pressing, and kneading on the buttock with mild force to release the spasmed muscles.
4) Apply rubbing techniques on the nerves and vessels until the is skin warm to promote local blood circulation and disperse bruising.
5) The patient sits on a stool and the Tui Na therapist sits behind the patient. Palpate the lumbar to find the original groove of the superior cluneal nerve and the movable, cord-like strip. Then push the strip slightly upward with one thumb. Apply kneading, rubbing, and pressing to assist the nerve to return to its original groove with the other thumb.
6) Apply pressing technique with an elbow on the buttock.

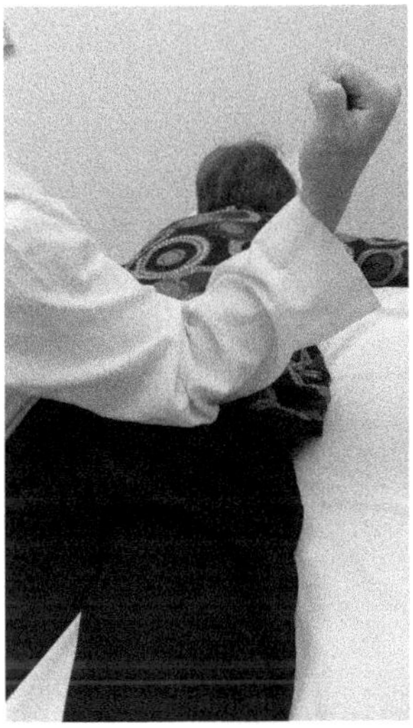
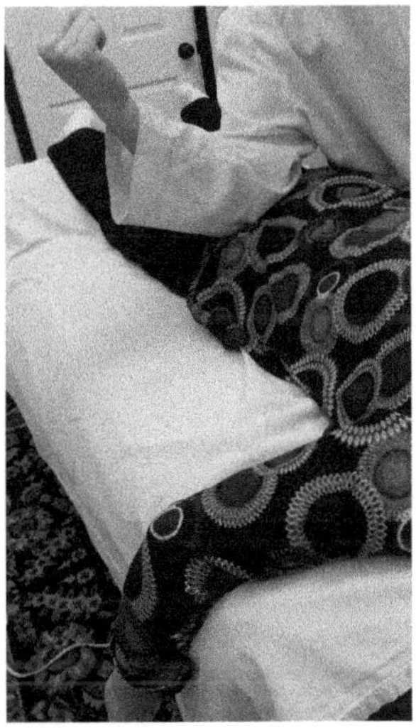
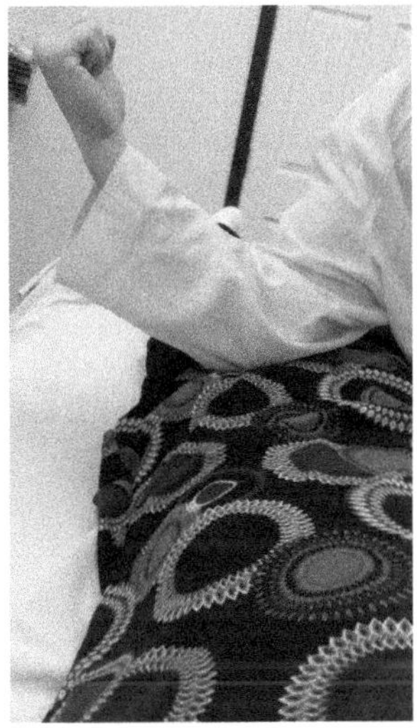

7) Apply stretching technique on the hip joint.

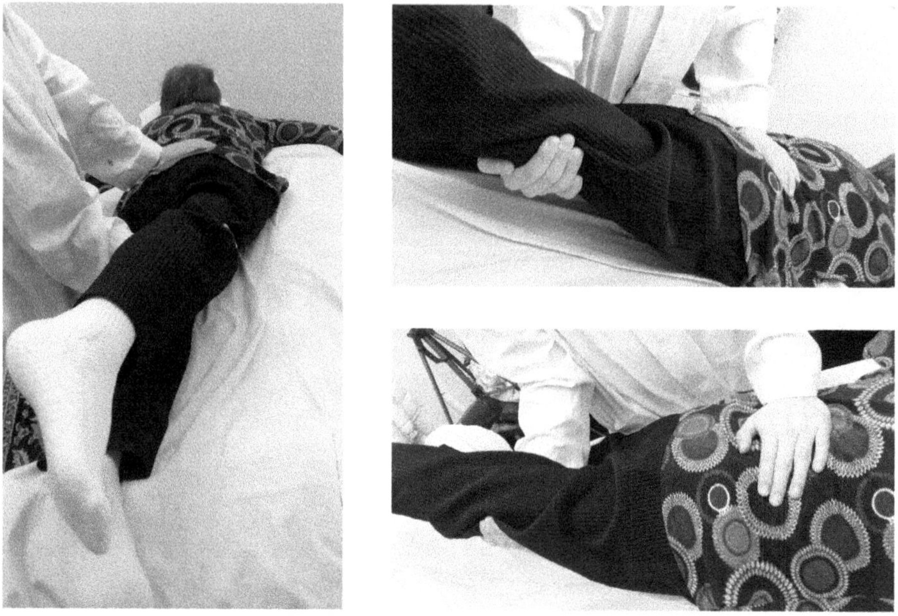

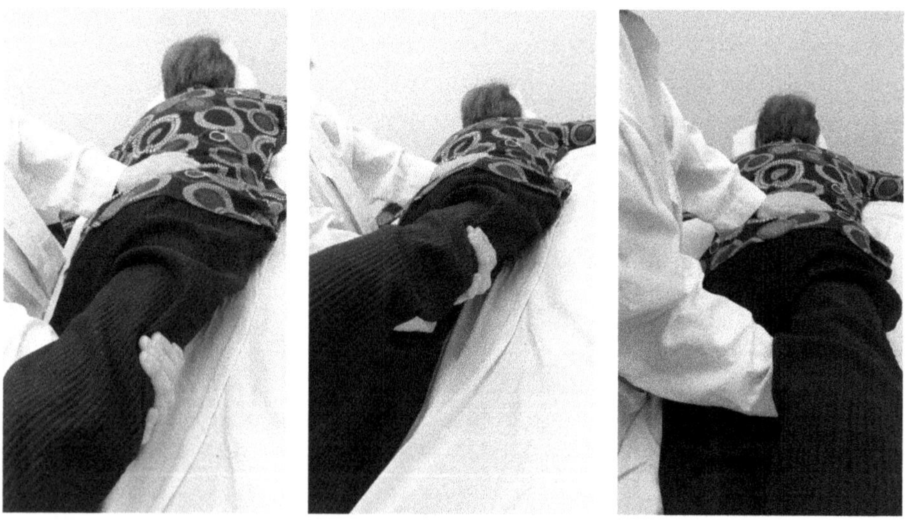

8) Apply herbal hot pad on the affected region.

Patient Advisory

1) Bed rest is necessary for the acute stage.
2) Avoid long period flexion and lumbar rotation.
3) Keep the affected region warm.
4) Avoid exposing the affected region when the weather is windy, rainy, and cold.
5) Diet should be light taste and consist of more green vegetables. Drink water regularly (about eight to ten cups daily). Increase fresh fruits, wheat, sunflower seeds, pork, and peanuts.
6) Routine stretching exercise is recommended after the acute stage; this lower body stretching exercise is helpful for pain recovery.
    a) Lie supine on a firm pad or carpet with both legs extended.
    b) Flex the affected knee and hip, place both hands on the patella.
    c) Continue to flex the leg and move the knee toward the opposite shoulder gently.
    d) Once it reaches the limit before feeling pain, hold it for five to thirty seconds. Be mindful, a gentle stretch is better than overstretch. Repeat five to ten times.
    e) Practice the same procedures on the other leg five to ten times.
    f) Practice the same procedures on both legs for five to ten times.
    g) Frequency is twice daily, once after waking up and once at night before sleep.

ADDUCTOR STRAIN 股内收肌群損傷

Introduction

Adductor strains are the common acute sports injuries amongst athletes such as soccer players who compete in sports that involve repetitive twisting, turning, sprinting, and kicking. These injuries also affect athletes who practice ice hockey, running, tennis, rugby, American football, basketball, breaststroke swimming, horse riding, and skiing.[24] Mostly, the injuries are due to excessive exertion during exercise, insufficient warm-up preparation, and improper exercise posture.

Anatomy

The muscles of the femoral adductor are composed of five muscles on the inner side of the femur. The superficial layer from the outside to the inside is the pectineus muscle, the adductor longus muscle, and the gracilis muscle. The deep side of the adductor longus muscle and the pectineus muscles are adductor brevis muscles, and the deep muscles of the muscles are the adductor magnus muscles, which are triangular.

Adductor Longus Muscle

It is a long and large flat muscle. It partially covers the adductor brevis and magnus. The muscle forms the medial border of the femoral triangle.

Attachments: It originates from the superior aspect of the pubic bone[25] under the pubic tubercle, and inserts at the medial ridge of linea aspera.

Nerve innervation: It is innervated by the anterior division (sometimes the posterior division) of the obturator nerve. The obturator nerve exits via the anterior rami of the spinal cord from L2, L3, and L4.[26]

Function: It adducts and medially rotates the thigh (References suggest it may also laterally rotate the thigh); it can also produce some degree of flexion/anteversion.

Adductor Brevis Muscle

It is a short muscle, lying underneath the adductor longus.

Attachments: It originates from the body of pubis and inferior pubic rami. It attaches to the linea aspera on the posterior surface of the femur, proximal to the adductor longus.

Nerve innervation: Obturator nerve (L2–L4).

Function: Adduction of the thigh.

Adductor Magnus Muscle

It is a broad triangular muscle forming a septum that divides the anterior (adductor) muscles from the posterior (ischiocondylar) aspect of the thigh. It is the largest muscle in the medial compartment. It lies deep to the adductor brevis and the adductor longus.[27]

Attachments: The anterior fibers originate from the inferior ramus of pubis and ischial ramus, and it attaches to the medial gluteal tuberosity, middle of linea aspera, medial supracondylar line, the posterior fibers arise from the ischial tuberosity. It attaches to the adductor tubercle of the medial condyle of the femur.[28]

Nerve innervation: The anterior muscle: Obturator nerve, L2,3,4; the posterior muscle: Tibial division of the sciatic nerve, L4.

Function: Adduction of the hip joint. The anterior fibers may assist with flexion, while the posterior fibers may assist in extension. Adductor magnus also stabilizes hips and lower limbs while walking and running to avoid excessive internal rotation.

Pectineus Muscle

The pectineus muscle is the most anterior adductor of the hip.

Attachments: It arises from the pectineal line of the pubis and to a slight extent from the surface of bone in front of it, between the iliopectineal eminence and pubic tubercle, and from the fascia covering the anterior surface of the muscle; the fibers pass downward, backward, and laterally to be inserted into the pectineal line of the femur which leads from the lesser trochanter to the linea aspera (Wikipedia 2019).

Nerve innervation: The femoral nerve, L2 and L3 (Wikipedia 2019). It may sometimes receive additional innervation from the obturator nerve called the accessory obturator nerve (20% of the population) (Wikipedia 2019).

Function: Adduction and external rotation of the thigh[29] but the primary function is hip flexion.

Gracilis Muscle

It is the most superficial muscle on the medial region of the thigh. It is thin, long, and slender, broad above, then narrows.

Attachments: It arises by a thin aponeurosis from the anterior margins of the lower half of the symphysis pubis and the upper half of the pubic arch.

The muscle's fibers run vertically inferiorly and eventually blend into a rounded tendon.[30] This tendon courses posterior to the sartorius tendon[31] and passes behind the medial condyle of the femur, curves around the medial condyle of the tibia where it becomes flattened, then inserts into the upper part of the medial surface of the body of the tibia, below the condyle. At its insertion, the tendon is situated immediately above that of the semitendinosus muscle, and its upper edge is overlapped by the tendon of the sartorius muscle, which it joins to form the pes anserinus. The pes anserinus is separated from the medial collateral ligament of the knee-joint by a bursa.

Nerve innervation: Obturator nerve

Function: Adduction and medially rotation, lateral rotation, hip flexion, and knee flexion.[32]

External Obturator Muscle

It is a flat, triangular muscle, which covers the outer surface of the anterior wall of the pelvis.

Attachments: It arises from the anterior surface of the obturator membrane[33] and surrounds the anterior bony margins of obturator foramen[34] from the inferior pubic ramus and the ramus of the ischium. It also arises from the medial two-thirds of the outer surface of the obturator membrane and from the tendinous arch which completes the canal for the passage of the obturator vessels and nerves.

The fibers springing from the pubic arch extend on to the inner surface of the bone, where they obtain a narrow origin between the margin of the foramen and the attachment of the obturator membrane.

The fibers converge and pass posterolaterally and upward and end in a tendon which runs across the back of the neck of the femur and the lower part of the capsule of the hip joint and is inserted into the posteromedial surface of the greater trochanter of the femur.[35]

Nerve innervation: Posterior division of obturator nerve innervates most of the obturator externus, L3, L4.[36]

Function: Lateral rotation of the hip and stabilization of the hip joint as a postural muscle.

Western Medicine Etiology and Pathology

The main functions of the adductor muscle group are first to stabilize the pelvis when standing and then to allow the hip joint to create adduction torque, which means under the torque, the hip is able to bring the lower extremity toward the midline. In some sports, such as horseback riding, hockey, football, rugby, skating, skiing, climbing, breaststroke swimming, and soccer, this torque plays an important role.

- Because of this torque, adductor injury is commonly the result of forced aggressive external rotation of an abducted leg. The athletes who play the above sports must externally rotate the hip suddenly from the inward direction. Such rotating movements often lead to injuring the insertion of the tendon of the adductor muscle to the bone[37] and may lead to partial or complete tearing of the muscle and tendon in severe cases. Tears can occur at any part of the muscle from muscle origin or insertion or the muscle-tendon junction to certain parts within the belly of the muscle(s). Most commonly, tears occur at the muscle-tendon junction. Uncommonly, the tendon injury occurs at the site of its bony attachment.[38]
- Rapid starts and stops while running and jumping are another chief cause of this injury. The constant change of load and over-stretching heavily stress the adductor tendons, leaving athletes more prone to adductor sprain and groin injuries.
- A sudden increase in physical training or a type of training such as the weekend-warrior.
- Long term overuse with repetitive abduction may lead to chronic injury to the adductor muscle group.[39]

Although there are several muscles located at the medial aspect of the thigh, the most common injury is the adductor longus.

Chinese Medicine Etiology and Pathology

Acute injury:

1) *Forceful contraction*: This is the most common groin injury in athletes who strain the adductor muscle group. This is one of the fundamental injury mechanisms of acute adductor longus injuries; the adductor attempts to decelerate an extending, abducting leg by using an eccentric contraction to adduct and flex the hip.
   a) Change of hip direction during hip extension and abduction with external rotation.
   b) Kicking and jumping that forcefully change the hip from extension to flexion and the leg from abduction to adduction with external rotation.

2) *Direct blunt trauma*: In the acute state, a direct contact injury to the soft tissues results in muscle hematoma.[40]
3) The theory of the etiology and pathology behind the circulation of Qi and blood in Chinese medicine is that Qi and blood flow are blocked by

the local hemorrhaged almost immediately after the violent direct-attack injury. Blockage causes pain in the local region; this is the diagnosis of Local Qi and Blood Stagnation.

Chronic injury:

1) Incomplete treatment of the acute state trauma becomes chronic pain when it receives improper treatment.
2) A healthy body is the result of a balance between Yin and Yang; Yin and Yang always support each other to keep the balance. Trauma or prolonged illness leads to consumption of Qi and blood and deficiency of liver and kidneys because the balance is disturbed. The Yin and Yang are no longer fit; the body becomes vulnerable to more injury or the wound or illness becomes chronic. When the body is weakened by an imbalance of Yin and Yang, climatic variations and environmental change may induce the invasion of exogenous pathogenic factors such as wind, cold, damp, and heat. When the body constitution is Yang deficient, Yang Qi is under a consequence of cumulative physical microtrauma such as the reoccurrence of injury in the same region or the same tissue, overuse of the adductor muscles, or repeated minor injury. Yang Qi is even more deficient from the result of over-draining of its energy. According to Sù Wèn 43 (痹論篇第四十三), "The presence of cold means a famine[41] of Yang Qi and a predominance of Yin Qi. Being affected[42] by the disease, cold manifests.[43] 其寒者, 陽氣少, 陰氣多, 與病相益, 故寒也。." When the constitution is cold, it is susceptible to the effects of wind, damp, and cold; the accumulation of these factors leads to stagnation of Qi and blood. This is diagnosed as wind damp cold Bi syndrome.

According to Sù Wèn 43 (痹論篇第四十三), "When a medley of the three Qi of Wind, Cold, and Damp reaches the body, they[44] form Bi obstructer.[45] In this condition, if the Wind prevails, the patient develops Wandering Bi Syndrome; if the Cold prevails, the patient develops Painful Bi Syndrome; if the damp prevails, the patient develops Fixed Bi Syndrome. 岐伯對曰: 風寒濕三氣雜至, 合而為痺也。其風氣勝者為行痺, 寒氣勝者為 痛痺, 濕氣勝者為著痺也。"[46] (Unschuld 2011).

Manifestations

- Patients is usually able to describe the history of injury.
- Sudden onset of pain, sometimes accompanied by the sensation of a "pop" in the inner thigh.
- Inability to continue activity after initial onset of pain.[47]
- Patients refuse palpation or pressure in the acute state.
- In the acute state, the affected region may have more pain and stiffness during the night or when getting up in the morning or getting up from a period of time of sitting.

- Patients may notice a crunchy sound or feeling when moving the leg.[48]
- Location of pain commonly is at the pubic area especially during abduction and lateral rotation of the hip and leg.
- Walking with a limp.
- In the chronic state of wind damp cold Bi syndrome, the pain is aggravated when the weather is rainy, cold, and windy.
- Patients complain of more pain when they are fatigued.

Physical Examination

- Begin by obtaining the patient's medical history:
  - After injury, history and physical exams are usually sufficient to establish the diagnosis.
  - Sports injury to the adductor muscles is from forced aggressive external rotation of an abducted leg. Sports involving this movement include ice hockey, soccer, running, tennis, American football, horseback riding, and skiing.
  - Careers that may be vulnerable to the injury include ice hockey and soccer players. Approximately 10% to 11% of all injuries are groin strains (Tyler et al. 2010).
  - Injury history that can irritate the adductor muscles includes patients who are not doing lower limb exercise regularly but are involved in excessive physical activities such as hiking on a challenging hiking trail on a weekend.
- Inspection:
  - Bruising.
  - Swelling may be present.
- Palpation:
  - Tenderness to palpation at the affected area.
  - Medial thigh is warm to touch in acute injury.
  - Patients usually refuse touching because of the severe pain.
- Range of motion.
  - The test reveals the ranges of motion of the leg joints are usually within normal ranges.
  - Pain may be induced when resisting the examiner's abduction when adducting the leg.

Differential Diagnosis

Epididymitis symptoms:

1) Pain or discomfort in the lower abdomen or pelvic area.
2) Testicle pain and tenderness, usually on one side, that usually comes on gradually.
3) Painful urination or an urgent or frequent need to urinate.

4) Discharge from the penis.
5) Blood in the semen.

Treatment

Treatment principles:

1) Relaxes tendons.
2) Opens meridians.
3) Moves blood, disperses bruise.
4) Releases adhesion.
5) Releases spasm.
6) Stops pain.

Herbal Treatments

Herbal formula:

- Shen Jin Huo Xue Tang 伸筋活血湯.

Formula ingredients:

Shen Jin Cao 伸筋草 9 g, Chuan Niu Xi 川牛膝 9 g, Zhi Gou Ji 制狗脊 9 g, Qin Jiu 秦艽 4.5 g, Dang Gui 當歸 9 g, Sang Ji Sheng 桑寄生 9 g, Mu Gua 川木瓜 9 g, Bai Shao 杭白芍 9 g, Xu Duan 續斷 9 g, Ru Xiang 乳香 6 g, Mo Yao 没藥 6 g, Du Zhong 杜仲 9 g, Gan Cao 甘草 6 g.

Ingredient explanations:

- Shen Jin Cao 伸筋草 expels wind, drains damp, relax tendons.
- Chuan Niu Xi 川牛膝 moves the blood, expels wind, drains dampness, tonifies the kidneys and liver, and strengthens the tendons and bones.
- Zhi Gou Ji 制狗脊 tonifies liver and kidneys, strengthens tendons and bones, expels wind damp, and invigorates blood flow.
- Qin Jiu 秦艽 expels wind damp and relaxes the sinews.
- Dang Gui 當歸 moves and tonifies blood, disperses cold, and stops pain due to blood stasis.
- Sang Ji Sheng 桑寄生 expels wind damp, tonifies liver and kidney Yin.
- Mu Gua 木瓜 promotes Qi and blood circulation and unblocks the channels, releases spasm.
- Bai Shao 白芍 nourishes blood, preserves Yin, calms liver Yang and liver wind, softens tendons, and alleviates spasm and pain.
- Xu Duan 續斷 nourishes liver and kidneys, strengthens bones and tendons, alleviates pain.
- Ru Xiang 乳香 moves blood, stops pain.
- Mo Yao 没藥 moves blood, reduces swelling, and stops pain.
- Du Zhong 杜仲 tonifies the kidneys and liver, strengthens bones and tendons, and lowers blood pressure.

- Gan Cao 甘草 releases spasms, alleviates pain, harmonizes the other herbs, and guides the herbs to all 12 meridians.

Formula action:

- For acute and chronic soft tissue injuries.

Acupuncture Treatments

Acupuncture point formula:

UB37 殷門Yinmen, UB40 委中Weizhong, UB57 承山Chengshan, SP12 沖門Chongmen and Ashi acupoints medial to UB36 承扶Chengfu; Ashi acupoints 阿是穴 along the adductor muscles.

Cupping Treatments

SP10 血海Xuehai.

Procedures:

1) The patient is wearing shorts and lies supine with the legs extended and cupping region facing upward; the therapist stands next to the knees, holding a cupping bowl in the left hands hand and a hemostat in the right.
2) Securely clamp a cotton ball into the hemostat, dab the cotton ball with isopropyl alcohol, squeeze off extra alcohol.
3) Light the cotton ball.
4) Hold the cupping bowl several inches above the skin, and place the lit cotton ball into the cupping bowl. Withdraw the cotton ball then quickly place the lip of the cupping bowl onto the acupuncture point SP10 血海 Xuehai.
5) Leave the cupping bowl on the skin for less than ten minutes.
6) To remove the cupping bowl, simply push a finger under the lip of the cupping bowl to allow air into the bowl and remove it.
7) Remind the patient that there might have some side effects during or after the treatment, including lightheadedness or dizziness, sweating, or nausea, and the skin around the rim of the cup may become irritated and may leave a circular mark, pain, and hematoma.

Moxibustion Treatments

Warming needling techniques on UB23 腎俞 Shenshu, UB25 大腸俞 Dachangshu.

Bleeding Treatments

SP1 隱白 Yinbai.

Electro-Acupuncture Treatments

UB37 殷門 Yinmen ~SP12 沖門 Chongmen 2/100 Hz for 20 minutes daily.

Tui Na Treatments

Tui Na techniques:

- Pressing, grasping, lifting, stretching, rotating, locking, pushing.

Tui Na procedures:

1) Finger pressing techniques:
   a) Patient is in a supine position; the therapist stands beside the ipsilateral side of the thigh.
   b) First search the medial side of the thigh to look for the tender points. Then apply the fingertips pressing techniques on the tender points. Always begin with light force, gradually increase the pressure until feeling a bouncing sensation, then hold that pressure.
   c) Continue with the pressure until the bouncing sensation slowly decreases, then increase pressure with the fingertips until the patient feels stabbing pain. This is beneficial for opening channels, regulating Qi, and blood flow.
2) Rotation techniques:
   a) The patient is in a supine position. The therapist stands next to the affected leg (the right, in this example).
   b) Place the left hand on the groin/hip region with the left thumb pressing on the tender point at about the inguinal area; the rest of the left hand presses the hip.
   c) Grasp the right ankle with the right hand, and lift and stretch the leg.
   d) Based on step (a), press the right hip joint with the left hand at the same time the right hand is lifting and stretching the right leg.

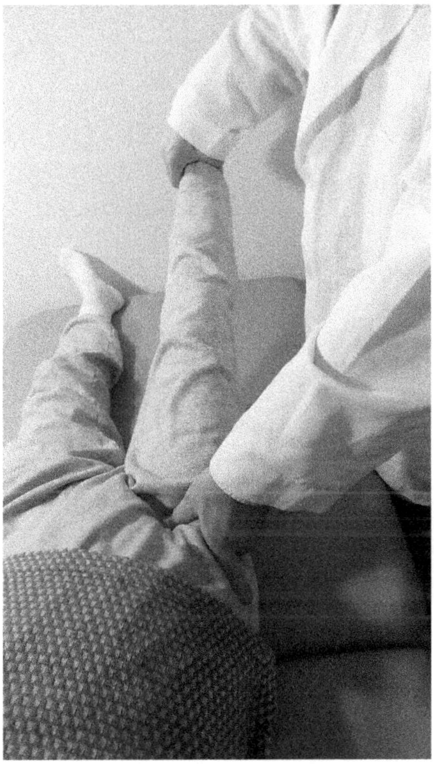

e) Apply rotating techniques with the left hand while still stretching and pressing. Start with small rotating, and reverse from the big ones. Total number of rotations is about ten times each direction.

3) Stretching techniques:
   a) While the patient is still in the supine position, the therapist stands near the affected leg.
   b) After the rotating techniques, the therapist places the patient's right ankle at the armpit and applies a locking technique by pressing the ankle toward the chest with the right arm.
   c) The therapist places the left hand at the patient's right popliteal fossa and flexes the patient's right knee and right hip by lifting the knee upward with the left hand and sending the knee toward the patient's chest.

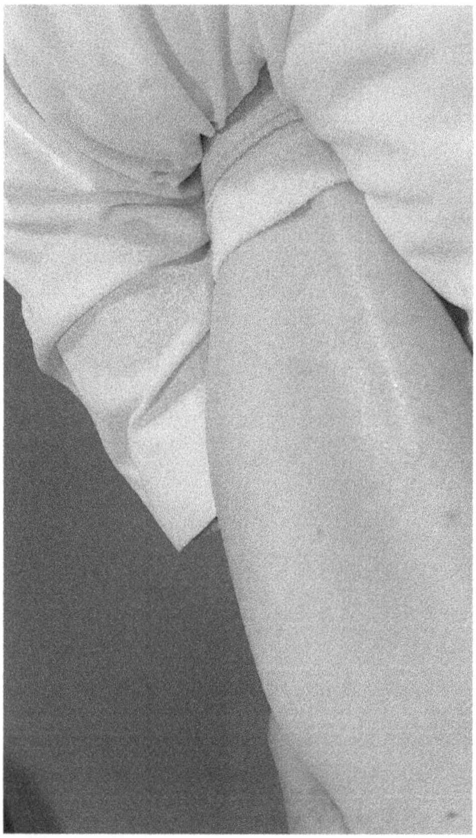

   d) The therapist applies pushing techniques with the right hand from the medial side of the knee toward the groin five to seven times.

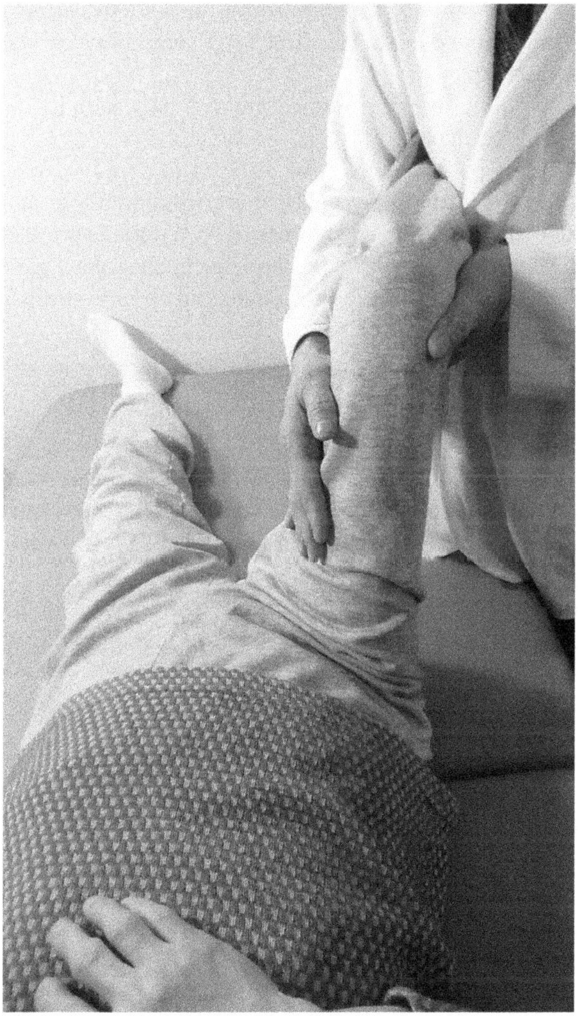

e) Straighten the patient's right leg on the treatment table.

Patient Advisory

1) Patients should be advised to rest the affected leg for two weeks after injury.
2) Exercise begins when pain is lessened. Training starts with some easy movements on a bed such as abducting the affected legs to stretch the adductors. Usually, exercise begins one week after injury to prevent muscle atrophy.
3) Crutches using may be used for several weeks for a severe strain.
4) Do your best to heal the acute strain, to prevent it from becoming chronic.

5) To prevent future strain, proper warming up and stretching before sport activity is highly beneficial. The stretching should not be just the leg, but also include the hip.
6) If there is any future injury, patients should stop the activity as soon as possible and seek treatment.[49]
7) In terms of diet, patients should be advised to avoid Fa Wu diet. Fa Wu foods are the foods that may cause more inflammation, as the foods to bring old injuries back. Some examples of Fa Wu foods are shellfish such as crabs and shrimp, fish, fried foods, lamb, duck, chocolate, garlic, and green onions. This concept applies to both acute and chronic strains.

SUBLUXATION OF SACROILIAC JOINT 骶髂關節紊亂症

Introduction

Subluxation or dislocation of the sacroiliac joint causes severe biomechanical dysfunction leading to inflammation of the surrounding soft tissues and nerves. It manifests similarly to many other pain disorders such as back, hip, and leg pain, and it often masquerades as sciatic pain or fibromyalgia.[50]

Sacroiliac joint pain from injury is commonly seen in patients involved in motor vehicle accidents, falls from ladders, and nurses or nursing aides who lift patients. The pain is also commonly seen in women after childbirth.[51]

Anatomy

The Sacroiliac joint is the articulation of the sacrum and the ilium. The bones are tethered together by strong ligaments just like other joints. There are two joints, one on each side of the sacrum. Both joints are generally considered as an immovable joint.

These joints are strong weight-bearing joints in the pelvis; they are as a single unit to offset the load of the upper body to the lower extremities and provide shock absorption for the spine to prevent spine injury.[52]

Western Medicine Etiology And Pathology

1) The sacroiliac joints are supposed to be immovable joints. Unexpected jarring or pounding-type accidents are the most common traumas leading to the subluxation. Examples are some types of sporting accidents, falling from a high place, tripping while hiking on a mountain trail, or lifting a heavy person with improper posture.

   Subluxation of the sacroiliac joint causes the bones to be slightly displaced. The joint is partially dislocated, which results in pain at the lower portion of the back and hip because it stresses and causes inflammation on the ligaments and other soft tissues that tie the joint together and puts pressure on the surrounding structures.[53] The inflammation causes swelling and puts pressure on the nerves.

Most sacroiliac joint injuries are caused by subluxation. In rare cases, the joints can become completely dislocated with extreme trauma such as injury from car accidents.[54] Trauma is the top leading cause of sacroiliac joint dislocation.

2) Poor posture while sitting at a desk working on a computer for hours or driving long distances causes pressure on the sacroiliac joints and strains the ligaments that hold the joint together.

Sleeping on a side, especially in a semi-prone position, causes compression in the sacroiliac joints.

Sitting with one leg crossed over the other and sitting with 4-figure position looks so comfortable, but these poor sitting postures also put pressure on the soft tissues surrounding the sacroiliac joints.[55]

3) The hormone Relaxin (Goldsmith and Weiss 2009) released in a pregnant woman's body allows ligaments to loosen, which is a normal part of pregnancy to prepare the pelvis for childbirth. The laxity of the ligaments allows the immovable joints to be overly mobile and leads to increase stress and abnormal wear on the surrounding soft tissues and bones, causing inflammation.

As the pregnancy progresses, the uterus becomes heavier, the increased weight is carried in the front of the body, the tendons and muscles at the lumbar and hip regions naturally lean backward and work much harder, resulting in strain on the sacroiliac joints.[56]

During pregnancy, the abdominal muscles become stretched and weaken as the fetus grows and they become less able to support the spine.

4) When a person has a leg length discrepancy, the person has one leg that is shorter or longer than the other. The abnormal alignment may end up causing stress and may cause malalignment of the sacroiliac joints, resulting in inflammation and pain.

5) When inflammation becomes chronic, the inflammatory types of arthritis can cause sacroiliac joint pain and may cause degenerative change leading to complete sacroiliac joints fusion.[57]

Chinese Medicine Etiology and Pathology

According to Chinese medicine, this is due to Qi stagnation and over-exertion causing deficiency. Meridians are blocked by the external injury or the blockage from catching wind, damp, and cold. This is a tendon Bi syndrome resulting Qi and blood stagnation.

Manifestations

- Pain is the main symptom of the disease. Most patients with dysfunction of sacroiliac joints suffer from pain after an injury or sitting a long period of time. Some may not have a history of injury.

- It is characterized by persistent dull achy pain, deep, diffusing, and constant. It is aggravated after exertion, prolonged positions such as sitting, standing, walking, or lying and released after rest. As it progresses, joint activity can be limited by pain, and pain can even occur during rest.
- Morning stiffness is also a complaint among older patients with dysfunction of sacroiliac joints. It is stiff at the beginning of movement and relieved by activities. Patients report that frequent position changes are needed to maintain any degree of comfort.
- Locations of pain are various, the low back near the Posterior Superior Iliac Spine (PSIS) region is most commonly in the acute state of pain. Buttock, hip, groin, and iliac crest are also regions where patients with dysfunction of sacroiliac joints complain of pain. Some also complain that the pain radiates to the buttock and the thigh.

There are two types in Chinese medicine:

1) Acute state:
   a) *Qi stagnation type*: There is a history of trauma. The pain intensity varies; sometimes it is severe, but sometimes it is mild. The location of the pain is wide with local swelling, and patients refuse pressure at the affected area. Patients have a difficult time with sitting, sleeping, and movement. Local swelling. Coughing may induce the pain. There is a history of trauma.
      - Tongue: Body is thin with thin coats.
      - Pulse: Wiry, tight, and rapid.
      - Traditional diagnosis: Qi and blood stagnation
   b) *Blood stasis type*: There is a history of trauma. Location of pain is unilateral. It is bruised, lumbar movement is limited, and the patient may have abdominal distension and constipation.
      - Tongue: Body is purple, dim, and bruised
      - Pulse: Choppy and tight.
2) Chronic state: Laxity tendons, sore and weak low back and knees, feeling cold in lower extremities. Fatigue when sedentary and long-distance walking. Preference pressure on the affected area. Pain is aggravated after exertion. Frequent urination, may be incontinent.
   - Tongue: Body is pale with thin coats.
   - Pulse: Deep and thin.
   - Traditional diagnosis: Liver and kidney deficiency.

Physical Examination

- Begin by obtaining the patient's medical history:
  - General health condition such as leg length discrepancy, pregnancy.
  - Injury history to the sacroiliac joint such as a car accident or falling from ladders.

- Careers that may be vulnerable to the injury such as nurses and nursing aides who lift patients, office workers who sit with poor postures such as crossed knees.
- Sports injury such as falling and directly attacking the sacroiliac joint.
- Inspection:
  - Patients often have to change postures quite frequently.
  - Coughing may induce the pain.
  - Bruising may present in acute injury.
- Palpation:
  - Acute injury patients refuse pressure on the sacroiliac joint while chronic pain patients prefer pressure.
  - Location of pain may be unilateral.
- Range of motion:
  - Lumbar movement is limited.

While one positive test raises suspicion, three or more positive tests would confirm the sacroiliac joint as a source of the lumbar pain.[58]

1) Seated flexion test (or Piedallu's sign): The patient is seated with the feet flat on the ground; the examiner sits or stands behind the patient, then locates the patient's iliac crest and rests the thumbs on it. Ask the patient to flex the back; the positive result presents when the affected side moves the thumb upward.
2) FABER test (Patrick's test): The patient lies supine, the tested leg is placed into flexion, abduction, and external rotation, and the ankle is resting on the contralateral thigh proximal to the knee. While stabilizing the opposite side of the pelvis at the anterior superior iliac spine, an external rotation, abduction, and posterior force are then lightly applied to the ipsilateral knee until the end of range of motion is achieved.

   The test result is positive when there is reproducing pain at the sacroiliac joint or the joint's range of movement is limited.[59]
3) Compression test:

   The patient lies on their side: The examiner stands behind the patient with the arms extended and hands overlapped over the upper part of the patient's iliac crest and exerts a downward force toward the floor. Increased pain in the sacroiliac joints indicates a positive sign.[60]
4) Distraction:
   - The patient lies supine with the hands placed at the low back.
   - The examiner crosses the arms with elbows straightened and places the hands on the anterior and medial aspects of the patient's left and right Anterior Superior Iliac Spine (ASIS).
   - Slowly and steadily presses both hands posteriorly (downward pressure).
5) Yaoman's test:
   - The patient lies prone.

- The examiner flexes the patient's test knee to 90 degrees and then hyperextends the same hip.
- The positive result is indicated by increasing pain localized to the sacroiliac joints; this indicated the pathology of the anterior sacroiliac ligaments.[61]

Differential Diagnosis

Ankylosing Spondylitis

Both sacroiliac joins dysfunction and ankylosing spondylitis have several similar symptoms. They are stiffness and pain in the lower back and buttocks. The pain is usually dull and diffuse, and it is worse in the morning and during the night, improved by a warm pad and light exercise.[62]

Ankylosing spondylitis cannot be improved by bed rest.[63]

Treatment

Treatment principle:

- *Acute state*: Moves Qi and blood, relaxes tendon, relocates joints.
- *Chronic state*: Tonifies liver and kidneys, warms meridians, releases adhesion.

Herbal Treatments

Qi Stagnation

Treatment principle:

- Moves Qi, opens meridians, stops pain.

Herbal formula:

- Ze Lan Tang 澤蘭湯.

Formula ingredients:

Ze Lan 澤蘭 9 g, Mu Dan Pi 丹皮 6 g, Niu Xi 牛膝 6 g, Tao Ren 桃仁 10 pieces, Hong Hua 紅花 1.5 g, Dan Gui Wei 當歸尾 15 g, San Qi 三七 3 g, Chi Shao 赤芍 4.5 g.

Ingredient explanations:

- Ze Lan 澤蘭 moves blood, promotes urination, and reduces swelling.
- Mu Dan Pi 丹皮 clears heat, cools blood, moves blood.
- Niu Xi 牛膝 expels wind dampness, tonifies liver and kidneys, directs herbs downward.

- Tao Ren 桃仁 breaks up stubborn blood stasis.
- Hong Hua 紅花 moves blood, releases bruise, opens meridians, stops pain.
- Dan Gui Wei 當歸尾 invigorates, harmonizes, and nourishes the Blood, and alleviates pain.
- San Qi 三七 stops bleeding, moves blood, and stops pain.
- Chi Shao 赤芍 moves blood, reduces swelling, and stops pain.

Modifications:

- For pain in sacroiliac joints, add Du Huo 獨活.
- For more blood stasis, add Ru Xiang 乳香, Mo Yao 没藥.

Taking instructions:

- Take with warm wine.

Formula action:

- Remove stubborn stagnation, promote new circulation.

Formula indication:

- Physical trauma, blood stasis, stabbing pain when rotating the joints.

Blood Stasis

Treatment principle:

- Move blood, disperse bruise, stop pain.

Herbal formula:

- Fu Yuan Huo Xue Tang 复元活血湯.

Formula ingredients:

Chai Hu 柴胡 9 g, Tian Hua Fen 天花粉 9 g, Dang Gui 當歸 9 g, Hong Hua 紅花 6 g, Gan Cao 甘草 6 g, Chuan San Jia 穿山甲 炮 6 g.

Da Huang 大黄 12 g, Tao Ren 桃仁 9 g.

Ingredient explanations:

- Chai Hu 柴胡 harmonizes Shao Yang disorders, relieves stagnation.
- Tian Hua Fen 天花粉 clears heat, expels pus, and reduces swelling.
- Dang Gui 當歸 moves and tonifies blood, disperses cold, and stops pain due to blood stasis.
- Hong Hua 紅花 moves blood, releases bruise, opens meridians, and stops pain.

- Gan Cao 甘草 releases spasms, alleviates pain, harmonizes the other herbs, and guides the herbs to all 12 meridians.
- Chuan San Jia 穿山甲 moves blood, reduces swelling, expels wind damp.
- Da Huang 大黄 moves blood and disperses bruising.
- Tao Ren 桃仁 breaks up stubborn blood stasis.

Cooking instructions:

1) Place the herbs into an herb pot. Traditionally, a clay cooker is used. Gently rinse the herbs once with a strainer. You may notice some red color flowers in the strainer; place them back in the herb pot.
2) Add water into the pot until it is just enough to cover the herbs, then add one more cup of water.
3) Presoak the herbs for 15 minutes.
4) Turn on the stove to high heat and bring the herbs to a boil. Then reduce the heat to medium low to simmer but with slight bobble. Turn off the heat when the decoction is reduced to one cup.
5) Drain the decoction out into a container.
6) Repeat steps two to five one more time, and drain each decoction to the same container.
7) Divide these two cup decoctions into two separated cups

Dose:

- Drink one cup of the decoction twice a day.

Formula action:

- Moves blood, disperses bruise, soothes liver, and opens meridians.

Formula indication:

- Physical trauma, blood stasis, severe pain.

Patten herbs:

- Die Da Wan 跌打丸, San Qi Pian 三七片, and Yun Nan Bai Yao 雲南白藥.

External Applications

Herbal formula:

- Shuang Bai San 双柏散.

Formula actions:

- Move blood, reduce swelling, stop pain.

Formula indications:

- Sprain, strain, swelling muscles, red and hot skin.

Formula ingredients:
Da Huang 大黃 10 g, Huang Ba i 黃柏 20 g, Ze Lan 澤蘭 10 g, Bo He 薄荷 10 g, Ce Bai Ye 側柏葉 10 g.

Ingredient explanations:

- Da Huang 大黃 moves blood and disperses bruising.
- Huang Bai 黃柏 clears deficient kidney Yin heat.
- Ze Lan 澤蘭moves blood, promotes urination, and reduces swelling.
- Bo He 薄荷 expels wind heat.
- Ce Bai Ye 側柏葉 cools blood, stops bleeding, clears damp heat, and dispels wind damp.

Methods: Grind the herbs to power and mix the power with honey 5 g and water 50 ml and cook to a thick paste. Then apply the paste 1.5 cm thick on the affected region.

Cautions: Avoid applying this formula on open wound; Pregnancy is contraindicated.

Acupuncture Treatments

Acute state: UB40 委中 Weizhong, UB31 上髎 Shangliao, UB32 次髎 Ciliao, UB33 中髎 Zhongliao, UB34 下髎 Xialiao, GB30 環跳 Huantiao, UB37 殷門 Yinmen, Ashi points.

- *For Qi stagnation type*: Add LI4 合谷 He Gu and LIV3 太沖Taichong, SJ5 外關 Waiguan, and GB34 陽陵泉 Yanglingquan.
- *For blood stasis type*: Add UB17 膈俞 Geshu, UB39 委陽 Weiyang, LI4 合谷 He Gu.
- *Chronic state*: UB31 上髎 Shangliao, UB32 次髎 Ciliao, UB33 中髎 Zhongliao, UB34 下髎 Xialiao, UB18 肝俞 Ganshu, UB23 腎俞S henshu, LIV5 蠡溝 Ligou, LIV8 曲泉 Ququan, GB34 陽陵泉Yanglingquan, GB39 懸鐘 Xuanzhong, KD1 涌泉 Yong Quan.

Cupping Treatments

UB31 上髎Shangliao, UB32 次髎Ciliao, UB33 中髎Zhongliao, UB34 下髎Xialiao.

Moxibustion Treatments

Warming needling techniques on UB23 腎俞Shenshu, UB25 大腸俞Dachangshu.

Bleeding Treatments

One Ashi point on UB31 上髎Shangliao, UB32 次髎Ciliao, UB33 中髎Zhongliao, UB34 下髎Xialiao.

Electro-Acupuncture Treatments

UB23 腎俞Shenshu ~ UB25 大腸俞Dachangshu 2/100 Hz for 20 minutes daily.

Tui Na Treatments

Tui Na techniques:

- Pressing, rolling, kneading, stretching

Tui Na procedures:

1) The patient is in a prone position; the therapist stands at the patient's left side.
2) The therapist applies rolling techniques on the sacroiliac joint and the ipsilateral buttock for two to three minutes.

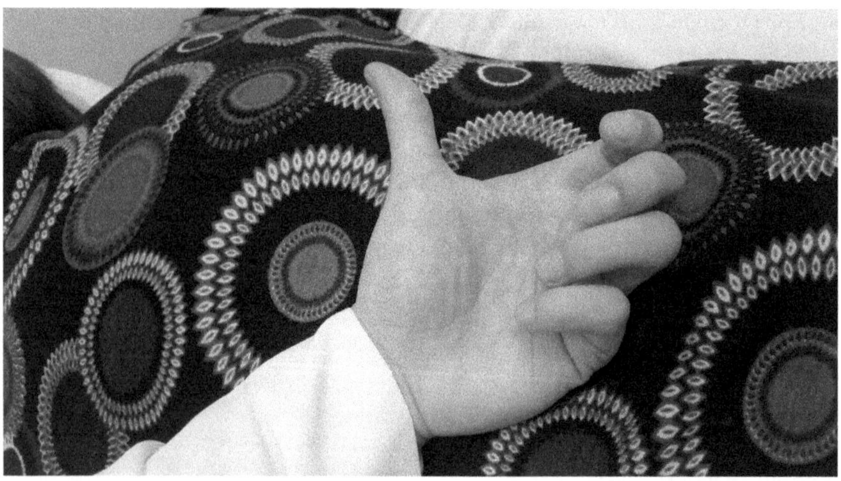

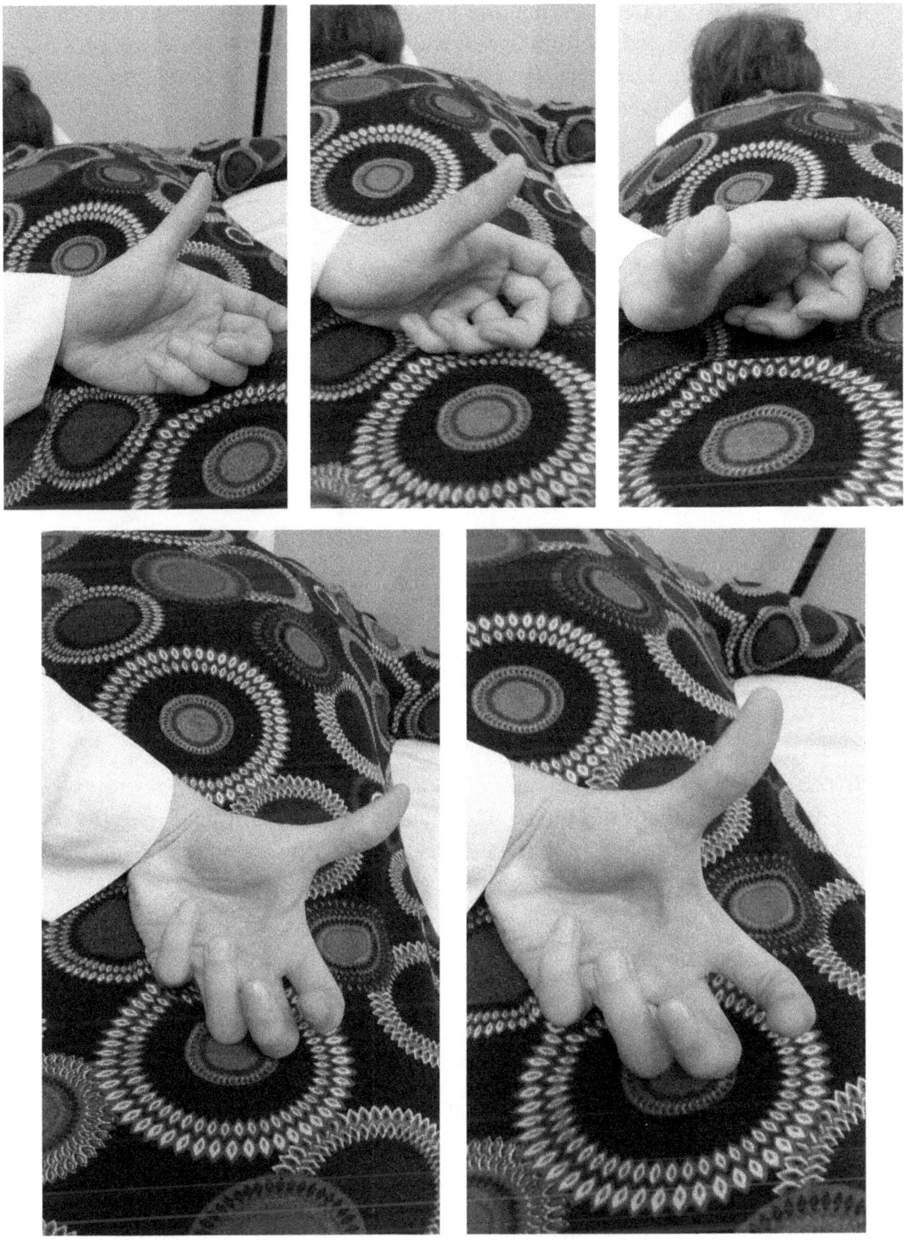

3) The therapist applies a thumb, pressing and kneading on the tender spots and on UB31 上髎Shangliao, UB32 次髎Ciliao, UB33 中髎Zhongliao, UB34 下髎Xialiao, GB31 風市Fengshi, UB54 秩邊Zhibian for two to three minutes each.

4) The therapist applies pressing and kneading techniques with the heel of a hand on the affected sacroiliac joint and buttock for two to three minutes or until it is warm. The therapist will apply the kneading and pressing on the ipsilateral leg as well if there is referent pain.

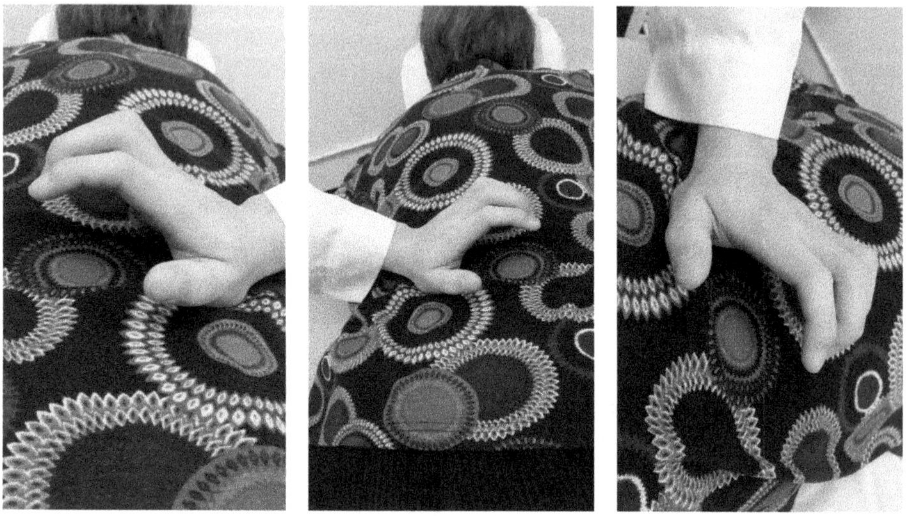

5) The patient is in a lateral recumbent position on the contralateral side at the edge of a treatment table. The unaffected leg is straight, and the affected knee and hip are bent toward the chest. The therapist stands in front of or behind the patient.

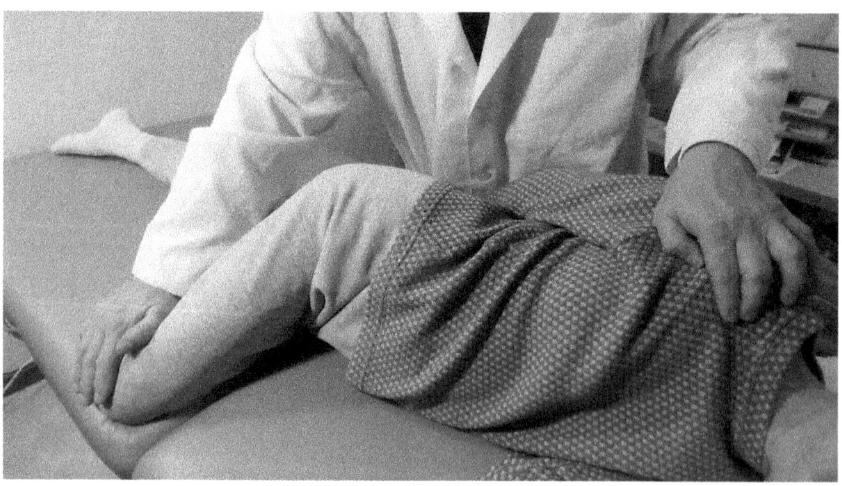

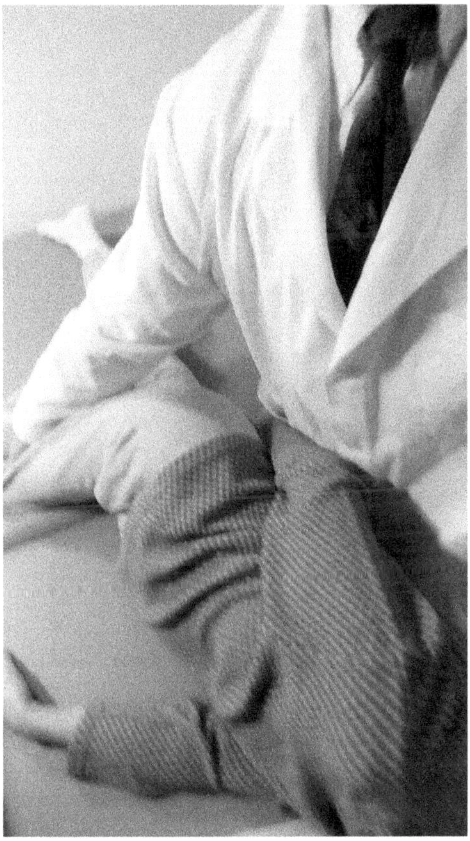

6) The therapist places a forearm at the patient's ipsilateral shoulder and stretches the shoulder posteriorly while placing a hand on the ipsilateral knee.
7) With both hands in the opposite directions, the sacroiliac joint is stressed to its full range of motion. At this point, a high-velocity, low-amplitude impulse can be applied. The therapist uses the hand which is on the patient's ipsilateral knee to apply a controlled, quick, gentle thrust to the joint.[64]
8) The patient is placed in a supine position; the therapist stands beside the patient. The therapist flexions the patient's hip and knee to the full range of motion, then quickly but gently extends the leg three to five times.
9) The patient is in a lateral recumbent position on the unaffected side. The unaffected leg is straight, and the affected knee and hip are bent toward the chest. The therapist stands on the back side of the patient.
10) The therapist places one hand on the affected sacroiliac joint and grasps the patient's ipsilateral ankle with another hand.

11) The hand that grasps the patient's ankle stretches the ipsilateral leg backward and at the same time presses the affected sacroiliac joint forward.

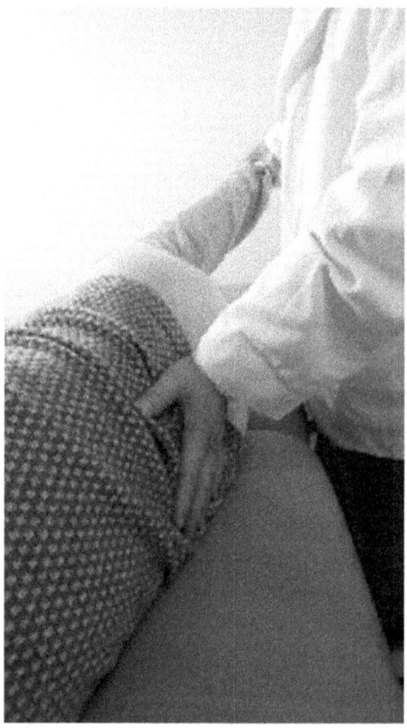

12) Patient is prone; the therapist stands at the side.
13) The therapist places one hand on the affected sacroiliac joint and lifts the ipsilateral leg with another hand. Both hands exert on the affected joint and leg toward in opposite directions gradually to the maximum range of motion. Then gently thrust the joint.
14) The therapist applies the hypothenar on the affected sacroiliac joint with kneading and pressing techniques until warm or hot.

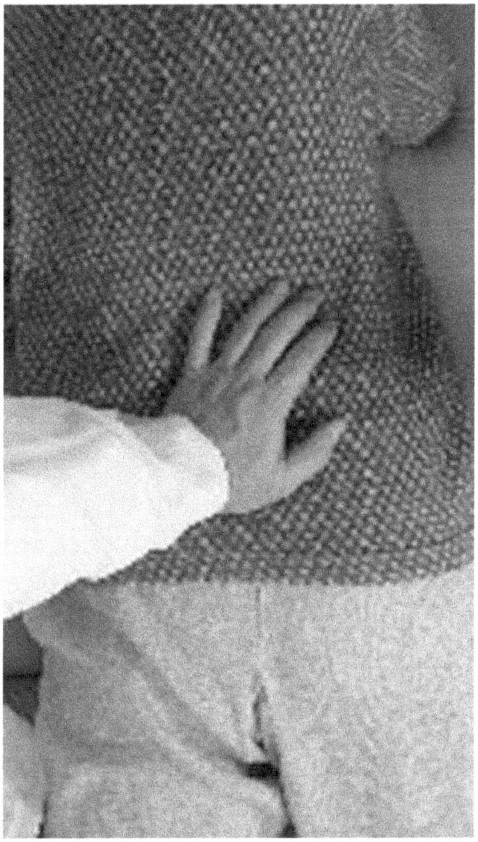

Patient Advisory

1) The puerperal woman should be advised to avoid over-exertion too soon after pregnancy.
2) To avoid habitual subluxation, patients are advised after each treatment to avoid over-exertion and long periods of standing.
3) Keep the lumbar warm.
4) Avoid sitting for too long a period.
5) Appropriate exercise such as walking or slow pace running is recommended to strengthen the joints.

To assist healing, foods that are high in calcium and vitamins are recommended.

COCCYDYNIA 尾骨痛

Introduction

Coccydynia is pain and tenderness at the coccyx; usually it is associated with inflammation. Injuries such as falls and childbirth can cause contusion of the sacrococcyx and inflammation of the adjacent ligaments or even dislocation and fracture of the coccyx. The risks of developing coccydynia include obesity and female sex. Coccydynia is five times more prevalent in women than men.[65]

Anatomy

The coccyx is the lowest region of the vertebral spine. It is located inferior to the sacrum. It is curved anteriorly with its apex facing downward and caudally.

It is an inverted triangular bone with the wider base at the top and the narrower apex at the bottom. It consists of three to five coccygeal vertebrae with semi-fused fibrocartilaginous joints between the sacrum and coccyx and between coccygeal segments.

The sacrococcygeal joint is where the coccyx articulates to the apex of the sacrum. Both are articulated to each other via the sacrococcygeal ligament. Clinically, both traumatic and nontraumatic coccydynia can occur at this joint.

In some cases, the coccyx has a slight amount of movement because the first vertebra is separated from the rest. It is articulated with the sacrum by an amphiarthrodial joint in early age and in more than half of adults, it may fuse with the sacrum later in life (Brouhard 2020).[66]

There are a number of soft tissues inserted in the coccyx. Levator ani at the apex is a group of three muscles including the pubococcygeus, the iliococcygeus, and the puborectalis. Levator ani forms the greater part of the pelvic floor.[67] The coccygeus muscle arises from the ischium across the anterior aspect of the coccyx and the gluteus maximus across the posterior aspect.

Contraction of these muscles may be the source of coccyx pain or the pain may be idiopathic in western medicine.

Functions: Even its size is small, the coccyx has several important functions. Along with being the insertion site for multiple muscles, ligaments, and tendons to form a strong pubic floor, it also serves as one leg of the tripod, along with the two the ischial tuberosities, that provides weight-bearing support to a person in the seated position,[68] with an increased weight load on the coccyx when a person leans back.[69] The coccyx also provides positional support to the anus.

Western Medicine Etiology and Pathology

The most common factors of coccydynia include:

1) Childbirth:

Giving birth is one of the most common causes of coccydynia. The hip and coccyx gain more laxity toward the end of pregnancy which allows the muscles and ligaments around the coccyx to overstretch.

On the other hand, giving birth can cause trauma that injures or even fractures the coccyx.[70]

2) Coccyx injury:

Injury can occur on the coccyx if there is a hard impact to the base of the spine. The trauma may cause dislocation or fracture in severe injuries. For example, falling backward when one tries to sit down but somebody has pulled away the chair directly attacks the coccyx. Horse riding sports, skiing, and skating are also high risk for coccydynia.

3) Overweight:

Being overweight can put excess pressure on the coccyx when sitting.

4) Underweight:

The buttock, especially the coccyx region, may not have enough fat to prevent tissues having friction on the coccyx.

5) Repetitive strain injury:

Some sports require leaning forward such as cycling or rowing may develop coccydynia. Continually leaning forward and stretching the base of the spine may cause muscles and ligaments around the coccyx to become strained and stretched. The muscles may no longer be able to maintain the coccyx in the correct position, resulting in pain and discomfort.

6) Poor posture:

Sitting with wrong posture for a long period of time, such as driving long distance, can increase pressure on the coccyx. The worst sitting posture is sitting leaning back, which puts pressure directly onto the coccyx.

Chinese Medicine Etiology and Pathology

Qi and Blood Stagnation

Physical trauma such as a direct blow to the coccyx, falling on the buttock, sitting or leaning backward for a long period of time, trauma during childbirth delivery, and straining the muscles and ligaments around the coccyx causes blockage in the meridians so Qi and blood cannot flow through, resulting coccyx pain.

Wind Damp Cold Bi Syndrome

In the case of underlined deficiency of Yin and Yang, the blood and Qi circulation to the region of coccyx has been poor. The pain occurs when patients with a deficiency condition are invaded by damp, cold, and wind from the environment, such as sitting on a cold wet floor or injury while skiing. With the deficiency, the coccyx region houses wind, damp, and cold, and these factors cause stagnation of Qi and blood, resulting in pain.

Manifestations

Pain occurring in the coccyx region is the main complaint. During the office visit, patients have difficulty sitting during the appointment, or some patients may sit on one side of the buttocks to avoid pressure on the coccyx.

1) Patients may have an obvious history of injury in the coccyx region. It can also appear over time with no apparent cause.
2) Bruise signs may be visible in the coccyx region, if the pain is from an acute traumatic injury.
3) Pain at coccyx may occur suddenly and immediately after an injury.
4) The sensation of pain may be a dull, throbbing ache or may be a sharp, stabbing pain depending on the type and chronicity of the injury.
5) Severe localized pain may be felt immediately when the coccyx is touched.
6) The pain may radiate to the hips, thighs, and rectum. It can radiate to the low back and down the legs.
7) The pain can occur in any sitting position and even sleeping. Some female patients may experience pain during sexual intercourse.
8) The pain may be exacerbated by exercise, direct contact with the coccyx, coughing, and defecation, especially if constipation is present. The pain is usually relieved temporarily by sitting on a coccyx cushion (sometimes called a wedge seat cushion).

Physical Examination

- Begin by obtaining medical the patient's history:
  - Injury history to the patella from a fall backward that directly hit the coccyx.
  - Medical history that may cause coccyx injury such as childbirth or being overweight and underweight.
  - Careers that require long periods sitting, such as a bus driver.
  - Sports activity that requires leaning forward, such as cycling.
- Inspection:
  - Bruise signs may be visible in the coccyx region.
- Palpation:
  - Severe localized pain when the coccyx is touched. Palpation at the border between sacrum and coccyx may find a severe tender pain.

  - Local swelling may be also detected around the coccyx.
  - Palpation may also be used to identify potential coccygeal spicules (bone spurs), cysts, or tumors.[71]

- Intrarectal exam:
  - The patient lies on the left side, bends both knees and hips, and bring the legs up to the chest. The physician slips a lubricated index into the rectum through the anus and palpates and exams the coccyx. The physician places the thumb outside of the coccyx and both index and thumb hold and exam the range of motion of the coccyx by shifting the coccyx forward and backward. The test is positive if there is pain when shifting.
  - If there is no pain presentation when shifting, the pain may be a referred pain from somewhere else, such as from another part of the spine.
  - Then the index finger exams the soft tissues around the sacrococcygeal joint for cysts, spurs, and tumors.

Differential Diagnosis

Sacral Chordoma

Coccydynia and sacral chordoma have quite similar symptoms; they both have the symptoms of constant severe pain in the coccyx, and it may be worse at night and may radiate to the groin.

Sacral chordoma has symptoms of bladder or bowel obstruction. An ultrasound scan may show a possible collection. A soft tissue mass may be palpable. An MRI or CT may show a tumor near the sacrum and coccyx.[72]

Treatment

Treatment principle:

- Moves blood, expels wind, damp, and cold, tonifies liver and kidneys, and stops pain.

Herbal Treatments

Herbal formula for topical use:

- San Qi Yu Tong Gao 三七瘀痛膏.

Formula ingredients:

San Qi 三七 5 g, Bai Jie Zi 白芥子 10 g, Hong Hua 紅花 10 g, Tao Ren 桃仁 10 g, Da Huang 大黃 10 g, Xue Jie 血竭 10 g, Ru Xiang 乳香 10 g, Mo Yao 沒藥 10 g, Dang Gui Wei 當歸尾 10 g, Bing Pian 冰片 3 g.

Ingredient explanations:

- San Qi 三七 moves blood, stops bleeding, reduces swelling, stops pain.
- Bai Jie Zi 白芥子 expels phlegm, promotes Qi movement, alleviates pain.

- Hong Hua 紅花 moves blood, releases bruise, opens meridians, stops pain.
- Tao Ren 桃仁 breaks up stubborn blood stasis.
- Da Huang 大黃 moves blood and disperses bruising.
- Xue Jie 血竭 moves blood, stops bleeding, and stops pain.
- Ru Xiang 乳香 moves blood and stops pain.
- Mo Yao 沒藥 moves blood, reduces swelling, and stops pain.
- Dang Gui Wei 當歸尾 moves and harmonizes the blood.
- Bing Pian 冰片 clears heat and opens orifices.

Procedure:

- Grind the herbs to power, mix the power of the above herbs with wine to form a paste, warm it, and apply the paste on the sacrum and coccyx region about 2 mm thick for ten hours each day over the course of ten days as one course or treatment, then rest for two to three days before the next course of treatment.

Herbal formula:

- Shu Jin Huo Xie Tang 舒筋活血湯.

Formula ingredients:

Qiang Huo 羌活 6 g, Fang Feng 防風 9 g, Jing Jie 荊芥 6 g, Du Huo 獨活 9 g, Dang Gui 當歸 12 g, Xu Duan 續斷 12 g, Qing Pi 青皮 5 g, Niu Xi 牛膝 9 g, Wu Jia Pi 五加皮 9 g, Du Zhong 杜仲 9 g, Hong Hua 紅花 6 g, Zhi Ke 枳殼6 g.

Ingredient explanations:

- Qiang Huo 羌活 expels wind-cold-dampness, unblocks painful obstruction, alleviates pain, and guides Qi to the Tai Yang and Du channels.
- Fang Feng 防風 releases the exterior, expels external wind, expels wind-dampness.
- Jing Jie 荊芥 stops bleeding, dispels wind, relieves muscle spasm.
- Du Huo 獨活 expels wind damp, stops pain.
- Dang Gui 當歸 moves and tonifies blood, disperses cold, and stops pain due to blood stasis.
- Xu Duan 續斷 nourishes the liver and kidneys, strengthens bones and tendons, and alleviates pain.
- Qing Pi 青皮 soothes liver Qi and breaks up stagnant Qi, dries dampness, and transforms phlegm.
- Niu Xi 牛膝 expels wind dampness, tonifies the liver and kidneys, and directs herbs downward.
- Wu Jia Pi 五加皮 expels wind and damp, nourishes and warms the liver and kidneys, and strengthens bones and tendons.
- Du Zhong 杜仲 tonifies the kidneys and liver, strengthens bones and tendons, and lowers blood pressure.

- Hong Hua 紅花 moves blood, releases bruise, opens meridians, and stops pain.
- Zhi Ke 枳殼 removes stagnant food, promotes Qi movement, reduces distention, and pressure.

Contraindication:

- Pregnancy.

Dose: 250 ml BID.

Wash formula:

- Hai Tong Pi Tang 海桐皮湯.

Formula ingredients:

Hai Tong Pi 海桐皮 6 g, Tou Gu Cao 透骨草 6 g, Ru Xiang 乳香 6 g, Mo Yao 沒藥 6 g, Dang Gui 當歸 4.5 g, Chuan Jiao 川椒 9 g, Chuan Xiong 川芎 3 g, Hong Hua 紅花 3 g, Wei Ling Xian 威靈仙 2.4 g, Bai Zhi 白芷 2.4 g, Gan Cao 甘草 2.4 g, Fang Feng 防風 2.4 g.

Ingredient explanations:

- Hai Tong Pi 海桐皮 expels wind, drains damp, and reduces swelling.
- Tou Gu Cao 透骨草 moves blood, reduces swelling, clears heat, cools blood, drains damp heat, dispels wind and damp, relaxes muscles and tendons, and stops pain.
- Ru Xiang 乳香moves blood and stops pain.
- Mo Yao 沒藥 moves blood, reduces swelling, and stops pain.
- Dang Gui 當歸 moves and tonifies blood, disperses cold, and stops pain due to blood stasis.
- Chuan Jiao 川椒 warms middle Jiao, disperses cold, dispels damp, and stops pain.
- Chuan Xiong 川芎 moves blood and Qi, expels wind, and stops pain.
- Hong Hua 紅花 moves blood, releases bruising, opens meridians, and stops pain.
- Wei Ling Xian 威靈仙 expels wind-damp, stops pain.
- Bai Zhi 白芷 expels wind, drains dampness, dispels cold, and stops pain.
- Gan Cao 甘草 releases spasms, alleviates pain, harmonizes the other herbs, and guides the herbs to all 12 meridians.
- Fang Feng 防風 releases the exterior, expels external wind, and expels wind-damp.

Instructions:

- Broil the herbs in water, transfer the herb liquid to a bucket, then instruct the patient to sit in the bucket for 20 minutes once or twice daily.

Formula action:
Move blood, disperse bruising, open channels, stop pain.

Acupuncture Treatments

Acupuncture point formula:

- Ba Liao Xue 八髎穴 (UB31 上髎Shangliao, UB32 次髎Ciliao, UB33 中髎Zhongliao, UB34 下髎Xialiao).
- UB31 上髎Shangliao is the first sacral foramen, UB32 次髎Ciliao is the second sacral foramen, UB33 中髎Zhongliao is the third sacral foramen, and UB34 下髎Xialiao is the fourth sacral foramen.

Procedure:

- Select the ipsilateral Ba Liao Xue 八髎穴, insert 1.5 cun needles into the sacral foramen. Retain the needles for 20 minutes, and manipulate the needles every ten minutes. Treatment is once a day.

Modifications:

- For blood stasis, add UB17 膈俞Geshu, and UB39 委陽Weiyang.
- For liver and kidney deficiency, add UB23 腎俞Shenshu, KD1 涌泉Yong Quan, KD3 太溪Taixi, and KD6 照海Zhaohai; SP6 三陰交Sanyinjiao.
- For severe pain, add LI4 合谷He Gu, SI3 後溪Houxi, and UB62 申脈Shenmai.

Cupping Treatments

Ba Liao Xue 八髎穴 (UB31 上髎Shangliao, UB32 次髎Ciliao, UB33 中髎Zhongliao, UB34 下髎Xialiao).

Moxibustion Treatments

Warming needling techniques on Ba Liao Xue 八髎穴.

Bleeding Treatments

Seven-star needle lightly taps Ba Liao Xue 八髎穴until slightly red.

Electro-Acupuncture Treatments

Ba Liao Xue 八髎穴2/100 Hz for 20 minutes daily.

Tui Na Treatments

Tui Na techniques:

- Kneading, stretching, slippery, lifting

Tui Na procedures:

1) With the patient in a prone position, place a pillow in between the hip and treatment table; the therapist stands next to the hip.
2) The therapist applies kneading techniques on the coccyx and sacrum regions with both thumbs. Repeat about 10 to 20 times.

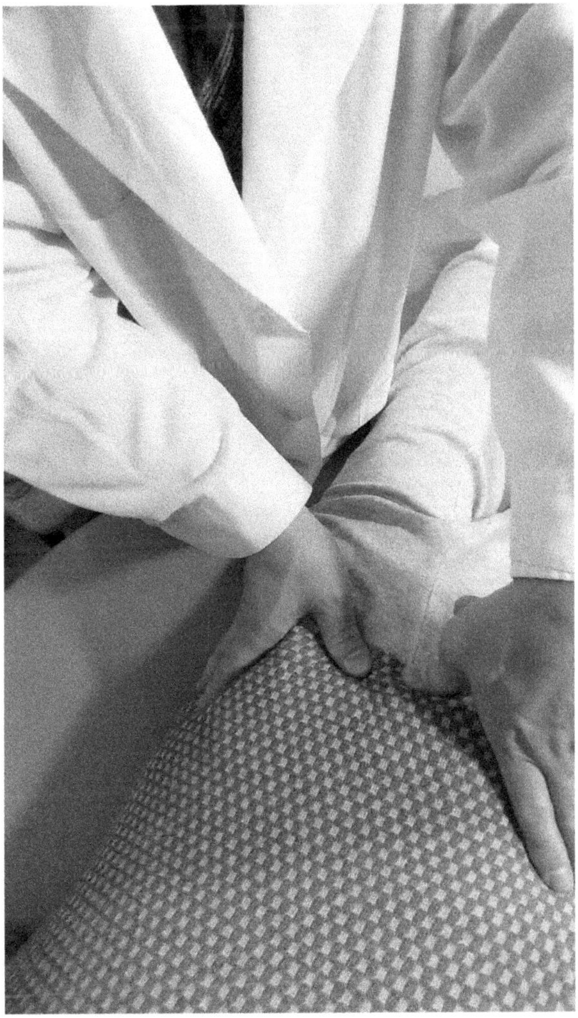

3) With an assistant stretching the patient's legs by pulling the ankles downward, the therapist places an arm on the patient's quadriceps tendons above the knees and lifts the legs backward while kneading the patient's sacrum with a thenar on another hand. Repeat five to six times.

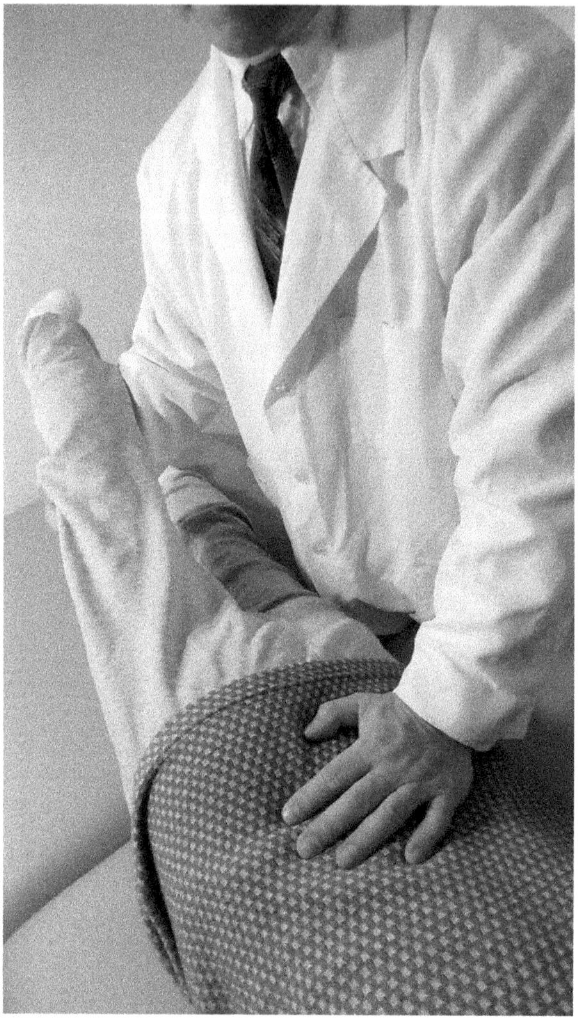

4) Change the patient's position to be supine. The therapist stands near the hip and the assistant stands near the feet holding the ankles.
5) The therapist places one hand on the sacrococcygeal joint and presses the patient's knees with another hand.
6) With the assistant stretching the patient's legs by pulling the patient's ankles downward, the therapist presses the hand on the patient's knees downward while pressing and lifting the patient's sacrum upward. These two hands must be coordinated applying force at the same time and ending the procedures by letting the patient's coccyx slip off the therapist's hand.
7) Advise the patient to rest in bed at home as much as possible.

Patient Advisory

To obtain better treatment outcome and to prevent future trauma or pain reoccurrences, patients are advised to strictly follow the recommendations. This home care is aimed at managing the pain and avoiding further irritation to the coccyx.

1) Applying ice or heat to the sacrococcygeal joint:

   Patients often raise a question to therapists about the use of the ice or heat. These two temperatures are opposite and can have opposite treatment outcomes. These two should be applied according to Chinese medicine etiology. Ice is good to use for Qi and blood stasis while heat is good to use for wind damp cold Bi syndrome.

   Using them correctly offers more rapid pain relief and long-term healing effects. Patients are advised to apply a hot pack or ice pack for 15–20 minutes each time, two to three times a day for the first few days after the injury.
2) *Spending less time sitting*: Since sitting for long periods is a causative factor, sitting causes more pain and slows the healing. Patients should avoid spending long hours sitting.
3) *Increasing physical activity level*: Patients who sit at desks using a computer or sitting in a classroom at school should consider taking a break of about 15 minutes each hour and spending these 15 minutes on beneficial physical activities such as walking.
4) *A coccyx cushion* (sometimes called a wedge seat cushion): This kind of cushion has a hole in the middle or at a side of the cushion that it is for the coccyx to rest in it while sitting. The hole helps to alleviate pressure on the coccyx and promotes an opportunity for better blood and Qi circulation.
5) *Sitting posture*: Patients are taught to correct the causative sitting postures. For example, the patient who used to sit and lean back, putting excessive pressure on the coccyx, will be taught to sit upright or lean forward and alleviate the body weight away from the spine and coccyx.
6) *Massage the affected area*: Patients learn how to massage the coccyx and the muscles around the coccyx. Gentle and correct methods of massage can help to promote the local Qi and blood circulation particularly if the pain is due to sitting for long periods or local muscle injuries.
7) *Prevent injury*: Patients will be recommended to avoid wearing high-heeled shoes, especially during snow or rain.
8) *Prevent catching cold*: Avoid sitting on the cold, hard floor.
9) *Avoid constipation*: Patients should avoid constipation by changing the diet to be high fiber, because constipation or dry stool may worsen the coccydynia. If high fiber food, water intake, and many other constipation remedies don't seem to work, try laxatives.
10) *Stretching*: Gently stretching the ligaments surrounding the coccyx and sacrum can reduce stress in the affected area and can promote circulation of Qi and blood flow.

11) *Seeking western medical service*: If patients also suffer from incontinence on both bowel movements and urination, and if the treatment outcome from alternative medicine is not satisfied, patients should be referred to seek further examination from western medicine to rule out other causes, such as cauda equina injury. Fracture of coccyx may irritate the urinary bladder and the rectum.

NOTES

1. https://www.hindawi.com/journals/aorth/2018/6578097/
2. https://www.kenhub.com/en/library/anatomy/piriformis-muscle
3. https://www.kenhub.com/en/library/anatomy/piriformis-muscle
4. https://www.kenhub.com/en/library/anatomy/piriformis-muscle
5. https://www.kenhub.com/en/library/anatomy/piriformis-muscle
6. https://www.kenhub.com/en/library/anatomy/piriformis-muscle
7. https://www.kenhub.com/en/library/anatomy/piriformis-muscle
8. http://www.restoretherapeutics.com/Injuries-Conditions/Lower-Back/Lower-Back-Issues/Piriformis-Syndrome/a~54/article.html
9. https://jaoa.org/article.aspx?articleid=2093614
10. https://www.webmd.com/pain-management/guide/piriformis-syndrome-causes-symptoms-treatments#1
11. https://www.medicalnewstoday.com/articles/318587.php
12. https://www.webmd.com/pain-management/guide/piriformis-syndrome-causes-symptoms-treatments#1
13. https://www.verywellhealth.com/intermediate-piriformis-syndrome-stretching-routine-4022709
14. https://josr-online.biomedcentral.com/articles/10.1186/s13018-014-0139-7
15. http://journalofprolotherapy.com/management-cluneal-nerve-referred-pain-prolotherapy/
16. https://www.sciencedirect.com/topics/neuroscience/superior-cluneal-nerves
17. https://www.sciencedirect.com/topics/neuroscience/superior-cluneal-nerves
18. https://www.sciencedirect.com/topics/neuroscience/superior-cluneal-nerves
19. https://www.sciencedirect.com/topics/neuroscience/superior-cluneal-nerves
20. https://www.verywellhealth.com/thoracolumbar-fascia-297293
21. https://www.verywellhealth.com/thoracolumbar-fascia-297293
22. https://link.springer.com/chapter/10.1007/978-3-319-27482-9_51
23. https://josr-online.biomedcentral.com/articles/10.1186/s13018-014-0139-7
24. http://www.med66.com/new/40a188aa2009/20091120liuhon164543.shtml
25. https://www.healthline.com/human-body-maps/adductor-longus-muscle#1
26. https://en.wikipedia.org/wiki/Adductor_longus_muscle
27. https://www.physio-pedia.com/Adductor_Magnus
28. https://www.kenhub.com/en/library/anatomy/the-hip-adductors
29. https://www.anatomynext.com/pectineus/
30. https://www.kenhub.com/en/library/anatomy/gracilis-muscle
31. https://www.kenhub.com/en/library/anatomy/gracilis-muscle
32. https://en.wikipedia.org/wiki/Gracilis_muscle
33. https://en.wikipedia.org/wiki/External_obturator_muscle
34. https://rad.washington.edu/muscle-atlas/obturator-externus/
35. https://rad.washington.edu/muscle-atlas/obturator-externus/

36. https://www.orthobullets.com/anatomy/10074/obturator-externus
37. https://www.physio-pedia.com/Groin_strain
38. https://www.wosm.com/wosm-resources/orthopaedic-conditions/adductor-muscle-strain/
39. https://www.sciencedirect.com/topics/neuroscience/hip-adductors http://www.med66.com/new/40a188aa2009/20091120liuhon164543.shtml
40. https://www.physio-pedia.com/Groin_strain#cite_note-.27Gino.27-3
41. http://dict.cn/famine
42. https://static1.squarespace.com/static/59dbb748e5dd5b5a1b458923/t/5bc24cc971c10bab9f2c8222/1539460298422/Suwen+43_.pdf
43. https://www.wosm.com/wosm-resources/orthopaedic-conditions/adductor-muscle-strain/
44. https://physioworks.com.au/injuries-conditions-1/adductor-tendinopathy
45. https://www.wosm.com/wosm-resources/orthopaedic-conditions/adductor-muscle-strain/
46. https://www.practicalpainmanagement.com/pain/spine/disordered-sacroiliac-joint-pain
47. https://www.practicalpainmanagement.com/pain/spine/disordered-sacroiliac-joint-pain
48. https://www.physio-pedia.com/Sacroiliac_joint https://www.ncbi.nlm.nih.gov/books/NBK507801/
49. https://stretchcoach.com/articles/sacroiliac-joint-pain/
50. https://www.livestrong.com/article/121609-causes-sacroiliac-subluxation/
51. https://www.livestrong.com/article/121609-causes-sacroiliac-subluxation/
52. https://www.acog.org/Patients/FAQs/Back-Pain-During-Pregnancy?IsMobileSet=false
53. https://www.livestrong.com/article/121609-causes-sacroiliac-subluxation/
54. https://si-bone.com/providers/diagnostic-resources/provocative-tests
55. https://www.physio-pedia.com/FABER_Test
56. http://morphopedics.wikidot.com/sacroiliac-joint-dysfunction-special-tests
57. http://morphopedics.wikidot.com/sacroiliac-joint-dysfunction-special-tests
58. https://www.spondylitis.org/Ankylosing-Spondylitis/Symptoms
59. https://baike.baidu.com/item/骶髂关节炎
60. https://binderchiropractic.com/spinal-adjustment
61. https://www.ncbi.nlm.nih.gov/pubmed?term=24688338 PMCID:PMC3963058
62. https://www.healthline.com/human-body-maps/coccyx#1
63. https://en.wikipedia.org/wiki/Levator_ani
64. https://www.ncbi.nlm.nih.gov/pmc/articles/PMC3963058/ https://pdfs.semanticscholar.org/1d6b/aa586f41a26fd5d13d416018f99d49db0173.pdf?_ga=2.112633668.1386928549.1571237138-190541184.1571024840
65. https://www.uptodate.com/contents/coccydynia-coccygodynia
66. https://www.medicalnewstoday.com/articles/318064.php
67. https://www.spine-health.com/conditions/lower-back-pain/diagnosis-coccydynia-tailbone-pain
68. https://www.ncbi.nlm.nih.gov/pmc/articles/PMC3149455/

5 Knee Pain

ANTERIOR CRUCIATE LIGAMENT INJURY 前十字韌帶損傷

Introduction

Anterior cruciate ligament injuries most commonly occur during sports that involve sudden stops or changes in direction, pivoting, cutting, jumping,[1] and landing. Soccer, basketball, football, and downhill skiing[2] are examples of high-risk sports.

The anterior cruciate ligament is one of four ligaments that maintain knee joint stability. Even though it is the smallest of the four main ligaments in the knee, it is the most important one. It keeps the leg stable when twisting the femur and tibia. When these two bones rotate too far in the opposite directions or when the knee is flexed in the wrong direction, the anterior cruciate ligament can be torn or sprained, and the knee will wobble and shift around when walking.

Despite its size, often the tear of the anterior cruciate ligament may damage to other ligaments and knee structures as well, most commonly the medial collateral ligament and the cartilage of the knee.[3]

Anatomy

The bone structure of the knee joint is formed by the femur, the tibia, and the patella on the top as protection.

The stability of the knee joint is maintained by four ligaments that hold the bone together; the anterior cruciate ligament is one of these ligaments. It originates from the posteromedial corner of the medial aspect of the lateral femoral condyle in the intercondylar notch (Purnell 2008)[4] and ends at the anterior to the intercondyloid eminence of the tibia, being blended with the anterior horn of the medial meniscus.[5]

Even though it is the smallest, it is a key structure in the knee joint, as it prevents excessive tibial medial and lateral rotation, as well as varus and valgus stresses.[6]

Western Medicine Etiology and Pathology

A sudden change of direction while running is the main cause of anterior cruciate ligament injury. When an athlete plants the foot hard to change direction, the femur and the tibia move in opposite directions, resulting in tearing of the anterior cruciate ligament.

The sports activities below are common factors that cause anterior cruciate ligament injury include:

- Sudden slowing down while running.
- Sudden stopping while running.

- Sudden change of direction while running.
- Female patients have high rate in the injury of anterior cruciate ligament. When landing from jumping, females are more likely to use less hip and knee flexion at initial contact and increased knee valgus with a wide stance or decreased knee-flexion displacement (Bell et al. 2014).
- Pivoting the knee with the foot firmly planted.
- Landing from a jump with the knee rotated inward.
- A direct blow from the front side of the tibia.[7]

Chinese Medicine Etiology and Pathology

Acute State

Acute traumatic injury to the ligaments leads to blood stasis and swelling, which block the meridians and results in pain.

Chronic State

Ligament injury is considered "Jin Shang 筋傷" or "traumatic injury on tendons." In traditional Chinese medicine (TCM) theory, the liver governs the tendons. "Traumatic injury on tendons" is consuming liver energy which mostly consumes Yin and blood. As the injury becomes chronic, the liver Yin and blood are deficient. The kidneys follow because the liver "wood" and the kidney "water" always affect each other.

Edema

This type of swelling is considered "Feng Shui 風水" or Yang edema. The wind invades meridians when there is blood stasis and blocks the dispersing, descending functions of the lung. The water is then accumulated.

Manifestations

1) Many patients hear or feel a "pop" in the knee when an anterior cruciate ligament injury occurs.
2) The injured knee will usually be swelling within 4–12 hours after the injury.
3) The knee is unstable when used and painful to bear weight.
4) The knee may have a sensation of giving out or buckling and easily bend backward.
5) The knee joint may lose full range of motion.
6) Tenderness along the joint line.

Anterior cruciate ligament injuries are classified by grades 1, 2, and 3.

A first-degree sprain is a ligament that is stretched but with no fibers torn, while a second-degree sprain is a partially torn ligament. A third-degree sprain is a completely torn ligament.

Physical Examination

Physical examination of the anterior cruciate ligament injury:

- Begin by obtaining the patient's medical history:
    - Sports activities that often involve sudden stops or changes in direction while running, including direct contact and noncontact injuries.
    - Sports activities that cause direct contact injury, including a direct blow to the front side of the tibia.
    - Sports that may cause noncontact injury include soccer, football, and basketball.
    - An anterior cruciate ligament injury may also affect other knee structures such as the meniscus, patella cartilage, and fractures. A thorough physical examination should include the exams for other knee structures as well.
- Inspection:
    - The knee is unstable with movement, and it is painful to bear weight.
    - The knee may be easily bent backward.
    - Swelling may occur within 4–12 hours after the initial injury.
- Palpation:
    - Tenderness may be found along the joint line.
- Range of motion:
    - The knee joint may lose full range of motion.
    - *Anterior Drawer Test*: The patient is lying supine on the exam table, the knee flexes at 90 degrees, and the feet are placed flat on the bed to keep relaxed. The examiner pulls the tibia forward. If the tibia moves excessively forward (more than 5 mm), the test is positive, and it suggests a torn anterior cruciate ligament.
    - *Lachman Test*: The patient is either supine or prone; the examiner stands next to the injured knee. The affected knee is bent at about 30 degrees. The examiner secures the femur with one hand and pulls the tibia forward with the other hand. If the tibia moves anteriorly significantly more than the unaffected leg, the test is positive, and it suggests a torn anterior cruciate ligament.

Extra tests:

- X-rays or MRI scans are highly recommended to study the degree of torn ligaments and fractures.

Differential Diagnosis

The anterior cruciate ligament injury may need to be differentiated from an injury of the meniscus and an injury of both anterior and posterior cruciate ligaments.

A "pop" sound or feeling in the knee presents with the anterior cruciate ligament injury, while posterior cruciate ligament injury does not present a pop. Usually the injury of the anterior cruciate ligament is caused by sport activity, while posterior cruciate ligament injury is mostly caused by motor vehicle accidents. Symptoms present in mild injury of the anterior cruciate ligament, while in mild posterior cruciate ligament injury there might be no symptoms or the affected knee joint might only feel unstable.

Treatment

Treatment principle:

- Moves blood and Qi. The acute injury causes blood stasis in meridians.
- Cools blood, expels wind, disperses lung, circulates water. Chinese medicine is very helpful for improving knee edema from traumatic inflammation.

Herbal Treatments

Herbal formula:

- Tao Hong Si Wu Tang + Si Miao Yong An Tang Jia Jian 桃紅四物湯合四妙勇安湯加減.

Formula ingredients:

Tao Ren 桃仁 15 g, Hong Hua 紅花 12 g, Dang Gui 當歸 30 g, Chuan Xiong 川芎 10 g, Niu Xi 牛膝 10 g, Huang Qi 黄芪 40 g, Gui Zhi 桂枝 15 g, Jin Ying Hua 銀花 30 g, Xuan Shen 玄参 20 g, Mu Tong 木通 15 g, Chi Shao 赤芍 15 g, Gan Cao 甘草 6 g.

Ingredient explanations:

- Tao Ren 桃仁 breaks up stubborn blood stasis.
- Hong Hua 紅花 moves blood, releases bruising, opens meridians, stops pain.
- Dang Gui 當歸 moves and tonifies blood, disperses cold, and stops pain due to blood stasis.
- Chuan Xiong 川芎 moves blood and Qi, expels wind, and stops pain.
- Niu Xi 牛膝 expels wind-dampness, tonifies liver and kidneys, directs herbs downward.
- Huang Qi 黄芪 strengthens spleen, reduces edema.
- Gui Zhi 桂枝 releases the exterior, warms meridians, and stops pain.
- Jin Ying Hua 金銀花 clears heat, disperses wind heat, clears damp heat, cools blood.
- Xuan Shen 玄参 clears heat, cools blood, nourishes Yin.
- Mu Tong 木通 drains heat, promotes urination.
- Chi Shao 赤芍 moves blood, reduces swelling, stops pain.
- Gan Cao 甘草 releases spasms, alleviates pain, harmonizes the other herbs, and guides the herbs to all 12 meridians.

Formula action:

- Cools and moves blood, reduces swelling, stops pain.

Acupuncture Treatments

Local acupuncture points to promote local Qi and blood circulation:

- SP9 陰陵泉Yinlingquan, ST35 犊鼻Dubi, He Ding 鶴頂 (the location is above the knee, in the depression of the midpoint of the upper border of the patella), GB34 陽陵泉Yanglingquan.

Distal acupuncture points to regulate Zang Fu functions:

- UB13 肺俞Feishu, UB22 三焦俞Sanjiaoshu, Ren9 水分Shuifen, LU7 列缺Liequeuo, LI4 合谷Hegu, SJ5 外關Waiguan, LI10 手三里Shousanli, Ren17 膻中Danzhong, ST36 足三里Zusanli.

Moxibustion Treatments

Warming needling techniques on ST35 犊鼻Dubi.

Bleeding then Cupping Treatments

He Ding 鶴頂.

Electro-Acupuncture Treatments

He Ding 鶴頂~ ST36 足三里Zusanli 2/100 Hz for 20 minutes daily.

Tui Na Treatments

Tui Na techniques:

- Grasping, kneading, bending, pressing

Tui Na procedures:

Tui Na treatment is only appropriate for a first- or mild second-degree injury. A third-degree injury must receive Western medical attention; Chinese medicine is then used for recovery.

1) The patient lies in a supine position; the therapist stands next to the affected knee. The left knee is used in the example below.
2) Apply grasping and kneading techniques beginning from the medial side of the affected thigh, and follow with the same techniques on the popliteal fossa, the quadriceps muscle and tendon, and the lateral side of the thigh. Lastly, end the techniques on the knee joint. This procedure is to open channels, move Qi and blood, reduce swelling, and disperse bruising.

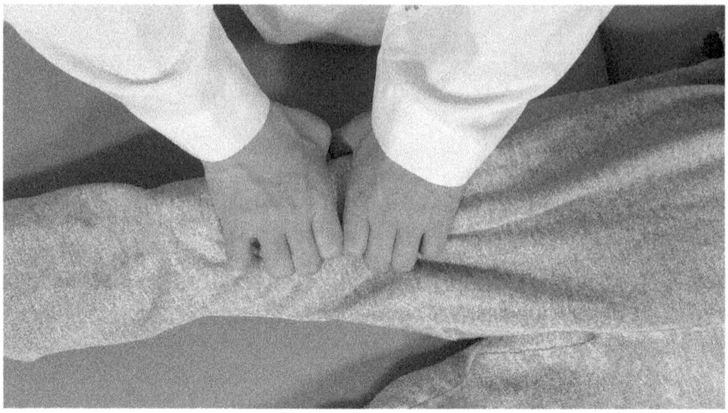

3) The left hand holds the ipsilateral ankle; the right hand helps the ipsilateral knee to bend and moves the ankle toward the contralateral hip. This step brings up more pain, showing on the tender spots. The right hand presses and kneads the painful spots for about one minute, then slowly extends the knee.
4) Repeat step three, but instead of moving the ankle toward the contralateral hip, now move the ankle toward the ipsilateral hip to bend the affected knee as much as possible so that the painful spots present. Press and knead the most painful spots with the right hand.

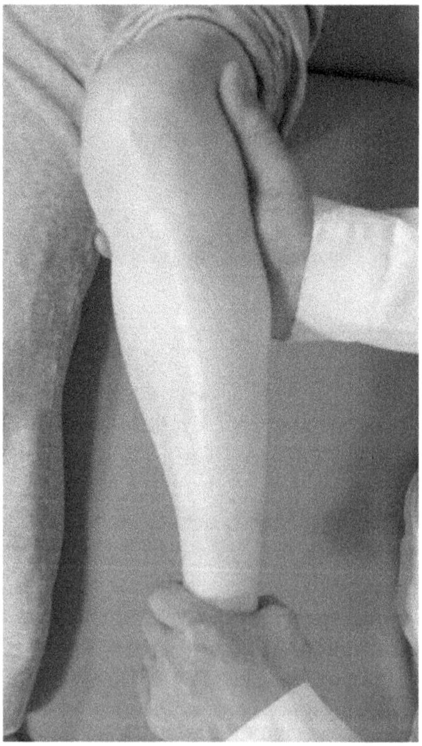

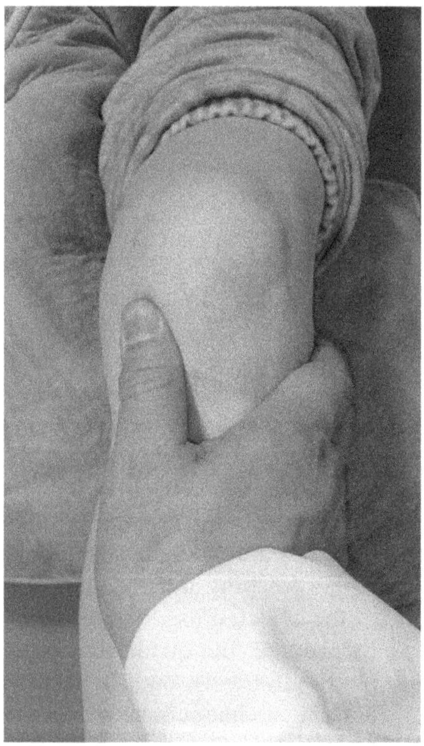

5) Slowly straighten the knee, then slowly dorsal flex the ipsilateral foot for one minute.

Cautions:

- Tui Na techniques should avoid passive movement techniques such as holding the ankle and shaking the leg forcefully.
- Chinese medicine is able to reduce swelling and stop pain quickly, but it should not be the main treatment modality for third-degree injuries.

Patient Advisory

- The affected leg should avoid bearing weight for at least four to six weeks.
- Try to avoid bending the affected knee.
- The quadriceps muscles should be trained during the recovery period.
- Patients should consider adding a brace to the injured knee and using a crutch to avoid putting weight on the injured leg.
- To prevent edema, the patient is advised to raise the affected leg higher than the heart while lying supine.
- For exercise in the acute stage, wear a brace on the injured knee. Patients may practice ankle pump exercise to maintain muscle tone in the thigh muscles, straighten the knee, and prevent thrombus build-up. The exercise procedure is below:
  a) Lie or sit with the knee fully extended.
  b) Point the toes as far away as possible.
  c) Planta flexion the ankle in a "pumping" movement.
  d) Raise the toes and dorsal flexion the foot toward the face.
  e) Repeat the movements for times.
- Dietary recommendation: Patients should be advised to avoid Fa Wu diet. Fa Wu foods are the foods that may cause more inflammation and bring back old injuries. Some examples of Fa Wu foods are shellfish such as crab and shrimp, fish, fried foods, lamb, duck, chocolate, garlic, and green onions. This concept applies to both acute and chronic strains.

POSTERIOR CRUCIATE LIGAMENT INJURY 後十字韌帶損傷

Introduction

The posterior cruciate ligament is one of the four main ligaments that stabilize the knee joint, and it is the primary posterior stabilizer among all four ligaments.

It is twice as strong and twice as thick as the normal anterior cruciate hence its strength against external force is twice that of the anterior cruciate ligament (Raj 2019).

Because it is the strongest and largest, it is less commonly injured. It accounts for 3% to 20% of all knee ligament injuries (Gliatis et al. 2019).

Injury to the posterior cruciate ligament may be the result of direct violent trauma such as a motor vehicle accident or acute sports injury to the knee from soccer, skiing, or football. But the injury with mild symptoms of pain and knee instability may go undiagnosed for months after the initial injury.

Posterior cruciate ligament injuries occur rarely in isolation and are more commonly a component of multi-ligament knee injuries.[8]

Anatomy

The posterior cruciate ligament is one of the two cruciate ligaments of the knee, and it is one of four stabilizers of the knee joint. Its functions are to prevent the femur from continuing to move anteriorly on the tibia and prevent the tibia from moving posteriorly on the femur. It also prevents hyper-extension and limits internal rotation, adduction, and abduction at the knee joint.[9]

The posterior cruciate ligament originates from the anterolateral aspect of the medial femoral condyle within the intercondylar notch[10] and inserts along the posterior aspect of the tibial plateau, approximately 1 cm distal to articular surface of the tibia (Logterman et al. 2018).[11] It crosses the anterior cruciate ligament to form an "x." The posterior cruciate ligament consists of two inseparable bundles: the anterolateral bundle for mid-flexion and internal rotation of the knee and the posteromedial bundle for extension and deep flexion of the knee.

Western Medicine Etiology and Pathology

Acute Injury

The most common mechanism of injury to the posterior cruciate ligament is a direct blow to the anterior aspect of the proximal tibia on a flexed knee with the ankle in plantarflexion. Because of its functions to prevent moving of the femur and tibia, posterior cruciate ligament injuries mostly occur with a sudden posterior tibial translation mechanism such as a motor vehicle accident that causes the knee to strike the kneebar of the dashboard and results in posterior translation of the tibia.

In sports activities, injury to the ligament usually occurs when the knee strikes the ground in a hyperflexed position or a posteriorly directed blow to the proximal tibia when the foot is flexed on the ground.[12] These injuries occur mostly during sports such as football, soccer, and skiing.

Chronic Injury

Chronic condition of posterior cruciate ligament injury is mostly from an undiagnosed acute injury. A secondary problem such as arthrosis in the medial and patellofemoral compartments and meniscal breakdown in the knee joint may occur.[13]

Chinese Medicine Etiology and Pathology

Acute State

Acute traumatic injury to the ligaments leads to blood stasis and swelling, which block the meridians so that Qi and blood are stagnant, resulting in pain. According to Chinese medicine, in *Plain Questions* or Sù Wèn 素問 chapter 39 (Unschuld and Tessenow 2011) on the topic of "Treatise on Lifting Pain" 《舉痛論篇第三十九》, the Yellow Emperor's personal physician Qi Bo 歧伯 explained that "When there is pain, there is no free flow, when there is free flow, there is no pain" (Unschuld and Tessenow 2011).

Chronic State

Ligaments injury is considered called "Jin Shang 筋傷" or "traumatic injury on tendons." In traditional Chinese medicine (TCM) theory, the liver governs the tendons. "Traumatic injury on tendons" consumes liver energy, which is mostly Yin and blood. As the injury becomes chronic, the liver is deficient, and the kidneys are then deficient because the liver "wood" and the kidney "water" always affect each other. With such deficiency, patients are vulnerable to any knee issue such as meniscal breakdown, patellofemoral syndrome, arthrosis and residual knee laxity, and loss of range of motion.[14]

Edema

This type of swelling is considered "Feng Shui 風水" or Yang edema. The wind invades meridians when there is blood stasis, the wind blocks the dispersing descending functions of the lung, and the water is then accumulated.

Manifestations

Posterior cruciate ligament injury can be graded according to severity. An MRI can assess for concomitant injuries such as meniscal, ligamentous, posterolateral joint capsule, popliteus, medial collateral ligament, and posterior oblique ligament injuries.

- *Grade 1*: Knee sprain with ligamentous partially torn the knee is still able to function and knee stability is preserved. One to five millimeter posterior tibial translation is present. The tibia remains anterior to the femoral condyles.
- *Grade 2*: Knee injury with ligament partially torn; the knee is unstable. Six to ten millimeter posterior tibial translation. Anterior tibia flush with femoral condyles.
- *Grade 3*: Knee injury with ligament completely torn accompanied by other ligamentous damage; the knee is unstable. Greater than 10 mm posterior tibial translation is present. Tibia posterior to femoral condyles (Raj 2019).

The typical presentations of acute posterior cruciate ligament injury are:

- Symptoms are often vague, depending on severity. Patients are often unsure there is an injury because they often are not even feeling or noticing the injury.
- Unlike anterior cruciate ligament injury, most people don't feel or hear a "pop" sound after posterior cruciate ligament injury and may be able to continue playing immediately after the injury.[15]
- Pain:

Mild to moderate, depending on the extent of the injury.

- The pain may be nonexistent if the ligament is mildly sprained.
- Mild pain at the posterior aspect of the knee when kneeling on the knee.
- There is a tenderness in the posterior aspect of the knee.
- Pain in the anterior aspect of the knee usually begins one to two weeks after the initial injury. Patients with mild knee pain may not seek medical attention immediately, and the type of knee pain is vague weeks later once they see a medical provider.

- Swelling:

  The knee swelling depends on the severity. If the torn ligament is grade 3, the swelling occurs almost right after the injury or within hours after the initial injury.
- Instability:

  Patients feels that the knee is loose, wobbly, unstable, and as if it is about to give out when walking or standing. Patients have a difficult time doing physical activities especially when bearing weight on the knee.

Physical Examination

Physical examination of the posterior cruciate ligament injury is as follows.

- Begin by obtaining the patient's medical history:
  - Injury history to the ligament may not be noticeable; patients may not be sure there was an injury. Most patients don't feel or hear a "pop" sound during the injury, and they may be able to continue the activity.
  - If there is pain, it might only be mild to moderate. It will occur at the posterior aspect of the knee in the kneeling position.
  - The pain may be in the anterior aspect of the knee one to two weeks later after the initial injury if the severity is grade three and the anterior cruciate ligament is involved.
- Inspection:
  - The knee swelling depends on the severity. Grade 1 injury doesn't manifest with effusion; in grade 3 injuries where the ligament is torn, the swelling occurs almost right after the initial injury or within hours. The swelling may make the knee stiff and may cause a limp.
  - The leg walks with give-out. Patients have unstable movements during physical activities, especially when bearing weight.
- Palpation:
  - Tenderness in the knee, specifically the back of the knee.
- Range of motion:
  - In grade 1, full range of motion is present.
  - In grade 3, the knee has a range of motion and difficulty with mobilization.

Posterior Drawer Test

The patient lies in a supine position with the hips flexed at 45 degrees and the knees flexed at 90 degrees. A posterior force is applied to the proximal tibia with one hand while stabilizing the distal femur with another hand. The test is positive if there is pain or an increase of tibia posterior translation of more than 5 mm, and the laxity will be apparent as compared to the contralateral knee. However, due to the pain and muscle spasm caused by knee flexion, the posterior drawer test is often difficult to perform, and some are misdiagnosed as a negative result.

As a result, I often modify the test procedures to minimize the pain if it is too painful to conduct.

The patient is in a supine position with the hips and knees relaxed completely and extended. I slowly raise the distal femur about 10 cm above the knee. When the posterior cruciate ligament is torn, the proximal tibia slips backward, and there is a visible sag under the inferior portion of the patella.

Posterior Sag Sign

The patient lies supine with hips flexed to 45 degrees and knees flexed to 90 degrees. There is a sag of the tibia compared to the contralateral side, which suggests a posterior cruciate ligament tear.

Lachman Test

The patient lies supine with the affected knee bent at 30 degrees; the examiner stands next to the injured knee. The examiner stabilizes the distal femur with one hand and presses the tibia posteriorly with another hand. If the tibia moves significantly more than the unaffected leg, the test is positive and suggests a torn posterior cruciate ligament.

Reverse Lachman Test

The patient lies prone with the knee flexed to 30 degrees; the examiner stabilizes the posterior distal femur then lifts the proximal tibia posteriorly. The Reverse Lachman's Test is positive if there is pain or an increase in tibia posterior translation of more than 5 mm.[16] The positive result indicates a strain or rupture of the posterior cruciate ligament.[17]

Extra Tests

X-rays or MRI scans are highly recommended to study the degrees of torn ligaments and fractures.

Differential Diagnosis

In an anterior cruciate ligament injury, there is a "pop" feeling in the knee; usually the cause is sporting activity. Posterior cruciate ligament injury is mostly caused by motor vehicle accidents, and there is no "pop" feeling in the knee. Symptoms present in a mild degree injury of anterior cruciate ligament, but there might be no symptoms or only instability in the affected knee joint in a mild degree injury of the posterior cruciate ligament.

Treatment

Treatment principle:

Acute stage:

- Moves blood and Qi. The acute injury causes blood stasis in meridians.
- Cools blood, expels wind, disperses lung, circulates water. Chinese medicine is very helpful in reducing knee edema from traumatic inflammation.

Chronic stage:

- Tonifies liver blood and Yin to strengthen the ligaments and tendons.
- Tonifies kidney Qi, Yin, and Yang to stabilize the knee joints and warm channels.

Herbal Treatments

Acute Stage

Herbal formula:

- Tao Hong Si Wu Tang + Si Miao Yong An Tang Jia Jian 桃紅四物湯合四秒勇安湯加減.

Formula ingredients:

Tao Ren 桃仁 15 g, Hong Hua 紅花 12 g, Dang Gui 當歸 30 g, Chuan Xiong 川芎 10 g, Niu Xi 牛膝 10 g, Huang Qi 黄芪 40 g, Gui Zhi 桂枝 15 g, Jin Ying Hua 銀花 30 g, Xuan Shen 玄参 20 g, Mu Tong 木通 15 g, Chi Shao 赤芍 15 g, Gan Cao 甘草 6g.[18]

Ingredient explanations:

- Tao Ren 桃仁 breaks up stubborn blood stasis.
- Hong Hua 紅花 moves blood, disperses bruise, opens meridians, stops pain.
- Dang Gui 當歸 moves and tonifies blood, disperses cold, and stops pain due to blood stasis.
- Chuan Xiong 川芎 moves blood and Qi, expels wind, and stops pain.
- Niu Xi 牛膝 expels wind-dampness, tonifies liver and kidneys, directs herbs downward.
- Huang Qi 黄芪 strengthens spleen, reduces edema.
- Gui Zhi 桂枝 releases the exterior, warms meridians, and stops pain.

- Jin Ying Hua 金銀花 clears heat, disperses wind heat, clears damp heat, cools blood.
- Xuan Shen 玄参 clears heat, cools blood, nourishes Yin.
- Mu Tong 木通 drains heat, promotes urination.
- Chi Shao 赤芍 moves blood, reduces swelling, stops pain.
- Gan Cao 甘草 releases spasms, alleviates pain, harmonizes the other herbs, and guides the herbs to all 12 meridians.

Formula action:

- Cools and moves blood, reduces swelling, stops pain.

Chronic Stage

Herbal formula:

- Du Huo Ji Sheng Tang 獨活寄生湯.

Formula ingredients:

Du Huo 獨活 9 g, Sang Ji Sheng 桑寄生 6 g, Du Zhong 杜仲 6 g, Niu Xi 牛膝 6 g, Xi Xing 细辛 3 g, Qin Jiao 秦艽 6 g, Fu Ling 茯苓 6 g, Rou Gui 肉桂 6 g, Fang Feng 防風 6 g, Chuan Xiong 川芎 6 g, Dang Shen 党参 6 g, Zhi Gan Cao 甘草 3 g, Dang Gui 當歸 6 g, Bai Shao 白芍 6 g, Shu Di Huang 熟地黄 6 g.

Ingredient explanations:

- Du Huo 獨活 expels wind-damp-cold from lower burner, bones, and sinews.
- Sang Ji Sheng 桑寄生 expels wind-dampness, tonifies liver and kidneys.
- Qin Jiao 秦艽 relaxes sinews, expels wind-dampness.
- Fang Feng 防風 expels wind-dampness.
- Chuan Xiong 川芎 nourishes the blood.
- Niu Xi 牛膝 expels wind-dampness, tonifies liver and kidneys, directs herbs downward.
- Du Zhong 杜仲 expels wind-dampness, tonifies liver and kidneys.
- Dang Gui 當歸 nourishes and moves the blood.
- Fu Ling 茯苓 tonifies spleen Qi, eliminates damp.
- Dang Shen 黨參 tonifies spleen Qi, eliminates damp.
- Shu Di Huang 熟地黃 nourishes and moves blood.
- Bai Shao 白芍 nourishes and moves blood.
- Xi Xin 細辛 expels wind-cold in the channels, sinews, and bones and stops pain.
- Zhi Gan Cao 炙甘草 tonifies Qi, harmonizes other herbs.
- Rou Gui 肉桂 warms and unblocks the channels, tonifies Yang Qi.

Formula actions:

- Dispels wind, cold, and damp.
- Relieves painful obstruction.
- Supplements Qi and blood.
- Tonifies the liver and kidneys.

Western uses: Chronic rheumatic arthritis.

- Osteoarthritis

Modifications:

- For severe pain, add Zhi Chuan Wu 制川烏, Zhi Cao Wu 制草烏, Bai Hua She 白花蛇, Xu Duan 川斷, Gou Ji 狗脊, Tu Si Zi 菟絲子, Hong Hua 紅花, and Di Long 地龍.
- For more cold symptoms, add Zhi Fu Zi 附子 and Gan Jiang 乾薑.
- For more damp symptoms, omit Shu Di Huang 熟地黃 and add Fang Ji 防己, Yi Yi Ren 薏苡仁, and Cang Zhu 蒼朮.

Contraindications:

- Damp heat excessive Bi syndrome.

Acupuncture Treatments

Acute Stage

Treatment principle:

- Moves blood and Qi.

Acupuncture point formula:

- LI4 合谷 Hegu, SP10 血海 Xuehai, GB34 陽陵泉 Yanglingquan, GB39 懸鐘 Xuanzhong, UB17 膈俞 Geshu, UB39 委陽 Weiyang, UB40 委中 Weizhong.

Cool blood, expel wind, disperse lung, circulate water:

- GB20 風池 Fengch, UB13 肺俞 Feishu, KD7 復溜 Fuliu, LIV2 行間 Xingjian.

Chronic Stage

Tonify liver blood and Yin:

- LIV5 蠡溝 Ligou, LIV8 曲泉 Ququan, UB18 肝俞 Ganshu, SP6 三陰交 Sanyinjiao.

Tonify kidney Qi, Yin, and Yang to stabilize the knee joints and warm channels:

- KD1 涌泉 Yongquan, KD 3 太溪 Taixi, KD6 照海 Zhaohai, UB23 腎俞 Shenshu.

Moxibustion Treatment

Warming needling techniques on ST35 犊鼻Dubi.

Bleeding then Cupping Treatment

He Ding 鶴頂 (the location is above the knee, in the depression of the midpoint of the upper border of the patella).

Electro-Acupuncture Treatment

He Ding 鶴頂 ~ ST36 足三里 Zusanli 2/100 Hz for 20 minutes daily.

Tui Na Treatments

Tui Na treatment is only appropriate for first and mild second degree injuries. Third-degree injury must receive Western medical attention; Chinese medicine is then used for recovery.

Tui Na techniques:

- Pressing, kneading, pushing, grasping

Tui Na procedures for acute stage:

1) The patient lies in a supine position with the knee flexed and supported by a pillow or a cushion; the therapist stands next to the affected knee. The left knee is used in the example below.
2) Apply pressing and kneading techniques with thumbs and other fingertips on "Eight Patella Blood Points."

   "Eight Patella Blood Points" are located at the edges of the patella. There are eight points: medial upper corner, medial lower corner, medial midpoint, lateral upper corner, lateral lower corner, lateral midpoint, mid-point of superior border, and midpoint of inferior border.

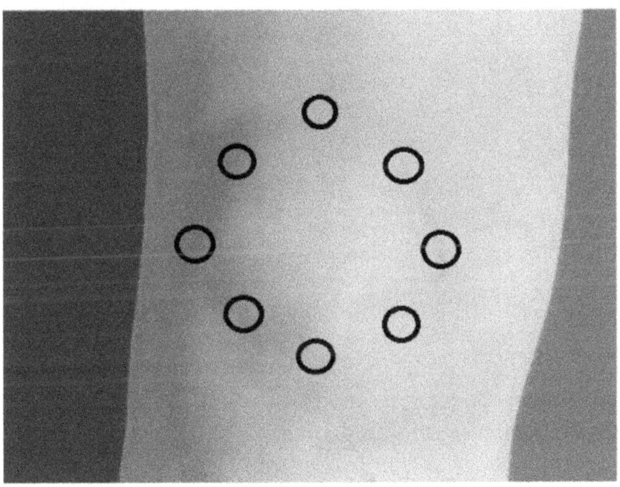

3) Apply pushing technique on the distal femur, directed from the knee joint toward the hip joint. the pushing areas are all aspects of lateral side on the IT band, the medial side on the adductor muscles, and the anterior aspect on the quadriceps tendon and muscles.
4) Apply grasping technique on the quadriceps muscles.

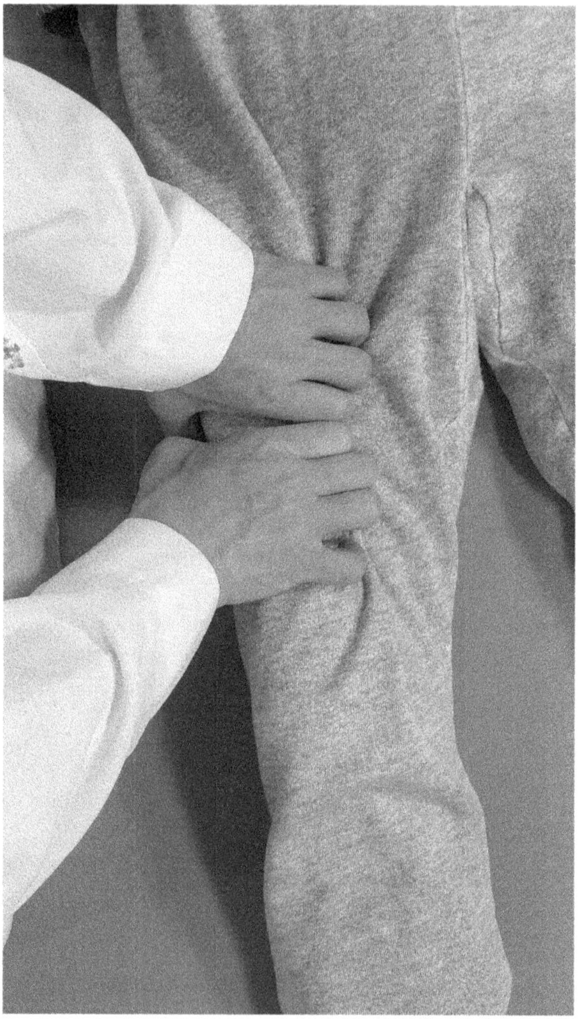

Patient Advisory

1) Before exercise, warm up the joints to prevent ligaments from over-stretching.
2) After acute injury, patients should rest the knee, apply ice packs, and apply an elastic bandage on the injured knee.

3) To prevent edema, the patient is advised to raise the affected leg higher than the heart while lying supine.
4) Wearing a knee brace for knee support is helpful for recovery.
5) Because the knee is unstable, patients are recommended to use a crutch to prevent falls.
6) Exercise to strengthen the leg muscles, especially the quadriceps muscles:
    a) The patient lies supine with the affected leg extended and the contralateral leg flexed.
    b) Slowly raise the leg until it is even with the contralateral thigh.
    c) Do not raise the leg higher than the contralateral thigh.
    d) Hold the affected leg until it is tremoring. Repeat 30–50 times.[19]
    e) Repeat on the other side.
7) In terms of diet, patients should be advised to avoid Fa Wu diet. Fa Wu foods are the foods that may cause more inflammation and bring back old injuries. Some examples of Fa Wu foods are shellfish such as crab and shrimp, fish, fried foods, lamb, duck, chocolate, garlic, and green onions. This concept applies to both acute and chronic strains.

MEDIAL COLLATERAL LIGAMENT INJURY 內側副韌帶損傷

Introduction

An injury to the medial collateral ligament is often called a medial collateral ligament sprain because the injury can either stretch the ligament or tear it. It is the most common knee ligament injury. It is often caused by the knee going into a valgus position so that the lateral side of the knee is attacked with a blunt force; football, hockey, wrestling, and martial arts are common high-risk sports that cause this injury. The ligament injury is also seen with non-sport related causes such as a wrong landing from a jump or a fall with the foot in the valgus position.

Anatomy

The medial collateral ligament is one of four major knee ligaments that help stabilize the knee joint. It is located on the medial side of the knee and connects the medial femoral condyle and the medial tibial condyle. There is a slight knee valgus physiologically. The ligament splits slightly at the knee joint, and some fibers also insert onto the medial meniscus. Its function is to resist valgus force which occurs if the tibia is forced outward in relation to the knee such as during a direct blow to the knee.

Western Medicine Etiology and Pathology

Clinically, medial collateral ligament sprains are the most common knee ligament injuries. Because there is a slight knee valgus physiologically, when an external valgus force is too great for the ligament to resist, the ligament is overstretched. This often occurs while the knee is slightly flexed.

The sporting injury can be caused by both direct contact sports and non-direct contact sports.

Injury from a direct contact sport can occur through a blunt force to the lateral side of the knee. Wrestling and American football are high-risk sports for direct contact injury.

Injury from a non-direct contact sport occurs when a knee twists, cuts, or stops suddenly while running or a landing wrong from a jump with the foot in valgus position. The injury increases stress to the medial collateral ligament, resulting in a sprain. The most common sports that might lead to the sprain of the medial collateral ligament are football, hockey, wrestling, and martial arts.

Because of the anatomy that the medial collateral ligament is related to the medial meniscus cartilage and the anterior cruciate ligament, these two structures may also be damaged in association with the medial collateral ligament injury. Due to the fact that some fibers of the medial collateral ligament insert on the medial meniscus, the anterior cruciate ligament ends at the anterior to the intercondyloid eminence of the tibia, and sprains may also associate with the medial meniscus tear.

When the sprain and inflammation become chronic, calcification along the insertion into the femur can occur.[20]

Chinese Medicine Etiology and Pathology

This is part of Bi syndrome. The patient's general health constitution is weak. Muscle incoordination or weakness predispose the patient to sprain or tear. Once the incident occurs, the injury causes blockages of blood and Qi in the meridians, and blockages manifest pain. According to Chinese medicine, in Plain Questions or Sù Wèn 素問 chapter 39 (Unschuld and Tessenow 2011) on the topic of "Treatise on Lifting Pain" 《舉痛論篇第三十九》, the Yellow Emperor's personal physician Qi Bo 岐伯 explained that "When there is pain, there is no free flow, when there is free flow, there is no pain" (Unschuld and Tessenow 2011).

Manifestations

- Patients have a history of valgus.
- An audible snap or a tearing sound at the time of injury may be noticed, depending on the severity of the injury.[21]
- Pain is the first symptom; it occurs almost immediately and is located at the injured area.
- Depending on the severity of the injury, the edema could be within the knee joint and may develop within minutes.
- The knee becomes weak and cannot bear weight; the patient has to limp in order to avoid putting pressure on the knee.

The medial collateral ligament injury is graded in three levels according to the degree of stretching and tearing which are the signs of severity and symptoms of the knee ligament sprain:

- *Grade 1*: The ligament has a slight stretch, but no tear. There may not be much swelling and pain, or there will be tenderness over the site of injury.
- *Grade 2*: There is a partial tear of the ligament. Knee pain occurs almost immediately; knee swelling and bruising are visible. There is some degrees of instability or sensation of the knee giving out.
- *Grade 3*: The ligament tears completely. Often there will be no pain, as all of the pain fibers are torn at the time of injury, or there is a severe pain that subsides quickly. The skin is bruised and swollen. The patient has to limp, as it is difficult for the knee to bear weight and it has a sensation of giving out. With these more severe tears, the meniscus and the anterior cruciate ligament are at risk of injury. In cases where there is an injury to these structures, hemarthrosis, or articular bleeding, warmth, excessive bruising,[22] the knee joint is instability, laxity, and feelings of catching or locking at a certain angle of movement are common.

Physical Examination

Physical examination of the patellar tendonitis is as follows.

- Begin by obtaining the patient's medical history:
  - Injury history to the ligament. The history should include direct contact and noncontact injuries.
  - Direct contact injury includes sports activities that are high risk for having a blunt force strike directly onto the lateral aspect of the knee or a history of injury from a motor vehicle collision such that the lateral side of the knee was hit directly.
  - Non-contact knee injury can be caused by sports activities that are high risks for valgus stress to the knee such as those that involve twisting, cutting, or landing awkwardly from a jump with the foot in valgus position.
- Inspection:
  - observation of the site of the affected knee to learn the location and the severity of the bruise and swelling.
  - The examination may also include the hip and the ankle to identify any other associated injuries.
- Palpation:
  - Touching the knee in various places to find the tenderness and pain, which are often located along the course of the medial collateral ligament.
- Range of motion:
  - Valgus Stress Test is proved positive by pushing on the lateral side of the affected knee to reproduce the valgus stress.
  - Lachman Test is also positive if there is an injury of the anterior cruciate ligament.
- Extra tests:
  - X-rays or MRI scans are highly recommended to study the degree of torn ligaments and fractures.

Differential Diagnosis

Pes Anserine Bursitis: The pes anserine bursa is distal and anterior to the medial collateral ligament; inflammation can cause acute or chronic pain that can mimic a medial collateral ligament injury. The patient may experience spontaneous anteromedial knee pain on climbing or descending stairs, and the knee is swollen or tender to touch.[23] The Lachman Test is negative.

Treatment

Treatment principle:

Acute stage:
- Moves blood and Qi. Acute injury causes blood stasis in meridians.
- Reduces swelling and stops pain. Chinese medicine is very helpful to reduce knee edema from traumatic inflammation.

Chronic stage:
- Tonifies liver blood and Yin to strengthen the ligaments and tendons.
- Tonifies kidney Qi, Yin, and Yang to stabilize the knee joints and warm channels.

Herbal Treatments

Acute Sprain

Herbal formula:

- Yun Nan Bai Yao Aerosol topical application 雲南白藥氣霧劑

Formula ingredients:

- The main ingredient in Yun Nan Bai Yao 雲南白藥 is San Qi 三七. San Qi 三七is used in the treatment of internal and external bleeding, and it also relieves any type of pain from injury.

The topical spray helps to promote blood circulation, stop pain, and reduce swelling. The spray is easier to apply on closed wounds than the Yun Nan Bai Yao powder form.

Dosage:

- The product comes in two bottles: a cream color Yunnan Baiyao Aerosol 雲南白藥氣霧劑 and a red color Yunnan Baiyao Aerosol Baoxianye 雲南白藥氣霧劑保險液.

Contraindication:

- Pregnancy.

Suggested usage:

- Shake the bottle several times before use. Hold the bottle vertically with the spray nozzle five to ten centimeters from the affected skin, and spray for about three to five seconds.

- Red color spray—Yunnan Baiyao Aerosol Baoxianye 雲南白藥氣霧劑保險液.
- For serious trauma, apply Aerosol Baoxianye 氣霧劑保險液 to the injured part first, then reapply it after one to two minutes later if the pain doesn't alleviate. Aerosol Baoxianye 氣霧劑保險液 should not be used more than three times a day. If pain continues, apply Yunnan Baiyao Aerosol 雲南白藥氣霧劑. Aerosol Baoxianye 氣霧劑保險液 should not be used on open wounds, cuts, and grazes.
- Cream color spray—Yunnan Baiyao Aerosol 氣霧劑保險液
- Externally used to spray on wounds, three to five times daily.

Herbal formula for Qi and blood stagnation type:

- Tao Hong Si Wu Tang + Si Miao Yong An Tang Jia Jian 桃紅四物湯合四秒勇安湯加減.

Formula ingredients:

- Tao Ren 桃仁 15 g, Hong Hua 紅花 12 g, Dang Gui 當歸 30 g, Chuan Xiong 川芎 10 g, Niu Xi 牛膝 10 g, Huang Qi 黄芪 40 g, Gui Zhi 桂枝 15 g, Jin Ying Hua 銀花 30 g, Xuan Shen 玄参 20g, Mu Tong 木通 15 g, Chi Shao 赤芍 15 g, Gan Cao 甘草 6 g.[24]

Ingredient explanations:

- Tao Ren 桃仁 breaks up stubborn blood stasis.
- Hong Hua 紅花 moves blood, releases bruising, opens meridians, stops pain.
- Dang Gui 當歸 moves and tonifies blood, disperses cold, and stops pain due to blood stasis.
- Chuan Xiong 川芎 moves blood and Qi, expels wind, and stops pain.
- Niu Xi 牛膝 expels wind damp, tonifies liver and kidneys, directs herbs downward.
- Huang Qi 黄芪 strengthens spleen, reduces edema.
- Gui Zhi 桂枝 releases the exterior, warms meridians, and stops pain.
- Jin Ying Hua 金銀花 clears heat, disperses wind heat, clears damp heat, and cools blood.
- Xuan Shen 玄参 clears heat, cools blood, and nourishes Yin.
- Mu Tong 木通 drains heat, promotes urination.
- Chi Shao 赤芍 moves blood, reduces swelling, stops pain.
- Gan Cao 甘草 releases spasms, alleviates pain, harmonizes the other herbs, and guides the herbs to all 12 meridians.

Formula action:

- Cools and moves blood, reduces swelling, stops pain.

Chronic Sprain

Herbal formula:

- Shu Jin Zhi Tong Liniment 舒筋止痛水.

Formula ingredients:

- San Qi 三七 18 g, San Leng 三棱 18 g, Hong Hua 紅花 30 g, Camphor 樟腦 30 g, Cao Wu 草烏 12 g, Chuan Wu 川烏 12 g, Wu Jia Pi 五加皮 12 g, Mu Gua 木瓜 12 g, Niu Xi 牛膝 12 g, Dang Gui We 當歸尾 18 g.

Ingredient explanations:

- San Qi 三七 Disperses bruise, stops bleeding, moves blood, stops pain.
- San Leng 三棱 moves blood, moves Qi.
- Hong Hua 紅花 moves blood.
- Camphor 樟腦 moves Qi, reduces swelling, stops pain.
- Cao Wu 草烏 expels wind, disperses blood and cold.
- Chuan Wu 川烏 expels exterior wind, drains interior damp, warms channels, stops pain.
- Wu Jia Pi 五加皮 expels wind, moves blood, tonifies liver and kidneys.
- Mu Gua 木瓜 releases spasm, relaxes tendon.
- Niu Xi 牛膝 tonifies kidneys, strengthens bones and tendons.
- Dang Gui Wei 當歸尾 moves blood.

Administration:

- Soak these herbs in 1500 ml of alcohol for a month. Topical application only.

Application:

- Rub the liniment on the affected medial collateral ligament two to three times daily.

Formula action:

- Relax tendon, move blood, and stop pain.

Acupuncture Treatments

Acute Injury

Acupuncture point formula:

- LI4 合谷 Hegu, SP10 血海 Xuehai, SP9 陰陵泉 Yinlingquan, ST34 梁丘 Liangqiu, GB34 陽陵泉 Yanglingquan, ST36 足三里 Zusanli, ASHI acupoints 阿是穴.

Chronic Injury

Acupuncture point formula:

- GB39 懸鍾 Xuanzhong, GB31 風市 Fengshi, LI4 合谷 Hegu, KD3 太溪 Taixi, LIV8 曲泉 Ququan.

Cupping Treatments

SP10 血海 Xuehai, LIV8 曲泉Ququan.

Moxibustion Treatments

Warming needling techniques on SP10 血海 Xuehai, LIV8 曲泉 Ququan, SP9 陰陵泉 Yinlingquan.

Bleeding Treatments

Ashi acupoints 阿是穴.

Electro-Acupuncture Treatments

SP10 血海Xuehai, LIV8 曲泉 Ququan2/100 Hz for 20 minutes daily.

Tui Na Treatments

Tui Na treatment is for grade 1 and 2 injuries.

Tui Na techniques:

- Kneading, pressing, pushing, bending.

Tui Na procedures:

1) The patient lies supine; the therapist stands next to the affected knee.
2) The therapist applies kneading techniques with the heel of a hand on the quadriceps muscle for three to five minutes.
3) The therapist applies finger pressing and kneading techniques on the acupuncture points SP10 血海 Xuehai, SP9 陰陵泉 Yinlingquan, and Ashi Points 阿是穴 on the affected medial collateral ligament for eight to ten minutes.
4) If the right knee is the injured one, the therapist clamps the patient's right ankle at the armpit, places the right hand at the affected leg's superior calf, puts the right thumb at the interior medial collateral ligament, and places the left hand on the quadriceps tendon.

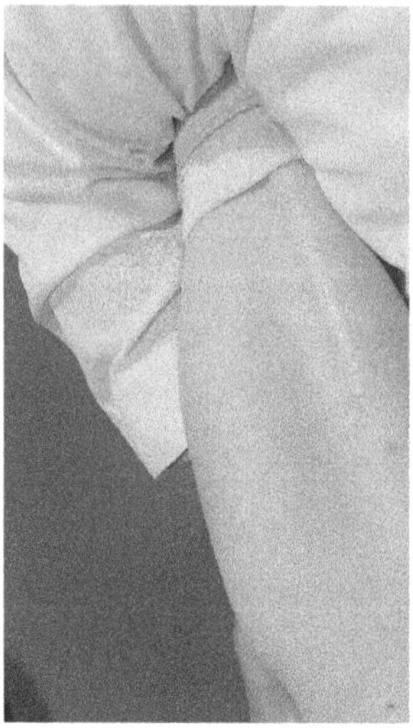

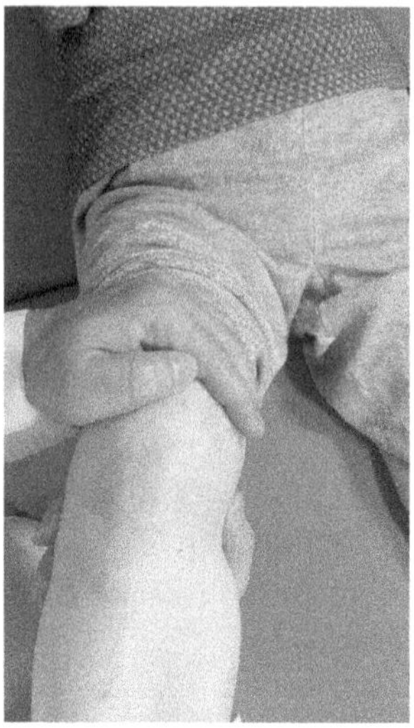

5) Flex the affected knee at the same time the right thumb pushes the medial collateral ligament superiorly; repeat five to seven times.

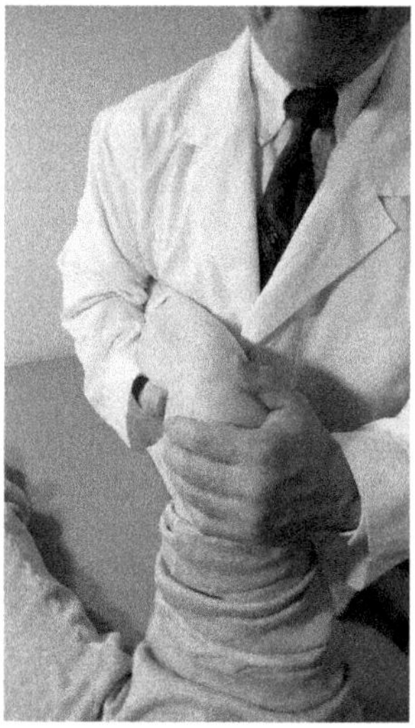

Patient Advisory

- Take a break from the activities that caused the injury.
- In the acute stage (about the first two to three days), an ice pack is helpful for swelling. Put the ice pack on the medial side of the knee every one to two hours, for ten to twenty minutes each time. Ensure that the ice pack is not directly on the skin by placing a thin cloth between the ice pack and the skin.
- When the knee is swelling, raise the leg higher than the heart when lying by putting a pillow or cushion under the knee.

- When the swelling stops, keep the knee warm to prevent cold invasion.
- Recommend adding a brace to the injured knee.
- Exercise the legs, and strengthen the soft tissues to prevent future knee injury:
  - The patient lies supine with the affected leg extended and the contralateral leg flexed.
  - Slowly raise the leg until it is even with the contralateral thigh.
  - Do not raise the leg higher than the contralateral thigh.
  - Hold the affected leg until it is tremoring. Repeat 30–50 times.
  - Repeat on the other side.
- Avoid bearing weight for four to six weeks.
- Recommend using crutches to avoid bearing weight and prevent falling.
- Patients should be advised to avoid Fa Wu 發物 diet. Fa Wu 發物 foods are the foods that may cause more inflammation and bring old injuries back. Some examples of Fa Wu 發物 foods are shellfish such as crab and shrimp, fish, fried foods, lamb, duck, chocolate, garlic, and green onions. This concept applies to both acute and chronic strains.

LATERAL COLLATERAL LIGAMENT INJURY 外側副韌帶損傷

Introduction

The lateral collateral ligament is one of four major ligaments that stabilize the knee joint.

An injury to the lateral collateral ligament could include straining, spraining, or partially or completely tearing any part of the ligament. The sprain is the most common form of lateral collateral ligament injury. Most injuries occur in combination with damage to the anterior cruciate ligament, meniscus, and posterior cruciate ligament. Sports that cause varus stress predispose patients to the injury. Varus force to the medial knee, poor landing after jumping, and sudden change of direction while running are common causes of varus stress to the knee; wresting, skiing, football, and soccer are high-risk sports.

Anatomy

The lateral collateral ligament originates from the lateral femoral epicondyle, joins the bicep femoris tendon, and then ends at the head of the fibula. It is a cord-like ligament about 66 mm long and 3.4 mm wide (Meister et al. 2000).

Western Medicine Etiology and Pathology

The main function of the lateral collateral ligament is to resist varus force and external tibial rotation.

Lateral collateral ligament sprain usually results from direct excessive varus stress to the knee, which puts pressure on the medial aspect of the knee that is too great for the ligament to resist and causes the lateral collateral ligament to overstretch or tear.[25]

Certain sports activities are high risk for the varus stress; they include twisting the knee whilst the foot is fixed, a cut that is a sharp change in direction, or a blunt force directly to the medial knee in a football tackle or a motor vehicle accident

The lateral collateral ligament injury rarely occurs in isolation; it is usually in conjunction with another knee injury such as anterior cruciate ligament injury.

With the anatomy structure such that the ligament ends at the fibular head, the peroneal nerve often is involved in the symptoms.

Chinese Medicine Etiology and Pathology

This is part of Bi syndrome. The patient's general health constitution is weak. Muscle incoordination or weakness predisposes them to the sprain or tear. Once the incident occurs, the injury causes blockages of blood and Qi in the meridians, and the blockages manifest pain. According to Chinese medicine, *in Plain Questions* or Sù Wèn 素問 chapter 39 (Unschuld and Tessenow 2011) on the topic of "Treatise on Lifting Pain" 《舉痛論篇第三十九》, the Yellow Emperor's personal physician Qi Bo 歧伯 explained that "When there is pain, there is no free flow, when there is free flow, there is no pain" (Unschuld and Tessenow 2011).

Manifestations

Depending on the severity of the injury, symptoms of a lateral collateral ligament injury can range from mild to severe and from sprain to tear.

- Mild pain or no symptoms if the ligament is mildly sprained.
- For a partial tear or a complete tear, a snap sound or a tearing sound may be noticeable.
- The knee may quickly swell.
- The knee is stiff.
- The knee can be locked when the meniscus is associated.
- The location of the pain or soreness is on the lateral collateral ligament, mostly at the inserting point on the fibular head.
- The knee feels like giving out.
- Drop foot-weak ankle dorsiflexion

The lateral collateral ligament injury is graded in three levels according to the degree of stretching and tearing and the severity and symptoms of the knee ligament sprain:

- *Grade 1*: The ligament has a slight stretch but not a tear. Even though the ligament may not be painful or the pain may be mild without swelling, a chronic condition may develop and lead to repeated ligamentous injury easily.
- *Grade 2*: The ligament has a partial tear. It is painful, and the pain increases when bearing weight. Movement is difficult. The knee is swollen, bruised, and unstable. Often, it has a sensation of giving out.

- *Grade 3*: The ligament is completely torn. The joint is unstable and feels like giving out, and it is difficult to bear weight. It is swollen and bruised. With these more severe tears, the meniscus and the anterior cruciate ligament are at risk of injury. In this case, there is an injury to these structures such as hemarthrosis, or articular bleeding, warmth, excessive bruising,[26] and the knee joint suffers from instability and laxity, and it feels like it is catching or locking at a certain angle of movement. Often there will be no pain, as all of the pain fibers are torn at the time of injury, or there is a severe pain that subsides quickly.

If the peroneal nerve is stretched or squeezed by the swelling of the surrounding tissues, the ipsilateral foot is numb and weak.[27] More severe tears by traction of the fibular attachment of the lateral collateral ligament result in fibular head avulsion fractures; patients lose sensation in the cutaneous distribution of the superficial and deep peroneal nerves, and the ankle dorsiflexion becomes weak, an issue called Foot Drop.

Physical Examination

Physical examination of the lateral collateral ligament injury.

- Begin by obtaining the patient's medical history:
  - Injury history to the ligament such as a history of a motor vehicle accident.
  - Sports activities that are high risk of varus stress such as a blunt force directly to the medial aspect of the knee while in a soccer tackle.
- Inspection:
  - Location of pain is at the lateral aspect of the knee.
  - The severity of the bruise.
  - The knee is swelling.
  - The knee is stiff with movement.
  - The joint is locked when the meniscus is involved.
  - The knee gives way when walking.
  - The examination may also include the hip and the ankle to identify any other associated injuries.
- Palpation:
  - Touching the knee in various places to find the tenderness and pain, which are often located along the course of the lateral collateral ligament.
- Range of motion:
  - Weak ankle dorsiflexion that the foot is in the "drop foot" position when the peroneal nerve is involved.
  - Varus Stress Test is positive when pushing on the medial side of the affected knee reproduces the varus stress.

- Lachman Test also is positive if there is an injury of the anterior cruciate ligament.
- Extra tests:
    - X-rays or MRI scans are highly recommended to study the degrees of torn ligaments and fractures.

Differential Diagnosis

If Iliotibial band syndrome is present, the location of the pain is at the lateral side of the knee, and the affected area could be swollen and warm. The pain is caused by repetitively flexing the knee during physical activities such as running, cycling, swimming, and climbing. The pain may radiate along the lateral side of the thigh to the hip causing hip pain; it may find some weakness or imbalance of the quadriceps muscles and hamstring muscles. Ober's Test is positive.

Treatment

Treatment principle:

- *Acute stage*:
    - Move blood and Qi. The acute injury causes blood stasis in meridians.
    - Reduce swelling and stop pain. Chinese medicine is very helpful to reduce knee edema from traumatic inflammation.
- *Chronic stage*:
    - Tonify liver blood and Yin to strengthen the ligaments and tendons.
    - Tonify kidney Qi, Yin, and Yang to stabilize the knee joints and warm meridians.

Herbal Treatments

Acute Sprain

Herbal formula:

- Yun Nan Bai Yao Aerosol topical application 雲南白藥氣霧劑.

Formula ingredients:

- The main ingredient in Yun Nan Bai Yao 雲南白藥 is San Qi 三七. San Qi 三七 is used in the treatment of internal and external bleeding. It also relieves any type of pain from injury.
- The topical spray helps to promote blood circulation, stop pain, and reduce swelling. The spray is easier to apply on closed wounds than the Yun Nan Bai Yao powder form.

Dosage:

- The product comes in two bottles: a cream color Yunnan Baiyao Aerosol 雲南白藥氣霧劑 and a red color Yunnan Baiyao Aerosol Baoxianye 雲南白藥氣霧劑保險液.

Contraindications:

- Pregnancy.

Suggested usage:

- Shake the bottle several times before use. Hold the bottle vertically with the spray nozzle five to ten centimeters from the affected skin, and spray for about three to five seconds.
- Red color spray—Yunnan Baiyao Aerosol Baoxianye 雲南白藥氣霧劑保險液.
- For serious trauma, apply Aerosol Baoxianye 氣霧劑保險液 to the injured part first, then reapply it after one to two minutes later if the pain doesn't alleviate. Aerosol Baoxianye 氣霧劑保險液 should not be used more than three times a day. If pain continues, apply Yunnan Baiyao Aerosol 雲南白藥氣霧劑. Aerosol Baoxianye 氣霧劑保險液 should not be used on open wounds, cuts, and grazes.
- Cream color spray—Yunnan Baiyao Aerosol 氣霧劑保險液.
- Externally used to spray on wounds, three to five times daily.

- Tao Hong Si Wu Tang + Si Miao Yong An Tang Jia Jian 桃紅四物湯合四秒勇安湯加減.

Formula ingredients:

Tao Ren 桃仁 15 g, Hong Hua 紅花 12 g, Dang Gui 當歸 30 g, Chuan Xiong 川芎 10 g, Niu Xi 牛膝 10 g, Huang Qi 黄芪 40 g, Gui Zhi 桂枝 15 g, Jin Ying Hua 銀花 30 g, Xuan Shen 玄参 20 g, Mu Tong 木通 15 g, Chi Shao 赤芍 15 g, Gan Cao 甘草 6 g.[28]

Ingredient explanations:

- Tao Ren 桃仁 breaks up stubborn blood stasis.
- Hong Hua 紅花 moves blood, releases bruise, opens meridians, stops pain.
- Dang Gui 當歸 moves and tonifies blood, disperses cold, and stops pain due to blood stasis.
- Chuan Xiong 川芎 moves blood and Qi, expels wind, and stops pain.

- Niu Xi 牛膝 expels wind dampness, tonifies liver and kidneys, directs herbs downward.
- Huang Qi 黃芪 strengthens spleen, reduces edema.
- Gui Zhi 桂枝 releases the exterior, warms meridians, and stops pain.
- Jin Ying Hua 銀花 clears heat, disperses wind heat, clears damp heat, cools blood.
- Xuan Shen 玄參 clears heat, cools blood, nourishes Yin.
- Mu Tong 木通 drains heat, promotes urination.
- Chi Shao 赤芍 moves blood, reduces swelling, stops pain.
- Gan Cao 甘草 releases spasms, alleviates pain, harmonizes the other herbs, and guides the herbs to all 12 meridians.

Formula action:

- Cools and moves blood, reduces swelling, stops pain.

Chronic Sprain

Herbal formula:

- Shu Jin Zhi Tong Liniment 舒筋止痛水.

Formula ingredients:

- San Qi 三七 18 g, San Leng 三棱 18 g, Hong Hua 紅花 30 g, Camphor 樟腦 30 g, Cao Wu 草烏 12 g, Chuan Wu 川烏 12 g, Wu Jia Pi 五加皮 12 g, Mu Gua 木瓜 12 g, Niu Xi 牛膝 12 g, Dang Gui Wei 當歸尾 18 g.

Ingredient explanations:

- San Qi 三七 disperses bruising, stops bleeding, moves blood, stops pain.
- San Leng 三棱 moves blood, moves Qi.
- Hong Hua 紅花 moves blood.
- Camphor 樟腦 moves Qi, reduces swelling, stops pain.
- Cao Wu 草烏 expels wind, disperses blood and cold.
- Chuan Wu 川烏 expels exterior wind, drains interior damp, warms channels, stops pain.
- Wu Jia Pi 五加皮 expels wind, moves blood, tonifies liver and kidneys.
- Mu Gua 木瓜 releases spasm, relaxes tendons.
- Niu Xi 牛膝 tonifies kidneys, strengthens bones and tendons.
- Dang Gui Wei 當歸尾 moves blood.

Administration:

- Soak these herbs in 1500 ml of alcohol for a month. Topical application only.

Application:

- Rub the liniment on the affected medial collateral ligament two to three times daily.

Formula action:
- Relax tendons, move blood, and stop pain.

Acupuncture Treatments

Acute Injury

LI4 合谷 Hegu, SP10 血海 Xuehai, SP9 陰陵泉 Yinlingquan, ST34 梁丘 Liangqiu, GB34 陽陵泉 Yanglingquan, ST36 足三里 Zusanli, ASHI acupoints 阿是穴 on the lateral collateral ligament.

Chronic Injury

Acupuncture point formula:

- GB39 懸鍾, GB31 風市, LI4 合谷, KD3 太谿, LIV8 曲泉.

Cupping Treatments

ST34 梁丘 Liangqiu, GB34 陽陵泉 Yanglingquan.

Moxibustion Treatments

Warming needling techniques on Ashi Points 阿是穴.

Bleeding Treatments

Ashi Points 阿是穴.

Electro-Acupuncture Treatments

ST34 梁丘 Liangqiu ~ GB34 陽陵泉 Yanglingquan 2/100 Hz for 20 minutes daily.

Tui Na Treatments

Tui Na treatment is for grades 1 and 2.

Tui Na techniques:

- Pressing, grasping, lifting, bending, kneading, rubbing.

Procedure:

1) Patient lies prone; the therapist stands next to the affected knee.
2) The therapist applies the thumb or other fingertips on the following acupuncture points with deep pressure for one minute per point: SP10 血海 Xuehai, GB34 陽陵泉 Yanglingquan, SP9 陰陵泉 Yinlingquan, UB40 委中Weizhong, LIV3 太沖Taichong, ST36 足三里 Zusanli.

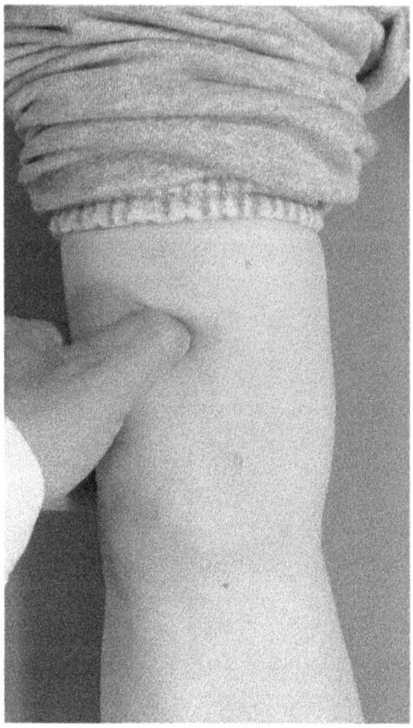

3) The therapist grasps and lifts the following muscles for several times-semitendinosus, semimembranosus, biceps femoris long and short head, and gracilis tendon.
4) The therapist flexes the knee joint at slow speed several times.
5) The therapist applies kneading techniques to the fibular head with thumbs several times.

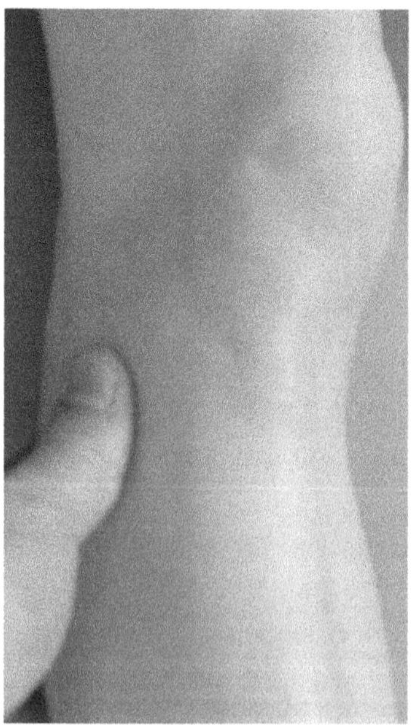

6) The therapist applies rubbing techniques with palms on the lateral collateral ligament for several times or until the skin is warm.

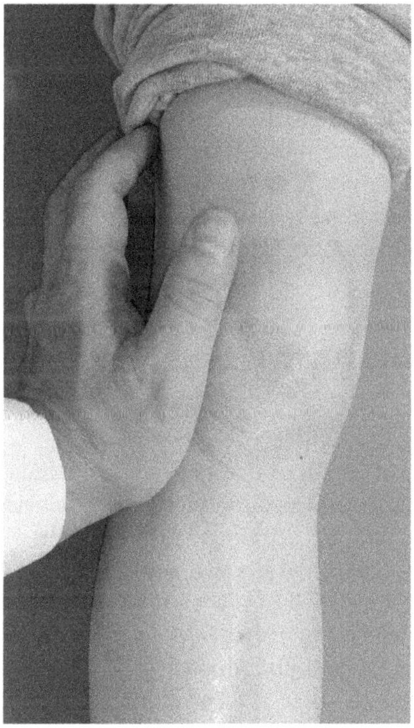

Patient Advisory

The goal is to heal the trauma soon and prevent reinjury.

- Take a break from the activities that caused the injury.
- In the acute stage, about the first two to three days, an ice pack is helpful for swelling. Every one to two hours, put the ice pack on the medial side of the knee for ten to twenty minutes. Ensure that the ice pack is directly put on the skin by placing a thin cloth between the ice pack and the skin.
- When the knee is swelling, raise the leg higher than the heart when lying by putting a pillow or cushion under the knee.
- When the swelling stops, keep the knee warm to prevent cold invasion.
- It is recommended to add a brace to the injured knee.
- Exercise the legs and strengthen the soft tissues to prevent future knee injury:

a) The patient lies supine with the affected leg extended and the contralateral leg flexed.
b) Slowly raise the leg until it is even with the contralateral thigh.
c) Do not raise the leg higher than the contralateral thigh.
d) Hold the affected leg until it is tremoring. Repeat 30–50 times.
e) Repeat on the other side.

- Avoid bearing weight for four to six weeks.
- Recommend using crutches to avoid bearing weight and prevent falling.
- Patients should be advised to avoid Fa Wu 發物 diet. Fa Wu 發物 foods are the foods that may cause more inflammation and bring old injuries back. Some examples of Fa Wu 發物 foods are shellfish such as crab and shrimp, fish, fried foods, lamb, duck, chocolate, garlic, and green onions. This concept applies to both acute and chronic strains.

MENISCAL TEARS 膝關節半月板損傷

Introduction

The meniscus is a piece of cartilage that provides a cushion between the femur and tibia. Each knee joint has two menisci. Meniscus tear injuries can happen at any age and often occur during athletic activities. Athletes, particularly those who play[29] contact sports such as football and hockey and sports that require sudden changes of direction and sudden stops such as soccer, tennis, and basketball are high risk for meniscus tears and can lead to meniscus injuries more easily than other non-contact sports or sports with less lower limb motion.

How well the knee heals after a torn meniscus depends on how good blood supply is. Each part of the meniscus has a different amount of blood supply; the better the blood supply the better the potential for recovery. The outer edge of the meniscus has better blood supply than the center of the meniscus (Healthwise Staff 2019). Unfortunately, blood supply to the knee cartilage decreases with age. In 1990, Renstrom and Johnson reported up to 20% of normal blood supply is lost by age 40 (Bhagia 2019).

Anatomy

The knee is one of the largest and most complex structures in the body.

The joint is formed by four bones, the femur, tibia, patella, and fibular.[30]

The menisci are located inside the knee joint and are not visible; each knee joint has two pieces, the lateral meniscus, which is located at the lateral side of the knee, and the medial meniscus, which is located at the inside of the joint. Both medial and lateral menisci are two large crescent-shaped discs that are superior to the tibia and inferior to the femur.

The medial and lateral menisci are fibrocartilaginous structures composed of 70% water and 60% to 70% collagen. Only the outer 10% to 30% of the meniscus is vascular, supplied primarily by the perimeniscular capillary plexus from the superior and inferior branches of the medial and lateral geniculate arteries (Rao et al. 2015).

The function of the meniscus is to protect the knee joint from the stresses and shocks placed on it from walking, running, climbing, bending, and the body's weight. The mechanism behind the functions sorbing ab weight loads is the ability of the meniscus to expand under compressive forces so that it increases the contact area and decreases the loads, stresses, and shocks to its elastic cartilage about 50–70%. It increases the contact area instead of a point to transmit the loads. In knee extension, the medial meniscus transmits up to 50% of a load applied to the medial side of the knee and the lateral meniscus transmits up to 70% of load applied to the lateral side of the knee of their compartmental loads. In knee flexion, their compartmental loads increase up to 85% and 90%, respectively. But the forces can increase to two to four times bodyweight when shifting from one meniscus to the other during activities such as walking, and the forces on the knee can increase up to six to eight times bodyweight while running or even 24 times bodyweight when landing from a jump (The Steadman Clinic 2020).

The medial meniscus is more commonly injured because it is less mobile than the lateral meniscus. The outer border of the medial meniscus is firmly attached to the joint capsule and deep fibers of the medial collateral ligament.[31] The lateral meniscus has no attachment to the lateral collateral ligament (The Steadman Clinic 2020) and has looser attachments with joint capsule.[32]

Western Medicine Etiology and Pathology

The knee is one of the most stressed joints in the body; it is vital for movement and vulnerable to injury.[33]

Meniscal injuries are in two categories—acute and degenerative tears.

An acute tear usually occurs when the knee is slightly flexed and the upper leg twists while the foot is planted, such as when landing in the wrong way after a jumping or a changing direction while running. Sports that might have such movements and put patients at high risk for meniscus tears include football, soccer, volleyball, basketball, tennis, and badminton.

Meniscus injuries may be associated with tears of the medial collateral ligament and cruciate ligaments.

Degenerative tears usually occur in members of the older population; sixty percent of patients are over the age of 65 (Wallis 2017). This is because the blood supply to the knee cartilage decreases with age, and up to 20% of normal blood supply is lost by age of 40 (Bhagia 2019). As the meniscus ages, it weakens and becomes less elastic (The Steadman Clinic 2020). If a person suffers from osteoarthritis, this person is at higher risk of injuring the knee and tearing the meniscus, as osteoarthritis is a health disorder caused by aging.[34]

Chinese Medicine Etiology and Pathology

Acute Meniscus Injury

According to Chinese medicine, injury of the meniscus is not an isolated condition; the patient's health constitution is predisposed to such an injury. Liver Yin or liver Blood or both are deficient, and the liver is no longer able to govern the tendons. The meniscus does not have enough blood supply, the quality of the meniscus becomes poor, and the elastic cushion of the menisci which is composed of 70% water and 60–70% collagen decreases and becomes weak. Traumatic factors such as a tackle, a cut in a changing direction while running, or a wrong landing from a jump may not have been the first time such trauma has occurred, but it is the one that happened to cause the injury that led to the presentation of the manifestations.

Chronic Meniscus Injury

Kidney and liver deficiency cause osteoarthritis. Kidney energy is no longer enough to govern the bone, and because the joints and bones surrounding the menisci are debilitated, they put stress on the menisci. Stress causes less blood supply to maintain elasticity and cushioning functions and repair any damage to the weak meniscus. Degeneration of menisci causes the presentation of the manifestations.

Edema

This type of swelling is considered "Feng Shui 風水" or Yang edema. The wind invades meridians when there is blood stasis, and the wind blocks the dispersing descending functions of the lung. Water is then accumulated.

Manifestations

When a meniscus tear occurs:

- A pop feeling or sound presents.
- Pain in the knee joint can be constant or may come and go and get worse when bearing weight.
- The knee swells. It usually takes a few hours to develop, and it may be difficult and painful to fully extend or straighten the knee. The swelling can be intermittent.
- The knee is difficult to move, and the patient has to limp in order to avoid putting pressure onto the knee.
- The knee has a sensation of locking or catching when bending the joint. Usually it flexes between 15 and 30 degrees, but it cannot flex or extend from that position.[35]
- There is a sensation of giving out while walking.

Physical Examination

There is no one reliable test specifically for a meniscus torn. Many other anatomic structures around the knee joint can mimic the positive result because they can be painful during the rotation of the knee.

Strobel and Stedtfeld state that examiners should perform three to five meniscus tests on every patient, since each test is inconclusive to confirm tears of the meniscus. They concluded the percentage of accuracy of each test. Apley's compression test is 46% accurate, the Steinmann I sign is 42% accurate, and the McMurray test is 35% accurate. The most reliable test among the 20 tests that Strobel and Stedtfeld listed is the joint line tenderness test, which is the palpatory tenderness on the meniscus at the tibiofemoral joint line which is 60–80% accurate (Strobel and Stedtfeld 1990).

These tests are for consideration:

1) The joint line tenderness test

Procedures to detect tears of the meniscus:

a) The patient lies supine; the examiner stands beside the affected knee. First, the examiner palpates the joint for warmth and areas of tenderness, assesses the stability of the ligaments, and tests the range of motion of the knee joint and the power of the quadriceps and hamstring muscles.

The medial and lateral joint lines have to be palpated separately. Systematically palpate the joint line as follows:

a) Passively flex the knee 90 degrees.
b) Identify the hollows that lie over the joint line at the sides of the patella ligament.
c) Identify the soft hollow between the prominence of the femur and tibia.
d) Palpation of medial joint line starts from the medial border of the patella ligament toward the posterior aspect of the knee.
e) Palpation of the lateral joint line starts from the lateral border of the patella ligament toward the posterior aspect of the knee.
f) Record the point of maximal tenderness.[36]

2) McMurray test

Procedures to detect tears of the meniscus:

a) The patient lies supine with hip and knee flexed at 90 degrees; the examiner stands beside the affected knee.
b) The examiner stabilizes the affected knee in one hand, locates the tibiofemoral joint line, and places a finger on the medial joint line.
c) The examiner grasps the patient's heel with the other hand.
d) The examiner maximally flexes the patient's knee.
e) The examiner rotates the tibia internally while applying a varus force.
f) The examiner rotates the tibia externally while applying a valgus force.
g) The test is positive when the patient has a sensation of clicking in the tibiofemoral joint line.

Clicking indicates medial meniscus injury when it occurs with internal rotation and varus force.[37]

3) Steinmann I Sign

Procedures to detect tears of the meniscus:

a) The patient lies supine with the leg near extension; the examiner stands beside the affected knee.
b) The examiner rotates the tibia externally and internally and slowly flexes the knee.
c) The test is positive for tears in the medial meniscus when the pain manifests if the tibia is rotated laterally.
d) The test is positive for tears in the lateral meniscus when the pain manifests if the tibia is rotated medially.[38]

4) Steinmann II Sign

Procedures to detect tears of the meniscus:

a) The patient lies supine with the knee slightly flexed; the examiner stands beside the affected knee.
b) Palpate the tibiofemoral joint line and locate tenderness.
c) The test is positive if the tenderness moves posteriorly with increasing flexion.
d) The test is positive if the tenderness moves anteriorly when the knee is extended.[39]
e) If the tenderness remains during the knee flexion or extension, the test is negative. The tenderness may be due to a collateral ligament lesion.[40]

5) Ege's Test

Procedures to detect tears of the medial meniscus:

a) The patient stands with the feet eight to ten inches apart and toes pointed forward.
b) The patient externally rotates the knee by turning the feet outward.
c) The patient squats and slowly stands up.
d) The test is positive for tears of the medial meniscus when there is an audible click sound or pain in the area of the meniscus.
e) For testing the tears of lateral meniscus, repeat the above procedures except the knees are internally rotated.

Procedures to detect tears of the lateral meniscus:

a) The patient stands with the feet eight to ten inches apart and toes pointed forward.
b) The patient internally rotates the knee by turning the feet inward.
c) The patient squats and slowly stands up.
d) The test is positive for tears of the lateral meniscus when there is an audible click sound or pain in the area of the meniscus.[41]

Precautions

In the early stage of trauma, it is not meaningful to perform these tests for at least three weeks, because the injuries to the surrounding soft tissues have not been repaired after the knee injury. With or without tears of the meniscus, doing these tests at this time can cause pain as long as the knee flexes, contracts, and rotates. Even when the test results are positive, it is difficult to be sure that the pain and crepitus sounds are from the damage to the meniscus.[42]

Differential Diagnosis

Snapping Biceps Femoris Tendon

Symptoms include an audible and usually visual lateral snapping lateral to the knee joint, usually over the fibular head.[43]

The Steinmann II Sign is negative because the snapping spot doesn't move following the knee flexion and extension.

Treatment

Treatment principle:

- *Acute stage*:
  - Moves blood and Qi. The acute injury causes blood stasis in meridians.
  - Cools blood, expels wind, disperses lung, circulates water. Chinese medicine is very helpful to reduce knee edema from traumatic inflammation.
- *Chronic stage*:
  - Tonifies liver blood and Yin to strengthen the ligaments and tendons surrounding the meniscus.
  - Tonifies kidney Qi, Yin, and Yang to stabilize the knee joints and warm channels.

Herbal Treatments

Herbal formula:

- Gu Sui Bu Tang 骨碎補湯.

Formula ingredients:

Gu Sui Bu 骨碎补 30 g, Shu Di Huang 熟地黄 25 g, Shan Zhu Yu 山茱萸 15 g, Niu Xi 牛膝 25 g, San Leng 三棱 15 g, E Zhu 莪术 15 g, Wu Jia Pi 五加皮 15 g, Ji Xue Teng 雞血藤 25 g, Huang Qi 黄耆 50 g.

Ingredient explanations:

- Gu Sui Bu 骨碎补 tonifies kidney Yang, strengthens tendons and bones, moves blood especially due to trauma.
- Shu Di Huang 熟地黄 nourishes blood; nourishes liver and kidney Yin.
- Shan Zhu Yu 山茱萸 nourishes liver and kidneys, strengthens kidney Yang.

- Niu Xi 牛膝 expels wind dampness, tonifies liver and kidneys, directs herbs downward.
- San Leng 三棱 moves stubborn blood stasis, regulates Qi, and stops pain.
- E Zhu 莪术 moves blood, regulates Qi, and stops pain.
- Wu Jia Pi 五加皮 expels wind, damp, nourishes and warms liver and kidneys, strengthens bones and tendons.
- Ji Xue Teng 雞血藤 moves and tonifies blood, relaxes tendons.
- Huang Qi 黄耆 tonifies Qi and blood and relieves numbness and pain.

Acupuncture Treatments

Acupuncture point formula:

- SP10 血海 Xuehai, ST34 梁丘 Liangqiu, ST35 犢鼻 Dubi, Xi Yan 内膝眼, GB33 膝陽關 Xiyangguan, UB40 委中 Weizhong, UB39 委陽 Weiyang, GB34 陽陵泉 Yanglingquan, SP9 陰陵泉 Yinlingquan, ST36 足三里 Zusanli, ST40 豐隆 Fenglong, Ashi acupoints 阿是穴.

Cupping Treatments

Ashi acupoints 阿是穴.

Moxibustion Treatments

Warming needling techniques on Ashi acupoints 阿是穴.

Bleeding Treatments

Ashi acupoints 阿是穴.

Electro-Acupuncture Treatments

SP10 血海 Xuehai, SP9 陰陵泉 Yinlingquan 2/100 Hz for 20 minutes daily.

Tui Na Treatments

Tui Na techniques:

- Kneading, pressing, grasping, rubbing, pushing, stretching, bending, one-finger meditation

Tui Na procedures—relaxing soft tissues:

1) The patient lies supine; the therapist stands next to the affected knee.
2) The therapist applies kneading, grasping, rubbing, and pushing techniques around the knee, quadriceps muscles, adductors muscles, IT band, hamstring muscles, and calf muscles. Repeat the techniques on the muscles until warm.

Tui Na procedures—releasing the locked meniscus:

1) The patient lies supine with the hip and knee bent at 90 degrees. The therapist's assistant stands near the hip joint and stabilizes the proximal femur; the therapist stands near the knee and holds the ipsilateral ankle.
2) The therapist slightly stretches the tibia while slowly rotating it.
3) Maximally flex the knee.
4) Slowly extend the leg.

Tui Na procedures—after unlocking the knee:

1) *Relaxing tendons*: The patient lies supine. The therapist stands next to the affected knee. Apply kneading techniques with fingertips and finger pads on the quadriceps tendon, patella tendon, and the tendons near the popliteal fossa for five to ten minutes.
2) *Opening channels on the anterior aspect of the meniscus:*
   a) The patient is in a supine position; the therapist stands beside the affected knee.
   b) The patient's knee is in a slightly flexed position with a cushion underneath. The therapist applies One Finger Meditation techniques on the affected ST35 犢鼻 Dubi if the injury is on the anterior horn of lateral meniscus or on Xi Yan 內膝眼 if it is an injury on the anterior horn of the medial meniscus. Duration is about five to ten minutes.

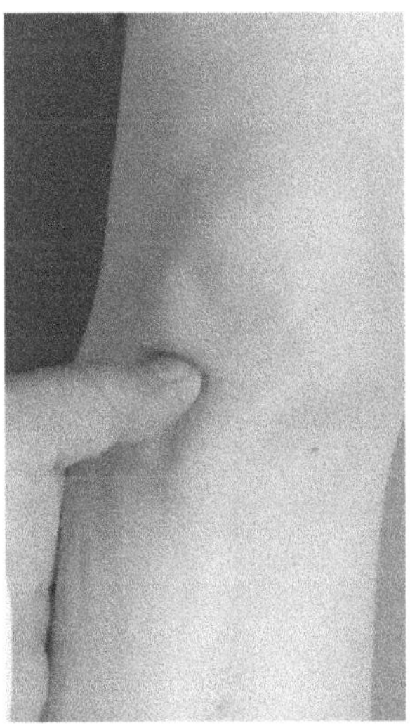
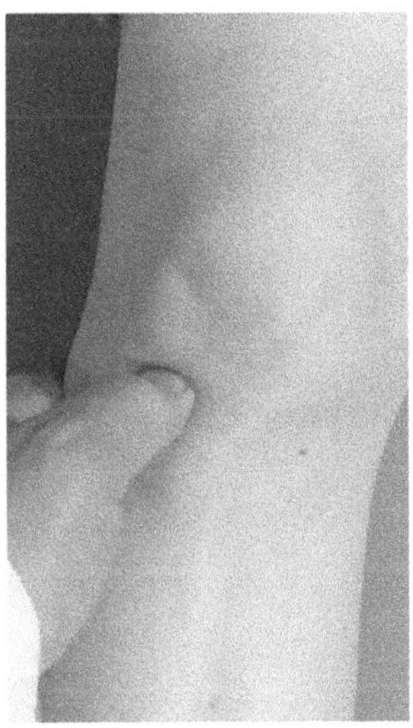
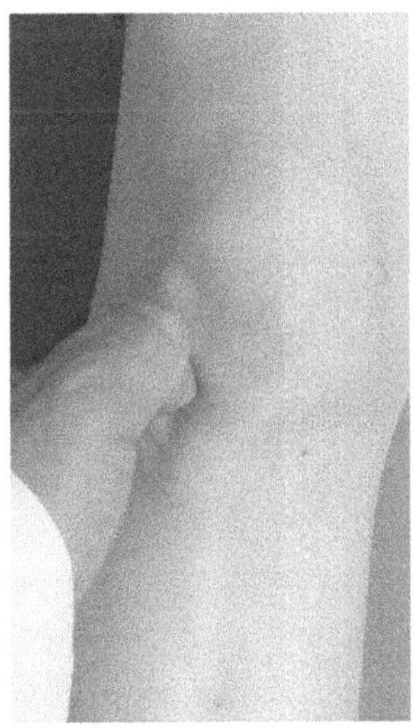

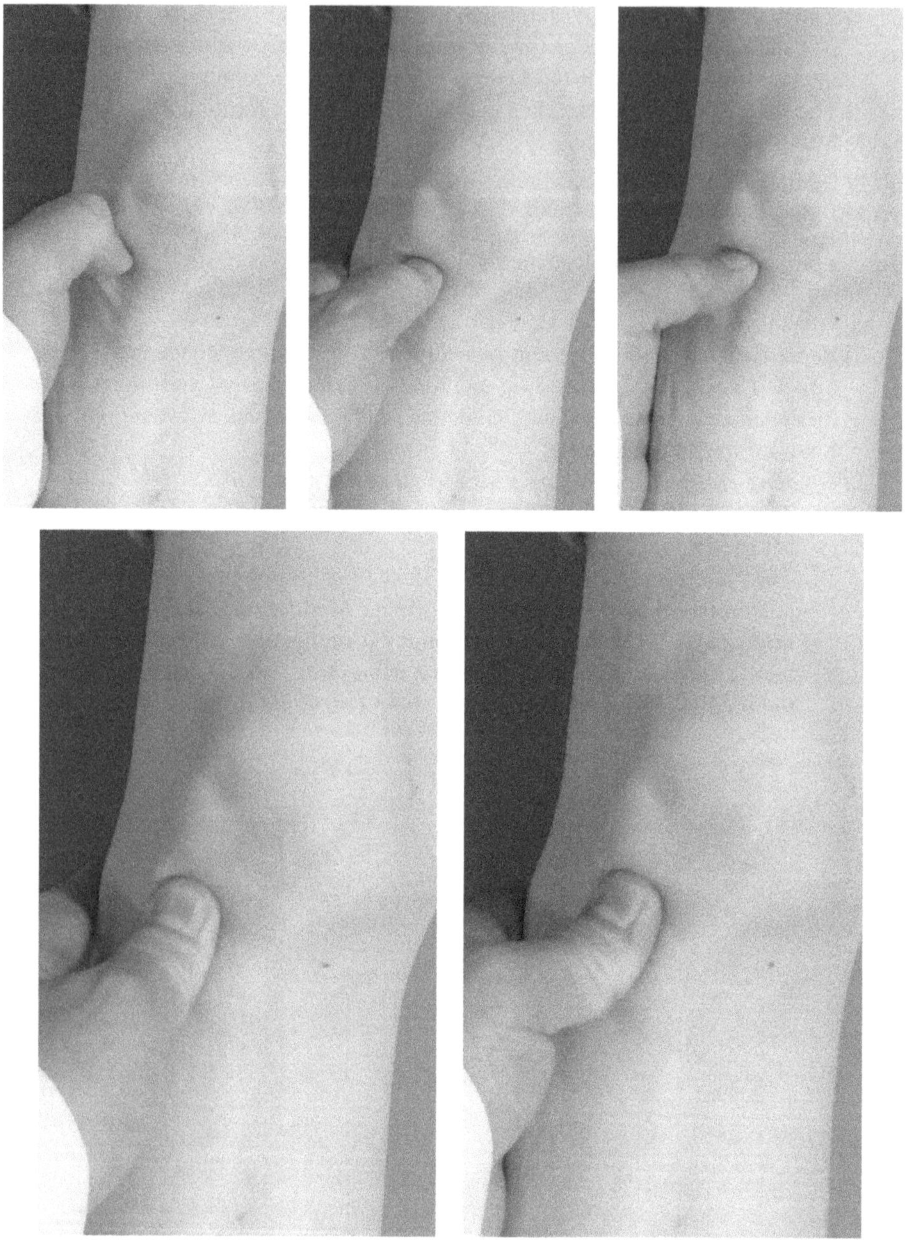

3) *Opening channels on the posterior aspect of the meniscus:*
   a) The patient is in a prone position; the therapist stands beside the affected knee.
   b) The patient's knee is slight flexed with a pillow or cushion supporting the ipsilateral foot.

c) The therapist applies One Finger Meditation techniques on the popliteal fossa and focuses on the affected aspect. Duration is about five to ten minutes.

4) *Pain reducing:*

a) The patient is in a supine position; the therapist stands next to the affected knee.

b) The therapist applies finger pressing techniques on the following acupuncture points: GB31 風市 Fengshi, SP10 血海 Xuehai, UB40 委中 Weizhong, SP9 陰陵泉 Yinlingquan, GB34 陽陵泉 Yanglingquan, Xi Yan 內膝眼. Strong pressing is considered for these points because they are for reducing Wind (GB31 風市 Fengshi), moving Blood (SP10 血海 Xuehai), draining Damp (SP9 陰陵泉 Yinlingquan), and local stimulation (UB40 委中 Weizhong and Xi Yan 內膝眼). When pressing UB40 委中 Weizhong, the patient is still in a supine position, and the therapist slides a hand and presses UB40 委中 Weizhong with a fingertip; the commonly used fingertip is the middle finger.

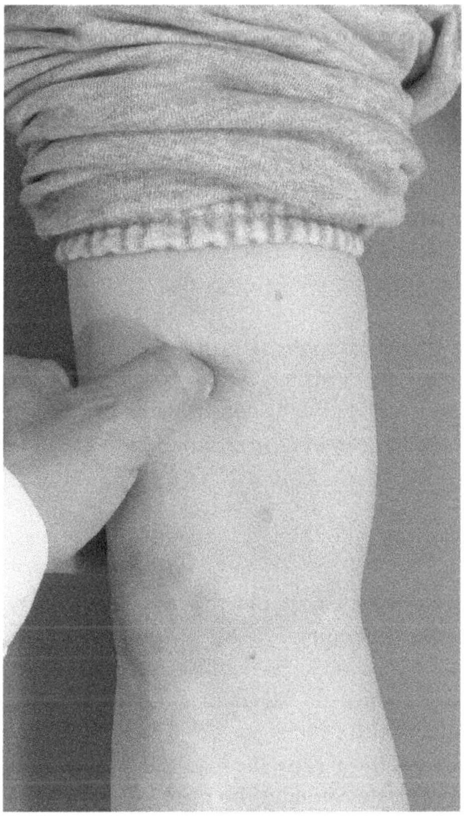

5) *Lubricating the joint:*

a) The patient is in a supine position; the therapist stands beside the affected knee.

b) The therapist places one hand on the affected knee, grasps the ankle with the other hand, and flexes the affected hip and knee to 90 degrees.

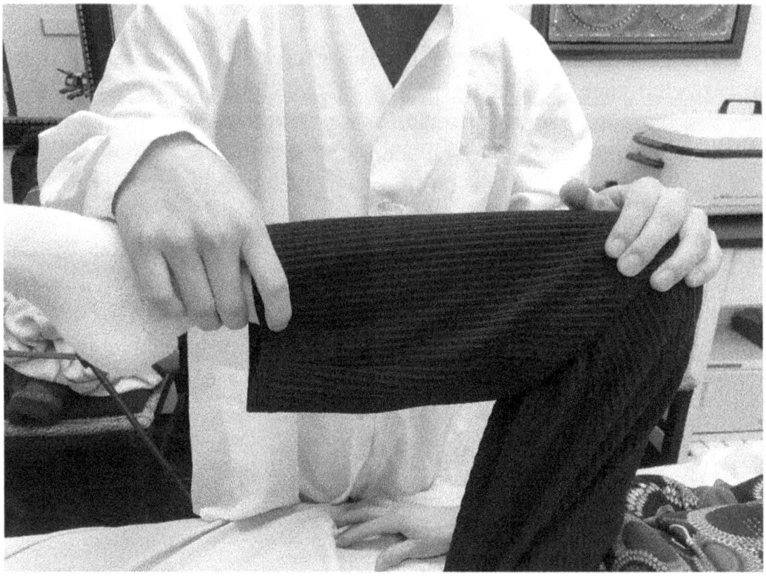

c) The therapist rotates the affected knee three to five times. The angles should start with small ones and move to the largest ones that the patient can tolerate to or to the maximal range of motion.
d) Then the therapist slowly extends the leg to place it on the treatment table to relax.
e) Finally, the therapist applies closing techniques to end the treatment. The techniques are pushing, rubbing, chafing, and rolling techniques. The duration is about five to ten minutes or until the skin is hot. The procedures are to promote the circulation of Qi and blood.

Patient Advisory

Acute condition:

- After acute injury, patients should rest the knee. Take a break from the activities that cause the injury.
- Apply ice packs.
- Apply an elastic bandage on the injured knee. Wearing a knee brace for knee support is helpful for recovery.
- When the knee is swelling, raise the leg higher than the heart when lying by putting a pillow or cushion under the knee.
- Because the knee is painful, unstable, and prone to locking, patients are recommended to use a crutch to prevent falls.
- The affected leg should avoid bearing weight for at least four to six weeks.

- The quadriceps muscles should be trained during the recovery period.
- Chinese medicine is able to reduce swelling and stop pain quickly, but it should not be the main treatment modality.

Chronic condition:

- Exercise to strengthen the quadriceps muscles:
  a) The patient lies supine with the affected leg extended and the contralateral leg flexed.
  b) Slowly raise the leg until it is even with the contralateral thigh.
  c) Do not raise the leg higher than the contralateral thigh.
  d) Hold the affected leg until it is tremoring. Repeat 30–50 times.
  e) Repeat on the other side.
- Before exercise, warm up the knee joints to loosen the tendons and ligaments and promote circulation.
- Recommend patients to slightly bend the knees 15–30 degrees when standing.
- Patients should be advised to avoid Fa Wu 發物 diet. Fa Wu 發物 foods are the foods that may cause more inflammation and bring old injuries back. Some examples of Fa Wu 發物 foods are shellfish such as crab and shrimp, fish, fried foods, lamb, duck, chocolate, garlic, and green onions. This concept applies to both acute and chronic strains.

PATELLAR TENDONITIS 髕腱炎

INTRODUCTION

Patellar tendonitis is commonly called Jumper's Knee because it is most common in athletes whose sports involve frequent jumping such as basketball and volleyball players. But anyone can suffer from patellar tendonitis, including people who don't participate in jumping sports. It is an injury involving overuse and overstress on the patella tendon.

ANATOMY

The patellar tendon is located inferior to the patella. The patellar tendon originates in the patellar apex and runs inferiorly to attach to the tibial tuberosity on the anterior aspect of the tibia. Technically the patellar tendon is not supposed to be called a tendon. A tendon is connective tissue that connects a muscle to a bone, but the patellar tendon connects a bone (patellar) to a bone (tibial tuberosity). A ligament connects between bones, so the patellar tendon should be called the patellar ligament.

This ligament is approximately 30 mm wide by 50 mm long, with a thickness of 5 to 7 mm (Hsu 2020).

WESTERN MEDICINE ETIOLOGY AND PATHOLOGY

When a tendon is stressed, microtears may occur. Most tendon injuries are the result of gradual wear and tear to the tendon from overuse, overstretch, and aging. With making the same motions over and over in the jobs, sports, or daily activities, the tendons are overused and aging. Even though the tendons are supposed to withstand high and repetitive loading, when the load is more than the tendon to withstand, microtears occur.

The human body is a huge repair factory if we provide a suitable environment. This consists of resting physically and mentally and the assistance of medicine, acupuncture, herbs, dietary, surgery, emotional support, and exercise.

Small tears can be quickly healed if the tendons are repaired appropriately. Yet, when the repetitive activities continue to stress the tendon, the lesions on the tendon exceed the healing progress, the destructive tear continues to cause microtrauma, the cumulative microtrauma appears to exceed the tendon's capacity to heal itself, and the quality of the tendons and the blood supply to the tendons diminishes, making the tendons susceptible to tearing, the result is tendinosis.

The following risk factors are most commonly associated with patellar tendinitis:

- Obesity increases loads to the knee.
- Tight quadriceps muscles in which the muscle strength declines and becomes unable to work together with the quadriceps tendon and patellar tendon to straighten the knee, therefore interfering with running and jumping especially on uneven or rigid surfaces.
- Rapid increase in the amount of training.
- Change in training intensity.
- Shoes that do not have enough supportive padding.
- Lower limb biomechanics cause muscular imbalance. Some muscles in the legs are much stronger or weaker than others. The stronger muscles pull harder on the weaker ones. This uneven pull may result in microtrauma in the patellar tendon. Examples include genu varum, also called bow-leggedness, bandiness, bandy-leg; tibia vara, commonly called bowlegs; genu valgum, commonly called knock-knee; Pes planus, commonly referred to as flat feet, and Pes cavus, also known as high arch feet.
- Some internal medicine diseases may also disrupt blood flow to the knee such as[44] kidney failure, autoimmune diseases such as systemic lupus erythematosus, type 1 and type 2 diabetes, rheumatoid arthritis, psoriatic arthritis developed from psoriasis, multiple sclerosis, or Hashimoto's thyroiditis. Such illnesses disrupt blood flow to the knee which weakens the tendon. The patellar tendon is then degenerated. This is more commonly seen in older patients.

These are some of the common factors that cause overload and impede the healing of the tendons.

Chinese Medicine Etiology and Pathology

Liver Blood and Yin Deficiency

According to Chinese medicine, liver governs blood, and nourishes tendons so that they are healthy and in good working condition. When liver blood and Yin are deficient, liver is no longer able to govern tendons and support their funcion. The tendons are prone to microtrauma by the risk factors leading to the presentation of the manifestations.

Liver blood and Yin are insufficient to repair the repetitive microtrauma; over time degeneration occurs.

Manifestations

Patellar tendonitis usually presents the following manifestations:

- Anterior knee pain that may become worse with activity such as jumping, running, or even prolonged sitting. Pain intensity is usually associated with an increase in sports activity and worse when kneeling down or getting up from a squat. The pain is at first sporadic, occurring only after sports or other physical activity. As the tendon becomes more traumatized, the pain is progressively worse and can interfere with any daily physical activity such as walking, using stairs, and even getting up from sitting for a long period and being unable to straighten the leg. The pain may be worse at night, and the patient may feel stiffness in the morning when getting up.
- Localized tenderness over the patellar tendon.
- It may manifest as red, swollen, and warm to touch.
- Crepitus sounds during leg activity.
- The affected tendon may appear thickened in comparison to the unaffected side.

Physical Examination

- Begin by obtaining the patient's medical history:
  - Injury history to the patellar tendon.
  - Career or sports activity that may be factors of patellar tendonitis.
  - General health condition such as obesity.
  - A related medical condition such as kidney failure, autoimmune disease such as systemic lupus erythematosus, type 1 and type 2 diabetes, rheumatoid arthritis, psoriatic arthritis developed from psoriasis, multiple sclerosis, or Hashimoto's thyroiditis might predispose the patient to patellar tendon injury.
- Inspection:
  - Shoe size and padding.
  - Muscular imbalance of the legs. Examples include genu varum and genu valgum, Pes planus and Pes cavus.
  - Crepitus sounds during leg activity.
  - It may manifest as red and swollen.
- Palpation:
  - The affected tendon may appear thickened in comparison to the unaffected side.
  - It may warm to touch.
  - Localized tenderness over the patellar tendon,

- Range of motion:
  - Pain in the patellar tendon may be reproduced with resisted knee extension. The patient sits on the edge of the exam table with the affected leg flexed at 90 degrees; the therapist sits in front of the affected leg with both hands stabilizing the ipsilateral tibia. The patient extends the knee while being resisted by the therapist's hands. The test is positive when the pain is reproduced.
  - Pain in the patellar tendon may be reproduced by performing single leg declining squats. The patient stands only on the affected leg and slowly squats down by flexing the knee. The test is positive when the pain at the patellar tendon reproduces at the knee flexion from 135 degrees to 90 degrees.[45]

Differential Diagnosis

Patellofemoral Pain Syndrome

Pain in the anterior patellar could mimic the location of patellar tendonitis. The patients who suffer from the patellofemoral pain syndrome also experience more diffuse kneecap pain when in sustained knee flexion such as when sitting in a chair.

Treatment

Treatment principle:

- Tonifies liver blood and Yin.
- Moves blood and stops pain.

Herbal Treatments

Herbal formula:

- Shu Jin Huo Xue Pian 舒筋活血片.

Formula ingredients:

Zhi Gou Ji 狗脊 15 g, Sang Ji Sheng 桑寄生 15 g, Zi Ran Tong 自然銅 15 g, Ze Lan 澤蘭 10 g, Hong Hua 紅花 5 g, Wu Jia Pi 五加皮 10 g, Shen Jin Cao 伸筋草 10 g, Luo Shi Teng 絡石藤 10 g, Zhi Xiang Fu 香附 10 g, Ji Xue Teng 雞血藤 15 g.

Ingredient explanations:

- Zhi Gou Ji 狗脊 15 g tonifies liver and kidneys, strengthens tendon and bone, expels wind-dampness, and invigorates blood flow.
- Sang Ji Sheng 桑寄生 15 g expels wind-dampness, tonifies liver and kidney Yin.
- Zi Ran Tong 自然銅 15 g dispels blood, stops pain.
- Ze Lan 澤蘭 10 g moves blood, promotes urination, and reduces swelling.
- Hong Hua 紅花 5 g moves blood, releases bruising, opens meridians, stops pain.

- Wu Jia Pi 五加皮 10 g expels wind and damp, nourishes and warms liver and kidney, strengthens bones and tendons.
- Shen Jin Cao 伸筋草 10 g expels wind, drains damp, relax tendons.
- Luo Shi Teng 絡石藤 10 g dispels wind-dampness, cools blood, reduces swelling, strengthens tendons and bones.
- Zhi Xiang Fu 香附 10 g spreads and regulates liver Qi and alleviates pain.
- Ji Xue Teng 雞血藤 15 g moves and tonifies blood, relaxes tendons.

Formula action:

- Relaxes tendon, moves blood, and disperses bruising.

Acupuncture Treatments

Acupuncture point formula:

ST35 犢鼻 Dubi, ST36 足三里 Zusanli, GB34 陽陵泉 Yanglingquan, ST34 梁丘 Liangqiu, SP10 血海 Xuehai, KD1 涌泉 Yong Quan, KD3 太溪 Taixi, GB39 懸鐘 Xuanzhong. Warming needle techniquc is used on each point.

Formula action:

- Move Qi, tonify liver and kidneys.

Treatment frequency: Daily.
Treatment duration: 30 minutes.

Cupping Treatment

He Ding 鶴頂 (above the knee, in the depression of the midpoint of the upper border of the patella).

Moxibustion Treatment

Warming needling techniques on Nei Xi Yan 膝眼 (in the depression on both sides of the patellar ligament when the kneels flexed), ST35 犢鼻 Dubi.

Bleeding Treatment

Ashi Points 阿是穴 (it is a tender point that is not included in the classic acupuncture points).

Electro-Acupuncture Treatment

SP10 血海 Xuehai, He Ding 鶴頂 2/100 Hz for 20 minutes daily.

Tui Na Treatments

Tui Na techniques:

- Pressing, rubbing, grasping, kneading, “holding a globe”

Tui Na procedures:

1) The patient lies supine with the affected knee bent at about 30 degrees; the therapist stands beside the affected knee.
2) The therapist applies finger pressing techniques on acupuncture points SP10 血海 Xuehai, ST34 梁丘 Liangqiu, Nei Xi Yan 膝眼 (in the depression on both sides of the patellar ligament when the kneels flexed), ST35 犊鼻 Dubi, ST36 足三里 Zusanli and UB40 委中 Weizhong. Pressing these points helps release pain.
3) The therapist applies rubbing techniques with palms on the anterior patellar.
4) The therapist applies grasping techniques on the patellar tendon by placing the thumb on the lateral side of the tendon and the four fingers on the medial side.

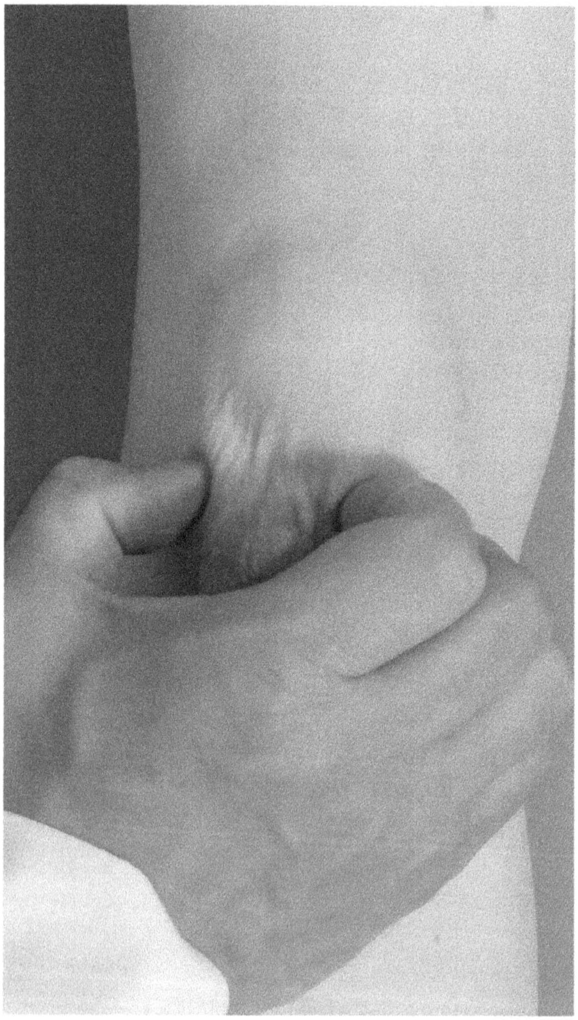

5) The therapist stabilizes the tibia and asks the patient to flex and extend the knee. With the resistant pressure, the patient practices the knee flexion and extension about three to five times.
6) The therapist applies finger kneading on Ashi acupoints 阿是穴 (it is a tender point that is not included in the classic acupuncture points) around the knee about 20 times going from light force to heavy force and another 20 times going from heavy force to light force.
7) With the patient's knee straightened, the therapist applies pressure with a palm on the affected patellar and kneads the patellar for about two to three minutes or until there is a warm sensation under the patellar.

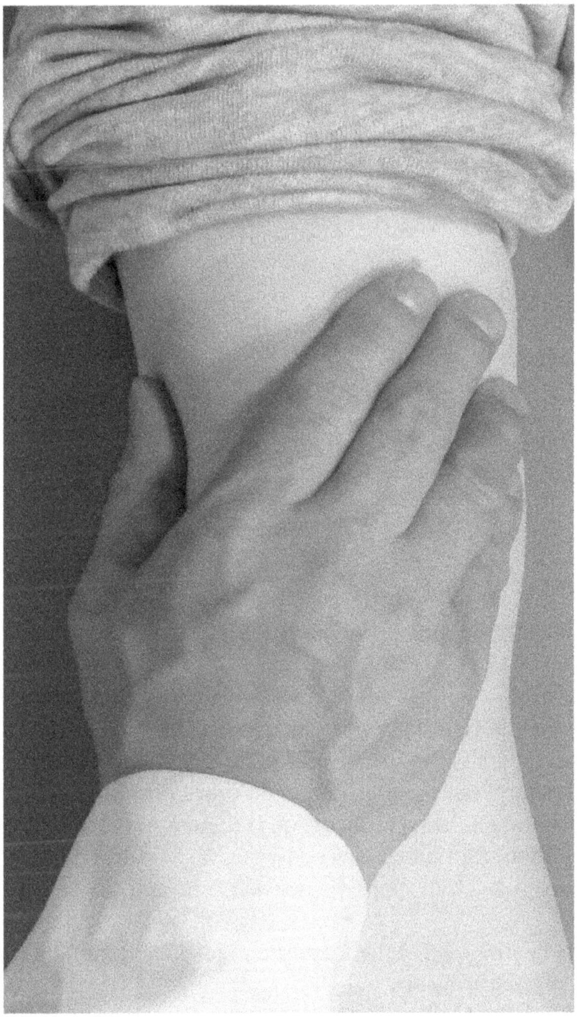

8) The patient bends the affected knee to 90 degrees; the therapist places both palms at the medial and lateral sides of the knee with the fingers interlocked.

This technique is called "holding a globe." Then the therapist presses and kneads the knee with both hands until the entire knee is warm.

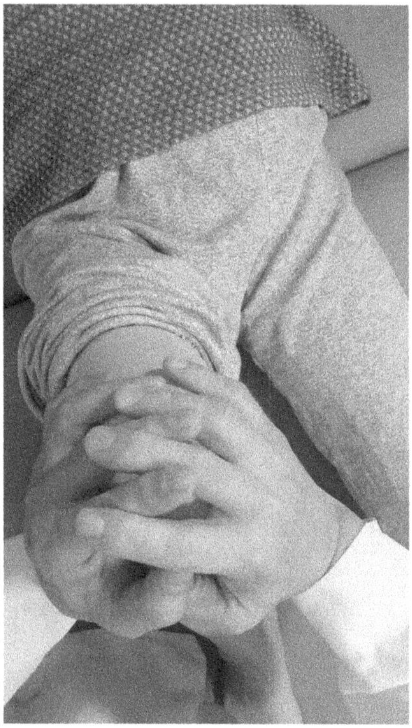

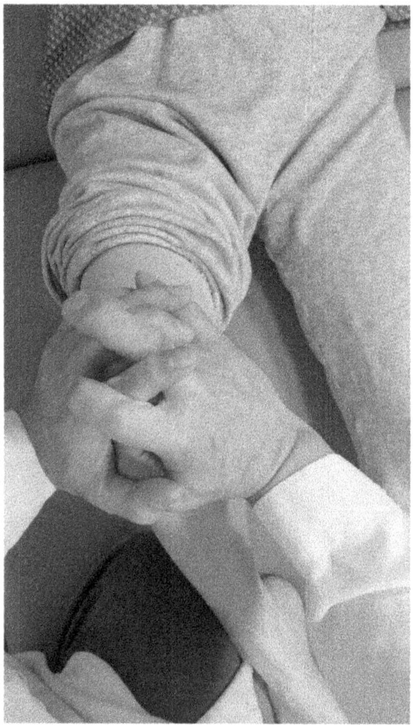

Patient Advisory

The first recommendation for the patient is REST:

1) Let the painful knee rest, and avoid any activity that causes more micro-trauma so that there is an opportunity for the body to repair the knee.
2) Getting rest mentally and physically in the afternoon and the evening nourishes Yin. Afternoon, evening, and night times nourish Yin because they are the Yin times which help produce Yin. If there is too much mental and physical activity in the Yin times, this consumes Yin instead of producing Yin.
3) Falling asleep at 11pm helps to nourish Yin and assists "Liver Governs Blood."
4) Applying a heating pad at the knee promotes local circulation and prevents wind-damp-cold invasion.
5) To prevent future injury:
   - Warm up and stretch the knee and leg muscles before exercise.
   - Cool down and stretch again after exercise.

- Wear appropriate knee support when playing sports.
- Avoid jumping and landing on very rigid or uneven surfaces.
- Practice exercise to strengthen the quadriceps muscles:
  a) The patient lies supine with the affected leg extended and the contralateral leg flexed.
  b) Slowly raise the leg until it is even with the contralateral thigh.
  c) Do not raise the leg higher than the contralateral thigh.
  d) Hold the affected leg until it is tremoring. Repeat 30–50 times.
  e) Repeat on the other side.

6) Dietary suggestions:
- Avoid foods that may interfere or decelerate the healing like cold food such as ice cream, tofu, cold or icy water and raw food such as salads; also avoid deep-fried and fatty food such as chips.
- Foods that benefit Qi and blood circulation include cooked food, chives, grapes, onions, yams, vinegar, and brown sugar.[46]
- Foods that are rich in fiber such as avocado, beans, bananas, oranges, apples, mushrooms, spinach, mangoes, strawberries, and raspberries.
- Foods that are rich in protein such as eggs,[47] almonds, oats, milk, and broccoli.

PATELLA TENDON RUPTURE 髌腱断裂

INTRODUCTION

A rupture in the patellar tendon occurs when the tendon is partially or completely torn and separated from the kneecap. In the United States, patellar tendon ruptures tend to affect less than 0.5% of the population each year. Males are more commonly affected than females. These injuries are typically seen in males in their third or fourth decade of life who participate in sports which involve jumping and running (Hsu 2020). Females have greater ligamentous laxity and hormonal changes due to the menstrual cycle may be protective (Hsu 2020). A complete tendon rupture can be acute or chronic; diagnosis within six weeks of injury is considered acute and after six weeks is considered chronic.[48]

ANATOMY

The patella tendon is located inferior to the patella. The patellar tendon originates in the patellar apex and runs inferiorly to attach to the tibial tuberosity.[49] Technically, the patellar tendon is not supposed to be called a tendon, because a tendon is connective tissue that connects a muscle to a bone, but the patellar tendon connects a bone (patellar) to a bone (tibial tuberosity). A ligament connects bones, so the patellar tendon should be called the patellar ligament.

This ligament is approximately 30 mm wide by 50 mm long, with a thickness of 5 to 7 mm (Hsu 2020).

To straighten the knee, the patellar tendon needs to work together with the quadriceps muscle and quadriceps tendon. The strong contraction of these three straightens the knees to lift the body to a jump into the air, whereafter they allow a controlled knee flexion to absorb the impact force to the knee when landing.[50]

Western Medicine Etiology and Pathology

The rupture of the patellar tendon can be either partial or complete. In a partial tear, some of the tendon fibers are torn, but the soft tissue is not damaged. In a complete tear, the soft tissues are ruptured into two pieces.[51] The patellar tendon most commonly tears at the place where it connects to the patellar inferior pole (Woodmass et al. 2017), usually seen in traumatic cases. The strain at the tendon-inferior pole patella interface is three to four times higher than at the midsubstance of the tendon (Hsu 2020). In some severe cases, a small piece of bone can break off along with the tendon; a very strong force is required to result in this outcome.[52] When the tear is caused by a medical condition such as patellar tendonitis or patellar degeneration, the tear usually occurs in the middle of the tendon.[53]

Patellar tendon rupture has been found under many circumstances:

- When atheletes forcefully land from a jump under eccentric contraction, the anterior knee suffers a major impact.[54] This usually involves an awkward landing from a jumping position where the quadriceps muscle is contracting so the knee is being forcefully straightened.[55]
- Falls result in cuts to the tendon.
- Chronic tendinosis damages the tendon[56] and may also cause small tears.
- Long-standing chronic tendon degeneration causes tensile overload on the extensor mechanism; the quadriceps muscle suddenly contracts with the knee in a flexed position. The greatest amount of force on the patellar tendon is when the knee flexed more than 60 degrees (Hsu 2020).
- A medical condition weakens tendons such as chronic renal failure, rheumatoid arthritis, diabetes mellitus, or systemic lupus erythematosus.[57]
- Recent surgery on the knee weakens tendons.[58]
- Long term use of corticosteroids or use of anabolic steroids (Hsu 2020).[59]

Chinese Medicine Etiology and Pathology

Liver Yin and Blood Deficiency

According to Chinese medicine, knee injury is not an isolated condition. Commonly, the liver and kidney are involved. Ligament injury is considered "Jin Shang 筋伤" or "traumatic injury on tendons." According to the theory of traditional Chinese medicine (TCM), the liver governs the tendons. "Traumatic injury on tendons" consumes liver energy which is Yin and blood. As the injury becomes chronic, the liver is deficient, and the kidney is then deficient because the liver "wood" and the kidney "water" always affect each other. When they both are deficient, the tendons surrounding the knee joint are prone to injury. This is because liver Yin and liver blood are not able to govern the tendons when they become deficient. If the tendons lack blood and Yin to repair the prior small tears, a new traumatic factor can easily rupture the tendon. The injury on the patellar tendon that underlined liver Yin and blood deficiency is typically seen in middle-aged patients, especially those who participate in the physical activities of jumping and running (Hsu 2020).

Qi and Blood Stagnation

A direct blow or cut to the patellar tendon usually causes blockages to the flow of the local Qi and blood in meridians almost immediately after the violence.

MANIFESTATIONS

A person with patellar tendon rupture typically experiences:

- An audible "pop."
- Sharp pain in the anterior knee.
- Bruising.
- Knee rapid swelling.
- Knee giving way.
- Knee cannot support the weight while the knee is bent such as when getting up and down the stairs.
- Inability to fully extend and stand on that leg[60] with a complete tear of the patellar tendon.
- The patellar is shifted up into the femur with a complete tear of the patellar tendon.
- A gap, marking, or indentation at the site of the rupture under the patellar.
- Difficulty walking due to patella instability.

PHYSICAL EXAMINATION

Physical examination of the patellar tendon rupture:

- Begin by obtaining the patient's medical history:
  - Injury history to the anterior patella.
  - Surgery history to the knee such as total knee replacement or anterior cruciate ligament reconstruction.
  - General health conditions such as patellar tendonitis.
  - A related medical condition such as kidney failure, rheumatoid arthritis, diabetes mellitus, or systemic lupus erythematosus might predispose the patient to patellar tendon injury.
- Inspection:
  - Knee effusion or swelling.
  - Surrounding skin for the signs of direct trauma.
  - Hemarthrosis and ecchymosis.

- Compare the patellar height from the affected to the unaffected knee. The affected patellar is elevated when the tendon ruptures.
- Palpation:
  - A palpable indentation below the inferior pole of the patella.
  - Tenderness at the infrapatellar aspect of the knee.
- Range of motion:
  - Loss of active knee extension with a complete tear of the patellar tendon.
  - Unable to perform the active straight leg raise test.

Differential Diagnosis

Quadriceps Tendon Rupture

The difference is the location of the rupture. A quadriceps tendon rupture has a suprapatellar gap while a patellar tendon rupture has an infrapatellar gap.

Patella Fracture

Palpation of closed injuries may reveal crepitus, or fracture gap.[61] Patients who suffer from a patella fracture may also present with fractures to the ipsilateral acetabulum and lower extremity.[62]

Treatment

Treatment principle:

Treatment for a patellar tendon tear with Chinese medicine is good for the partial tear, and it is extremely useful for recovery after surgery for the complete tear. It is helpful for patients who suffer from a complete tear but are not surgical candidates due to medical comorbidities.

- o Moves Qi and blood, opens meridians, reduces swelling, and stops pain.
- o Tonifies liver and kidneys, strengthens tendons and bones.

Herbal Treatments

Acute Stage

Herbal formula:

- Huo Xue Zhi Tong Tang 活血止痛汤.

Formula ingredients:

- Dang Gui 當歸 6 g, Su Mu 蘇木 6 g, Luo De Da 落得打 6 g, Chuan Xiong 川芎 2 g, Hong Hua 紅花 1.5 g, Ru Xiang 乳香 3 g, Mo Yao 沒藥 3 g, San Qi 三七 3 g, Chi Shao 赤芍 3 g, Chen Pi 陳皮 3g, Zi Jing Teng 紫荆藤 9 g, Di Bie Cong 地鱉蟲 9 g.

Ingredient explanations:

- Dang Gui 當歸 moves and tonifies blood, disperses cold, and stops pain due to blood stasis.

- Su Mu 蘇木 moves blood, reduces swelling, stops bleeding.
- Luo De Da 落得打 reduces swelling, clears heat, drains damp.
- Chuan Xiong 川芎 moves blood and Qi, expels wind, and stops pain.
- Hong Hua 紅花 moves blood, releases bruise, opens meridians, stops pain.
- Ru Xiang 乳香 moves blood, stops pain.
- Mo Yao 沒藥 moves blood, reduces swelling, and stops pain.
- San Qi 三七 moves blood, stops bleeding, reduces swelling, stops pain.
- Chi Shao 赤芍 moves blood, reduces swelling, stops pain.
- Chen Pi 陳皮 dries damp, transforms phlegm
- Zi Jing Pi 紫荆皮 moves blood, reduces swelling, promotes urination.
- Di Bie Cong 地鱉蟲 breaks up stubborn blood stasis.

Formula action:

- Moves Qi and blood, drains damp, reduces swelling, stops pain.

Contraindications:

- Contraindicated during pregnancy.

Modifications:

- For chronic cold damp wind Bi syndrome, add Qiang Huo 羌活 for opening meridians so that Zi Jing Pi 紫荆皮 is able to expel and drain the wind, damp, and cold in the joints.
- For swelling, add Ze Lan 澤蘭, Mu Tong 木通, Che Qian Zi 車前子, and Jiang Huang 薑黃.
- For severe pain, add Yan Hu Suo 延胡索, Niu Xi 牛膝, and Mu Xiang 木香.
- For swelling, red, and hot sensation, add Mu Dan Pi 牡丹皮, Zhi Zi 梔子and Huang Bai 黃柏.

Intermediate Stage

Herbal formula:

- Shu Jin Huo Xue Pian 舒筋活血片.

Formula ingredients:

- Zhi Gou Ji 狗脊 15 g, Sang Ji Sheng 桑寄生 15 g, Zi Ran Tong 自然銅 15 g, Ze Lan 澤蘭 10 g, Hong Hua 紅花 5 g, Wu Jia Pi 五加皮 10 g, Shen Jin Cao 伸筋草 10 g, Luo Shi Teng 絡石藤 10 g, Zhi Xiang Fu 香附 10 g, Ji Xue Teng 雞血藤 15 g.

Ingredient explanations:

- Zhi Gou Ji 狗脊 tonifies liver and kidneys, strengthens tendons and bones, expels wind damp, and invigorates blood flow.
- Sang Ji Sheng 桑寄生 expels wind damp, tonifies liver and kidney Yin.

- Zi Ran Tong 自然銅 dispels blood, stops pain.
- Ze Lan 澤蘭 moves blood, promotes urination, and reduces swelling.
- Hong Hua 紅花 moves blood, releases bruise, opens meridians, stops pain.
- Wu Jia Pi 五加皮 expels wind and damp, nourishes and warms liver and kidneys, strengthens bones and tendons.
- Shen Jin Cao 伸筋草 expels wind, drains damp, relaxes tendons.
- Luo Shi Teng 絡石藤 dispels wind damp, cools blood, reduces swelling, strengthens tendons and bones.
- Zhi Xiang Fu 香附 spreads and regulates liver Qi and alleviates pain.
- Ji Xue Teng 雞血藤 g moves and tonifies blood, relaxes tendons.

Formula action:

- Relaxes tendons, moves blood, and disperses bruising.

Contraindications:

- Contraindicated during pregnancy.

Acupuncture Treatments

Acute Stage

Acupuncture point formula:

- ST35 犊鼻 Dubi, ST36 足三里 Zusanli, GB34 陽陵泉 Yanglingquan, ST34 梁丘 Liangqiu, SP10 血海 Xuehai with electro-acupuncture stimulation in the frequency of 2/100 Hz for 20 minutes per treatment.

Formula action: Moves blood.
Treatment frequency: Daily.
Treatment duration: 30 minutes.

Intermediate Stage

Acupuncture point formula:

LI4 合谷 He Gu, LIV8 曲泉 Ququan, ST36 足三里 Zusanli, ST35 犊鼻 Dubi, Nei Xi Yan 膝眼 (in the depression on both sides of the patellar ligament when the knees are flexed), KD3 太溪 Taixi, KD1 涌泉 Yong Quan, GB39 懸鐘 Xuanzhong, SP10 血海 Xuehai, Ren12 中脘 Zhongwan, Ren6 氣海 Qihai with warming needle technique for three to five cones per point.

Formula action: Moves Qi, tonifies liver and kidney.
Treatment frequency: Daily.
Treatment duration: 30 minutes.

Cupping Treatment

He Ding 鶴頂 (above the knee, in the depression of the midpoint of the upper border of the patella).

Bleeding Treatment

Ashi Points 阿是穴 (a tender point that is not included in the classic acupuncture points).

Tui Na Treatments

1) The patient lies supine with the affected knee flexed at about 30 degrees; the therapist stands beside the affected knee.
2) The therapist applies finger pressing techniques on acupuncture points SP10 血海 Xuehai, ST34 梁丘 Liangqiu, Nei Xi Yan 膝眼 (in the depression on both sides of the patellar ligament when the kneels flexed), ST35 犊鼻 Dubi, ST36 足三里 Zusanli, and UB40 委中 Weizhong. Pressing these points helps release pain.
3) The therapist applies rubbing techniques with palms on the anterior patellar.
4) The therapist applies grasping techniques on the patellar tendon while placing the thumb on the lateral side of the tendon and the four fingers on the medial side.

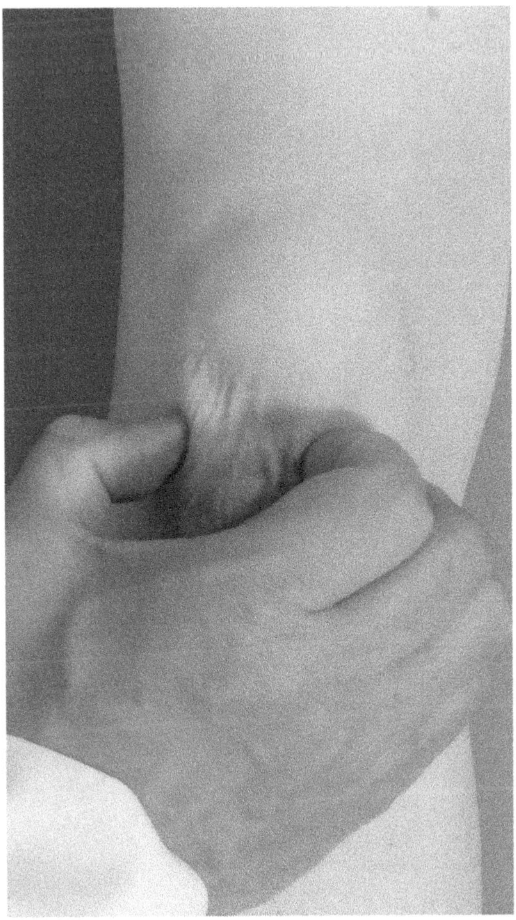

5) The therapist stabilizes the tibia and asks the patient to flex and extend the knee. With the resistant pressure, the patient practices the knee flexion and extension about three to five times.
6) The therapist applies finger kneading on Ashi acupoints 阿是穴 (it is a tender point that is not included in the classic acupuncture points) around the knee about 20 times, going from light force to heavy force and another 20 times going from heavy force to light force.
7) With the patient's knee extended, the therapist applies pressure with a palm on the affected patellar and kneads the patellar for about two to three minutes or until there is a warm sensation under the patellar.

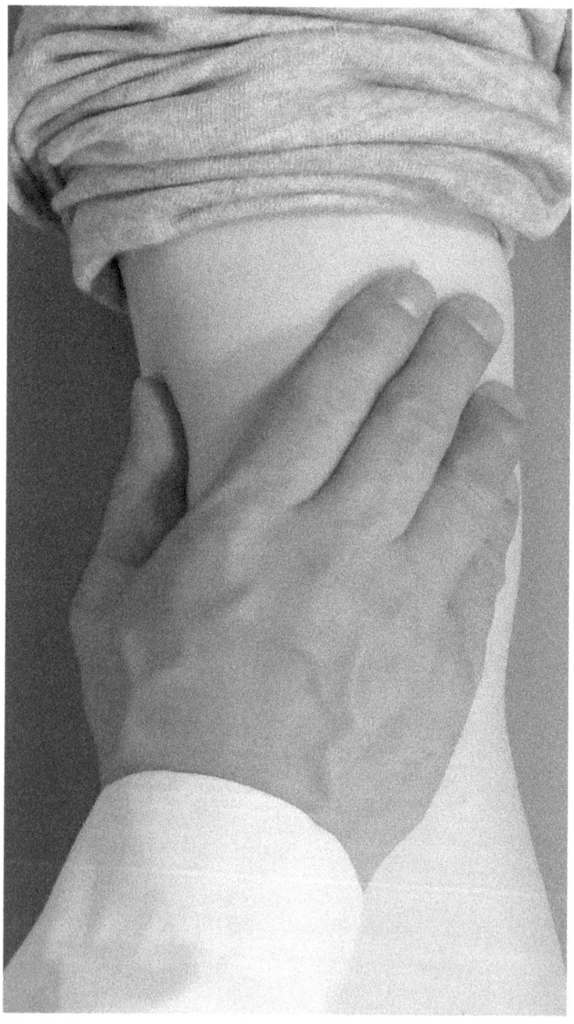

8) The patient flexes the affected knee to 90 degrees; the therapist places both palms at the medial and lateral sides of the knee with the fingers interlocked. This technique is called "holding a globe." The therapist then presses and kneads the knee with both hands until the entire knee is warm.

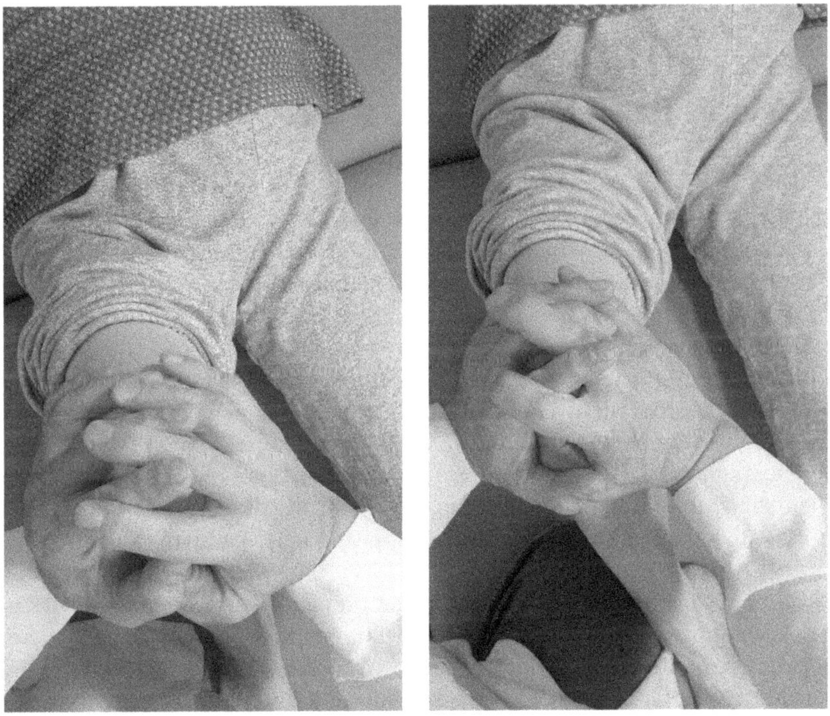

Patient Advisory

- Do exercise to strengthen the quadriceps muscle.
- During the acute stage or after surgery, avoid hard exercise such as running, jumping, getting up and down long stairs, and riding a bicycle.
- Place an ice pack on the affected knee when injury is fresh or after surgery for the duration of 15–30 minutes each time every hour for the first 24 hours.
- Use crutches, a cane, or a walker to support and assist walking, as these devices decrease stress and strain on the knee.
- Wear a knee brace, splint, or cast when the injury is acute and fresh and after surgery, as they help to keep the knee joint straight, relax the knee, decrease load on the patellar tendon, and heal the tendon.
- Dietary recommendation: High protein foods such as eggs and milk.
- Exercise recommendation: To prevent tendon adhesion, knee flexion and extension exercises are recommended:
  a) Lie prone on the bed with both legs extended.

b) Flex the affected knee to bring the heel as close to the buttock as possible.
c) Extend the leg to the start position.
d) Speed should be slow and gentle; stop when there is pain.
e) Repeat 10 to 20 times.

QUADRICEPS TENDON TEAR 股四頭肌損傷

Introduction

Clinically, a quadriceps tendon rupture is more commonly seen than a patellar tendon rupture.[63] The quadriceps muscle is essential for daily activities, such as climbing stairs or getting up from a chair. The quadriceps tendon, patellar tendon, and patella bone make up the extensor mechanism of the lower leg. Injuries to the quadriceps tendon are serious and can lead to disability. They can cause significant loss of time from sport and work, and the patient's knee loses the ability to extend if not treated appropriately. However, the degree that a quadriceps tendon rupture restrains lower leg extension is based on the severity of tendon damage. Minor tendon tears may have minimal impact on extensor function, while complete tendon tears may totally impede lower leg extension.[64] If the injury of the quadriceps tendon is diagnosed quickly and treated appropriately, patients can expect a full recovery.

According to studies, a quadriceps tendon rupture usually occurs in males older than 40, particularly those with chronic conditions that degenerate the intratendinous structure of the knee (Zuke et al. 2017). According to different studies, 30–76% of quadriceps ruptures occur in patients with underlying medical predispositions, including the following: hyperparathyroidism, chronic renal failure, obesity, rheumatoid arthritis, diabetes mellitus, long term immobilization, jumper's knee (Ristić 2013).[65]

Anatomy

The quadriceps femoris muscles are located at the anterior femur and are composed of four muscle groups:

- Vastus medialis.
- Vastus intermedius.
- Vastus lateralis.
- Rectus femoris.

The rectus femoris originates with its attachment to the ilium. It runs down the anterior aspect of the femur, and its muscle portion transitions to become a tendon which joins the other three tendons (vastus medialis, vastus intermedius, and vastus lateralis) to form a single tendon called the quadriceps muscle tendon. The quadriceps tendon is a trilaminar structure formed by the coalescence of all four muscle tendons just proximal to the suprapatellar margin. The most superficial layer of the tendon is the rectus femoris which continues and covers the patella. The middle layer is comprised of vastus medialis and lateralis. The deep layer is composed of the vastus intermedium.[66] The tendon crosses the knee joint and attaches to the tibial tubercle.

The quadriceps tendon is the extensor mechanism that allows humans to straighten the knee or perform a kicking motion. When the quadriceps muscle groups contract, force is transmitted through the quadriceps tendon, across the patella, and to the patellar tendon which then allows the knee joint to be extended.

Western Medicine Etiology and Pathology

In most cases, the rupture of the quadriceps tendon occurs because the tendon is predisposed to a rupture. A study found that quadriceps femoris muscle tendon ruptures usually occur in patients over 40 years of age and are much more frequent in males than in females (8:1) (Garner et al. 2015).

Most quadriceps tendon ruptures are the result of direct or indirect trauma and some medical problems.

- Contusion by direct contact to the anterior thigh. The rectus femoris muscle is the one most often contused since it is the closest to the surface of the anterior thigh. The contusion injury damages the blood vessels, resulting in bleeding within the compartment. When pressure inside the muscle compartment is high, it prevents adequate blood supply to the muscle. Contusion can be from a high energy accident such as a motor vehicle accident especially if the crash is a high speed or a low energy accident such as knee injury from a fall that results in cuts to the tendon.
- Tears occur under the condition that the quadriceps muscle makes a rapid, eccentric contraction. The quadriceps tendon rupture is often caused by a strong impact action when landing from a jump where the quadriceps muscle contracts with the knee in slight flexion and the foot is planted on the ground.
- Long-standing chronic tendon degeneration causes tensile overload on the extensor mechanism; the quadriceps muscle suddenly contracts with the knee in a flexed position. The greatest amount of force on the patellar tendon is when the knee flexed more than 60 degrees (Hsu 2020).
- Chronic tendinosis damages the tendon[67] and may also cause small tears.
- The quadriceps tendon is vulnerable to the rupture under some medical conditions that weaken the tendon such as chronic renal failure, systemic lupus erythematosus, gout, obesity, leukemia, rheumatoid arthritis, diabetes mellitus, hyperparathyroidism, and jumper's knee.
- The use of corticosteroids is linked to increased muscle and tendon weakness.
- Recent surgery on the knee that weakens tendons.[68]

Chinese Medicine Etiology and Pathology

Liver Yin and Blood Deficiency

According to Chinese medicine, knee injury is not an isolated condition. Usually, the liver and kidney are involved. They are deficient, and the tendons surrounding the knee joint are prone to be predisposed for pain. Liver Yin and liver Blood are not able to govern the tendons when they are deficient, and the tendons lack blood supply to repair the prior small tears. A new traumatic factor can easily rupture the tendon. The injury on the quadriceps tendon that underlined liver Yin and blood deficiency is typically seen in middle-aged patients who practice the physical activities of jumping and running (Hsu 2020).

Qi and Blood Stagnation

A direct blow or cut to the quadriceps tendon usually causes blockages to the flow of the local Qi and blood in meridians almost immediately after the violence.

Liver and Kidney Deficiency

Ligament injury is considered "Jin Shang 筋伤" or "traumatic injury on tendons." In traditional Chinese medicine (TCM) theory, the liver governs the tendons. "Traumatic injury on tendons" consumes liver energy, mostly Yin and blood. As the injury becomes chronic, the liver is deficient, and the kidney is then deficient as well because the liver "wood" and the kidney "water" always affect each other.

Manifestations

- Acute knee pain.
- A pop or tearing sensation at the time of injury.
- A palpable defect or a gap at the suprapatellar region.
- A visible hematoma around the knee.
- A large hemarthrosis.
- Ecchymosis around the knee.
- A freely mobile patella.
- Knee effusion at the suprapatellar region. It is sometimes difficult to assess the clinical diagnosis; the absence of the palpable gap because of the soft tissue effusion may limit the clinical evaluation.
- The quadriceps muscle becomes sensitive and cramps.
- Loss of stability. Some patients are still able to walk following the injury when it is a partial rupture. The patients have to limp in order to avoid putting pressure onto the knee.
- Some patients are not able to walk and straighten the knee when it is a complete rupture.
- The knee feels like giving way.
- Loss of knee extension.

Physical Examination

- Begin by obtaining the patient's medical history:

    - Injury history such as an awkward landing from a jump.
    - Surgery history to the knee such as total knee replacement or anterior cruciate ligament reconstruction.
    - General health conditions such as patellar tendonitis.
    - A related medical condition might predispose to the quadriceps tendon injury such as kidney failure, rheumatoid arthritis, diabetes mellitus, or systemic lupus erythematosus.
- Inspection:
    - Knee effusion or swelling.
    - Surrounding skin for the signs of direct trauma.
    - Hemarthrosis, hematoma, and ecchymosis.
    - A difficult time walking on the affected leg; the patient may still be able to walk but with elevation of the hip to accommodate the swing phase of walking.[69]
    - Locking of the knee and difficulty with climbing stairs.
- Palpation:
    - A palpable indentation or a gap at the suprapatellar region.
    - Tenderness at the suprapatellar aspect of the knee.
- Range of motion:
    - Loss of active knee extension with a complete tear of the quadriceps tendon.
    - Unable to perform active straight leg raise test
    - Resistant Maximal Hip Flexion Active Knee Extension Test is positive in partial tear.
        a) The patient lies supine with the hip in maximal flexion by using the arms to pull the femur toward the ipsilateral shoulder and the affected knee flexed. The therapist stands next to the affected knee and places a hand on the ipsilateral tibia to resist the patient's knee extension.
        b) The patient maximally extends the affected knee joint with resistance from the therapist. The therapist measures the range of knee extension angle.
        c) A positive test indicates a partial quadriceps tendon tear.
    - In incomplete ruptures, the patient may be able to fully extend the knee from the supine position but not from the flexed position.[70]
- Neurologic examination:
    - Decreased quadriceps motor function.
    - Absent patellar reflex.[71]

Differential Diagnosis

Patellar Tendon Rupture

The difference is the location of rupture. Quadriceps tendon rupture has a suprapatellar gap while patellar tendon rupture has an infrapatellar gap.

Patella Fracture

Palpation of closed injuries may reveal the crepitus and fracture gap.[72] Patients who suffer from a patella fracture may also present with fractures to the ipsilateral acetabulum and lower extremity.[73]

Patellar Tendonitis

If only tendonitis is present, no extension lag should be noted with any test position.

TREATMENT

Treatment principle:

Treatment for quadriceps tendon rupture with Chinese medicine is good for partial tears and extremely useful for recovery after surgery for the complete tear. It is helpful for patients who suffer from complete tears but are not surgical candidates due to medical comorbidities.

- Moves Qi and blood, opens meridians, reduce swelling, and stops pain.
- Tonifies liver and kidneys, strengthens tendons and bones.

Herbal Treatments

Early Stage

Herbal formula:

- Huo Xue Zhi Tong Tang 活血止痛湯.

Formula ingredients:

Dang Gui 當歸 6 g, Su Mu 蘇木 6 g, Luo De Da 落得打 6 g, Chuan Xiong 川芎 2 g, Hong Hua 紅花 1.5 g, Ru Xiang 乳香 3 g, Mo Yao 沒藥 3 g, San Qi 三七 3 g, Chi Shao 赤芍 3 g, Chen Pi 陳皮 3 g, Zi Jing Teng 紫荆藤 9 g, Di Bie Cong 地鱉蟲 9 g.

Ingredient explanations:

- Dang Gui 當歸 moves and tonifies blood, disperses cold, and stops pain due to blood stasis.
- Su Mu 蘇木 moves blood, reduces swelling, stops bleeding.
- Luo De Da 落得打 reduces swelling, clears heat, drains damp.
- Chuan Xiong 川芎 moves blood and Qi, expels wind, and stops pain.
- Hong Hua 紅花 moves blood, releases bruise, opens meridians, stops pain.
- Ru Xiang 乳香 moves blood, stops pain.
- Mo Yao 沒藥 moves blood, reduces swelling, and stops pain.
- San Qi 三七 moves blood, stops bleeding, reduces swelling, stops pain.
- Chi Shao 赤芍 moves blood, reduces swelling, stops pain.

- Chen Pi 陳皮 dries damp, transforms phlegm.
- Zi Jing Pi 紫荆皮 moves blood, reduces swelling, promotes urination.
- Di Bie Cong 地鱉蟲 breaks up stubborn blood stasis.

Formula action:

- Moves Qi and blood, drains damp reduces swelling, stops pain.

Contraindications:

- Contraindicated during preegnancy.

Modifications:

- For chronic cold damp wind Bi syndrome, add Qiang Huo 羌活 for opening meridians so that Zi Jing Pi 紫荆皮 is able to expel and drain the wind, damp, and cold in the joints.
- For swelling, add Ze Lan 澤蘭, Mu Tong 木通, Che Qian Zi 車前子, and Jiang Huang 薑黃.
- For severe pain, add Yan Hu Suo 延胡索, Niu Xi 牛膝, and Mu Xiang 木香.
- For swelling, redness, and hot sensation, add Mu Dan Pi 牡丹皮, Zhi Zi 梔子, and Huang Bai 黃柏.

Topical Application

Herbal formula:

- Xiao Zhong Hua Yu San 消肿化瘀散.

Formula ingredients:

Dang Gui 當歸, Chi Shao 赤芍, Sheng Di Huang 生地黄, Yan Hu Suo 延胡索, Xie Jie 血竭, Ru Xiang 乳香, Hong Hua 紅花, Da Huang 大黄, Jiang Huang 薑黃, Bie Jia 鱉甲, Qie Zi Gen 茄子根, Hong Qu 紅曲, Chi Xiao Dou 赤小豆.

Ingredient explanations:

- Dang Gui 當歸 moves and tonifies blood, disperses cold, and stops pain due to blood stasis.
- Chi Shao 赤芍 moves blood, reduces swelling, stops pain.
- Sheng Di Huang 生地黄 clears heat for heat symptoms due to Yin deficiency, cools the blood, nourishes Yin, generates fluids, and tonifies kidney Yin.
- Yan Hu Suo 延胡索 moves blood, stops pain.
- Xie Jie 血竭 moves blood, stops bleeding, stops pain.
- Ru Xiang 乳香 moves blood, stops pain.

- Hong Hua 紅花 moves blood, releases bruising, opens meridians, stops pain.
- Da Huang 大黄 moves blood and disperses bruising.
- Jiang Huang 薑黃 moves blood, opens meridians, expels wind, reduces swelling.
- Bie Jia 鱉甲 nourishes Yin, anchors Yang, moves blood, softens hardness and nodules.
- Qie Zi Gen 茄子根 moves blood, stops bleeding, stops pain, clears Heat.
- Hong Qu 紅曲 moves blood, tonifies spleen.
- Chi Xiao Dou 赤小豆 clears damp heat, reduces swelling, promotes urination.

Dose:

- All herbs are in equal doses.

Application procedures:

Method 1:
1) Grind the herbs to fine granules.
2) Make a paste by mixing the granules with vinegar.
3) Apply the paste on the quadriceps tendon.

Method 2:
1) Grind the herbs to fine granules.
2) Spread some Vaseline on a piece of gauze.
3) Sprinkle the granules on the Vaseline over about 0.2 cm, then apply the gauze to the torn quadriceps tendon.

Formula action: Moves blood, stops pain.

Indications: Contusion injury, joint dislocation, tendon injury, hemarthrosis in meridians.

Intermediate Stage

Herbal formula:

- Shu Jin Huo Xue Pian 舒筋活血片.

Formula ingredients:

- Zhi Gou Ji 狗脊 15 g, Sang Ji Sheng 桑寄生 15 g, Zi Ran Tong 自然銅 15 g, Ze Lan 澤蘭 10 g, Hong Hua 紅花 5 g, Wu Jia Pi 五加皮 10 g, Shen Jin Cao 伸筋草 10 g, Luo Shi Teng 絡石藤 10 g, Zhi Xiang Fu 香附 10 g, Ji Xue Teng 雞血藤 15 g.

Ingredient explanations:

- Zhi Gou Ji 狗脊 tonifies liver and kidnesy, strengthens tendon and bone, expels wind-dampness, and invigorates blood flow.
- Sang Ji Sheng 桑寄生 expels wind-dampness, tonifies liver and kidney Yin.
- Zi Ran Tong 自然銅 dispels blood, stops pain.
- Ze Lan 澤蘭 moves blood, promotes urination, and reduces swelling.
- Hong Hua 紅花 moves blood, releases bruising, opens meridians, stops pain.
- Wu Jia Pi 五加皮 expels wind and damp, nourishes and warms liver and kidneys, strengthens bones and tendons.
- Shen Jin Cao 伸筋草 expels wind, drains damp, relaxes tendons.
- Luo Shi Teng 絡石藤 dispels wind-dampness, cools blood, reduces swelling, strengthens tendons and bones.
- Zhi Xiang Fu 香附 spreads and regulates liver Qi and alleviates pain.
- Ji Xue Teng 雞血藤 moves and tonifies blood, relaxes tendons.

Later Stage

Herbal formula:

- Tao Hong Si Wu Tang 桃紅四物湯.

Formula ingredients:

Dang Gui 當歸 15 g, Shu Di Huang 熟地黃 15 g, Chuan Xiong 川芎 15 g, Bai Shao 白芍 15 g, Tao Ren 桃仁 15 g, Hong Hua 紅花 15 g.

Ingredient explanations:

- Dang Gui 當歸 moves and tonifies blood, disperses cold, and stops pain due to blood stasis.
- Shu Di Huang 熟地黃 nourishes blood, nourishes liver and kidney Yin.
- Chuan Xiong 川芎 moves blood and Qi, expels wind, and stops pain.
- Bai Shao 白芍 nourishes blood, preserves Yin, calms liver Yang and liver wind, softens tendons, and alleviates spasm and pain.
- Tao Ren 桃仁 breaks up stubborn blood stasis.
- Hong Hua 紅花 moves blood, releases bruising, opens meridians, stops pain.

Formula action:

- Tonifies and moves blood, breaks up stubborn blood stasis, regulates blood circulation.

Contraindications:

- Contraindicated during pregnancy.

Topical Application

Herbal ingredients:

- Gu Sui Bu 骨碎補 20 g, Ji Xing Zi 急性子 20 g, Niu Xi 牛膝 20 g, Tou Gu Cao 透骨草 20 g, Shen Jin Cao 伸筋草 20 g, Chuan Wu 川烏 20 g, Cao Wu 草烏 20 g, Wei Ling Xian 威靈仙 20 g, Hai Tong Pi 海桐皮 20 g, Ze Lan 澤蘭 20 g, Ru Xiang 乳香 20 g, Qiang Huo 羌活 20 g, Du Huo 獨活 15 g, Mo Yao 沒藥 15g, Hong Hua 紅花 15 g, Mu Gua 木瓜 15 g, Da Huang 大黄 15 g, Chuan Jiao 川椒 15 g, Sang Ji Sheng 桑寄生 30 g, Ji Xue Teng 雞血藤 30 g, salt 100 g, wine 100 ml.

Ingredient explanations:

- Gu Sui Bu 骨碎補 tonifies kidney Yang, strengthens tendons and bones, moves blood.
- Ji Xing Zi 急性子 breaks up stubborn blood stasis, softens hardness, removes toxins.
- Niu Xi 牛膝 expels wind dampness, tonifies liver and kidneys, directs herbs downward.
- Tou Gu Cao 透骨草 moves blood, breaks up stubborn blood stasis, reduces swelling, clears heat, cools blood, stops pain.
- Shen Jin Cao 伸筋草 expels wind, drains damp, relaxes tendons.
- Zhi Chuan Wu 川烏 expels wind-dampness, warms meridians, stops pain.
- Zhi Cao Wu 草烏 reduces swelling, stops pain.
- Wei Ling Xian 威靈仙 expels wind-dampness, stops pain.
- Hai Tong Pi 海桐皮 expels wind, drains damp, and reduces swelling.
- Ze Lan 澤蘭 moves blood, promotes urination, and reduces swelling.
- Ru Xiang 乳香 moves blood, stops pain.
- Qiang Huo 羌活 expels wind-cold-dampness, unblocks painful obstruction, alleviates pain, and guides Qi to the Tai Yang and Du channels.
- Du Huo 獨活 expels wind-dampness, stops pain.
- Mo Yao 沒藥 moves blood, reduces swelling, and stops pain.
- Hong Hua 紅花 moves blood, releases bruise, opens meridians, stops pain.
- Mu Gua 木瓜 promotes Qi and blood circulation, unblocks the channels, and releases spasm.
- Da Huang 大黄 moves blood and disperses bruise.
- Chuan Jiao 川椒 warms middle Jiao, disperses cold, dispels damp, and stops pain.
- Sang Ji Sheng 桑寄生 expels wind-dampness, tonifies liver and kidney Yin.
- Ji Xue Teng 雞血藤 moves and tonifies blood, relaxes tendons.
- Salt 100g induces the herbs to go downward to the kidneys, tonifies kidneys, tonifies Yin, cools heat, and harmonizes the dry hot herbs in the formula.

Application procedures:

1) Grind the herbs to powder and store them with the salt and wine in a fabric bag.
2) Steam the bag for ten minutes.
3) Apply the bag to the quadriceps tendon for an hour.
4) Repeat the procedure daily.

Acupuncture Treatments

Early Stage

Acupuncture point formula:

- GB31 風市 Fengshi, ST35 犊鼻 Dubi, ST36 足三里 Zusanli, GB34 陽陵泉 Yanglingquan, ST32 伏兔 Futu, SP10 血海 Xuehai with electro-acupuncture stimulation in the frequency of 2/100 Hz for 20 minutes per treatment.

Formula action: Moves blood.
Treatment frequency: Daily.
Treatment duration: 30 minutes.

Intermediate and Later Stage

Acupuncture point formula:
LI4 合谷 He Gu, LIV8 曲泉 Ququan, ST36 足三里 Zusanli, ST35 犊鼻 Dubi, KD7 復溜 Fuliu, KD3 太溪 Taixi, KD1 涌泉 Yong Quan, GB39 懸鐘 Xuanzhong, SP10 血海 Xuehai, Ren12 中脘 Zhongwan, Ren6 氣海 Qihai with warming needle technique for three to five cones per point.
Formula action: Moves Qi, tonifies liver and kidney.
Treatment frequency: Daily.
Treatment duration: 30 minutes.

Cupping Treatments

- He Ding 鶴頂 (the location is above the knee, in the depression of the midpoint of the upper border of the patella).

Bleeding Treatments

Ashi points 阿是穴.

Tui Na Treatments

Tui Na techniques:

- Pressing, kneading, rubbing, grasping, swing, pushing

Tui Na procedures:

1) The patient lies supine with the leg extended and a cushion placed below the knee; the therapist stands beside the affected knee.
2) Apply pressing technique with fingertips on ST36 足三里 Zusanli, SP10 血海 Xuehai, and UB40 委中 Weizhong with the greatest force that the patient is able to tolerate. Duration is one minute on each point.

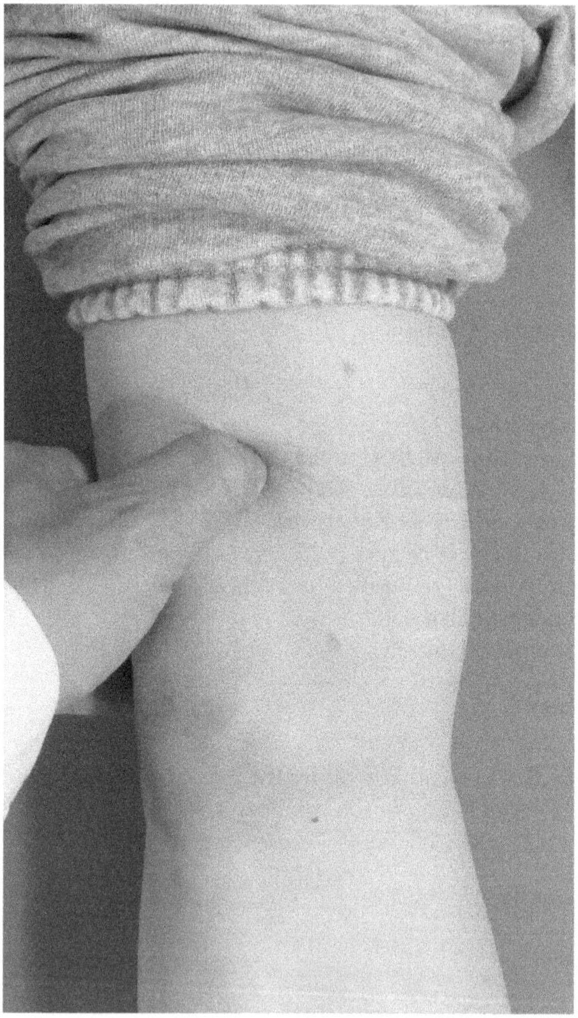

3) Apply pressing, kneading, and rubbing techniques on the tender spots around the knee especially around the quadriceps tendon for one minute on each spot. The force should begin light and grow heavier.

4) Apply the grasping technique on the quadriceps muscle for three minutes. Grasping on the quadriceps muscle should begin from proximal to distal toward the knee.

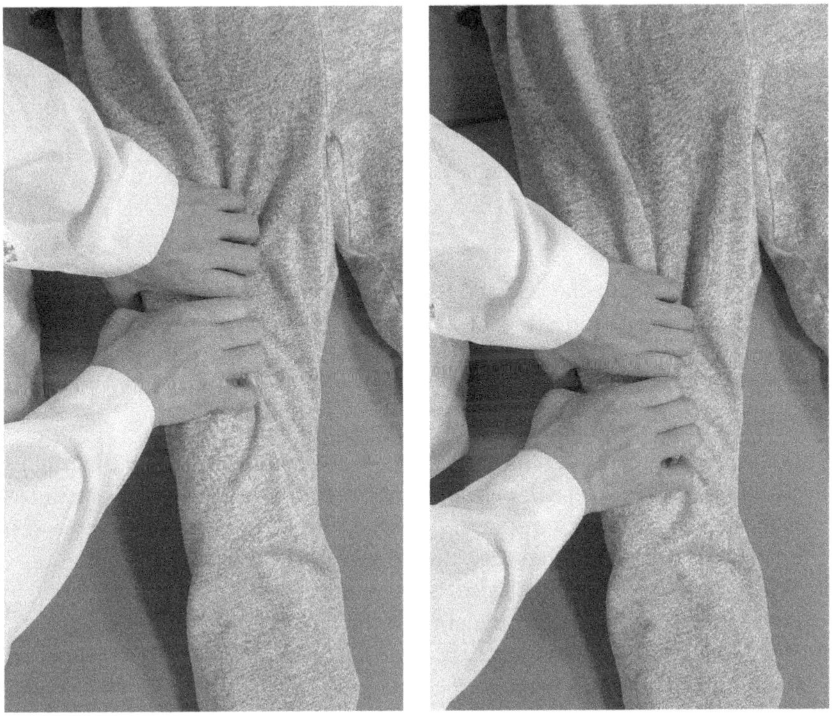

5) Apply the swinging technique on the knee. The therapist places a hand at the ipsilateral popliteal fossa and grasps the ipsilateral ankle with another hand, then flexes and extends the knee joint five to seven times. The range of motion should begin with small angles.

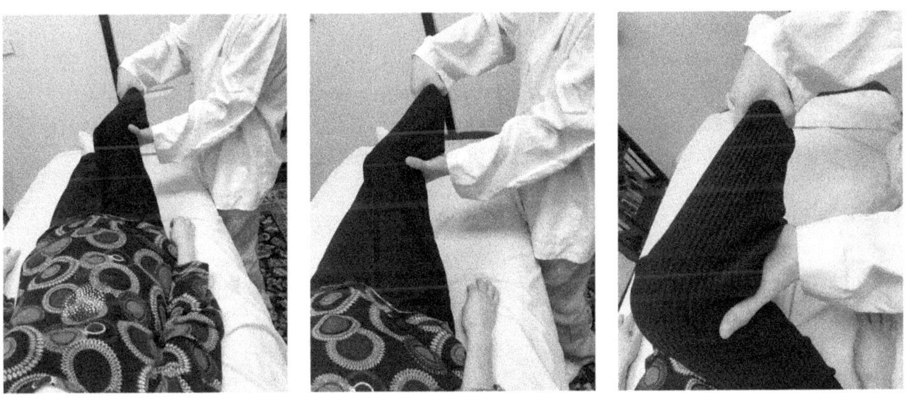

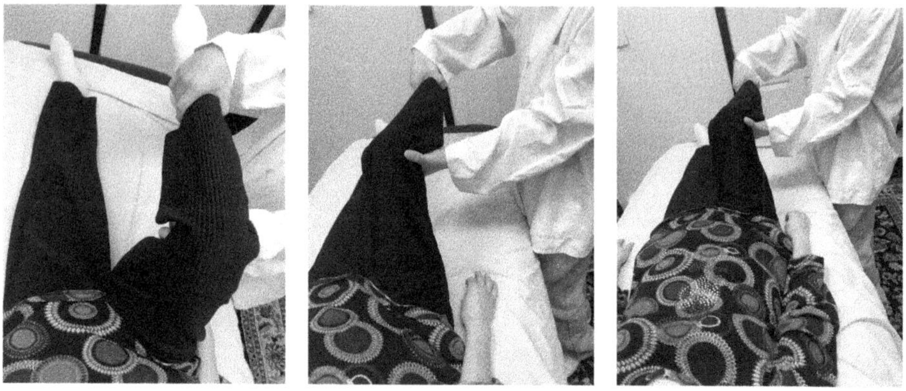

6) The patient extends the affected leg; the therapist lightly pushes the quadriceps muscle a few times to warm up.

Patient Advisory

- Take a break from the activities that caused the injury.
- In the acute stage, about the first two to three days, an ice pack is helpful for swelling. Put the ice pack on the medial side of the knee every one to two hours for ten to twenty minutes. Ensure that the ice is directly on the skin by placing a thin cloth between the ice pack and the skin.
- When the knee is swelling, raise the leg higher than the heart when lying by putting a pillow or cushion under the knee.
- When the swelling stops, keep the knee warm to prevent cold invasion.
- Recommend adding a brace to the injured knee.
- Exercise the legs and strengthen the soft tissues to prevent future knee injury:
  - The patient lies supine with the affected leg extended and the contralateral leg flexed.
  - Slowly raise the leg until it is even with the contralateral thigh.
  - Do not raise the leg higher than the contralateral thigh.
  - Hold the affected leg until it is tremoring. Repeat 30–50 times.
  - Repeat on the other side.
- Avoid bearing weight for four to six weeks.
- Recommend using crutches to avoid bearing weight and prevent falling.
- To prevent future injury, athletes should make sure that the quadriceps muscle has been adequately stretched and warmed up before exercise.

Dietary suggestions:

- Foods that benefit Qi and blood circulation such as grapes, onions, yams, vinegar, and brown sugar.
- Foods that are rich in fiber such as avocado, beans, bananas, oranges, apples, mangoes, strawberries, and raspberries.

- Foods that are rich in protein such as eggs, almonds, oats, milk, and broccoli.

DISPLACEMENT OF PATELLA 髕骨脱位

Introduction

Chronic or habitual patellar displacement is more common in females in the second and third decades of life. Athletes are at higher risk. But it is relatively equal in both sexes in acute onset.

It is approximately 2–3% of all knee injuries (Hayat 2020).

When the patella is completely slipped out of the patellofemoral groove of the femur, it is called patellar dislocation and is most commonly dislocated laterally.

When the patella is partially slipped out of the patellofemoral groove of the femur, it is called patellar subluxation or patella displacement. Clinically, lateral dislocation is most commonly seen.

Anatomy

The patella is located at the anterior aspect of the knee joint within the patellofemoral groove of the femur. It protects the anterior aspect of the knee joint from physical trauma.[74]

It is inferior to the quadriceps tendon and superior to the patellar tendon. It is the largest sesamoid bone in the body. A sesamoid is a bone embedded in a tendon that acts like a pulley for both quadriceps and patellar tendons. It provides a smooth surface for the quadriceps tendon to slide over. It enhances the leverage for the quadriceps tendon to exert on the femur in leg extension. It increases the tendon's ability to transmit muscular forces.

Western Medicine Etiology and Pathology

Patella dislocation and patella displacement are caused by high force impact on the patella or sudden forceful twisting of the knee. It is often caused by twisting or a direct blow to the patella.

Common factors that can predispose the dislocation of the patella:

- *Chronic patella dislocation or displacement*: The patient has previously suffered a patella displacement or dislocation. Prior injuries have weakened the knee joint. Pre-existing ligamentous laxity[75] predisposes the patient to later episodes.
- *Tight iliotibial band*: Usually it is more common in females. Women have been shown to generally have wider pelvises. Wider hips provide a larger area for a fetus to develop during pregnancy and cause an increase in the angle of the knee where the femur and tibia come together. This is called the Quadriceps Angle (Q-angle) (Jones 2013).

The normal Q-angle for males is usually 14 degrees and 17 degrees for females (Jones 2013). The wider pelvis and an increased Q-angle in females slant the femur inward.

- *Direct trauma to the knee*: A direct blow by an excessive force to the patella such as when the patella collides with another player in a soccer game can cause the patella to dislocate. Usually, it is to the medial aspect of the knee. The force able to blow the patella out of the patellofemoral groove is obviously much greater; such injury usually causes more severe damage than just displacing the patella. It usually restrains the ligaments and tendons and may cause fractures.
- *Indirect trauma to the knee*: A rapid change in direction by pivoting the femur in internal rotation and the tibia in external rotation while running at high speed in one direction with the foot planted and the knee in slight flexion is called valgus-flexion-external rotation.[76] The knee joint sustains a tremendous force; the patella may slide out of the patellofemoral groove.
- *Patella alta*: The patella locates abnormally higher toward the hip above the knee joint line. The patella must travel a greater distance before engaging fully in the patellofemoral groove[77] during knee flexion, resulting in patellar dislocation.
- *Patellar and trochlear dysplasia*: The femoral trochlea groove is abnormally shallow, and it doesn't fully contain the patella until the knee is in deep flexion. The patella is vulnerable to be dislocated in early knee flexion. Common high-risk sport activity is any during which the foot strikes the ground when running.[78]
- *Marfan Syndrome or Ehlers-Danlos*: These are secondary connective tissue disorders that increase the risk of displacement.
- *Muscular imbalance*: Particularly weakness of the vastus medialis oblique.
- *Congenital dislocations such as Down syndrome*: The small patella in combination with a hypoplastic condyle is the risk factor resulting congenital dislocation (Hayat 2020).

The direction of the dislocation or the displacement of the patella depends on the direction of the violence and the range of motion of the knee during the injury.

Patella dislocation is generally divided into five types:

1) Lateral dislocation or displacement is the most common seen dislocation, which is partly because of:
   - Imbalance between stronger lateral tissues such as lateral retinaculum and vastus lateralis and the weaker medial anatomic structures such as the medial patellofemoral ligament and vastus medialis.
   - Increased Quadriceps Angle (Q-Angle).
   - Femoral and tibial biomechanical rotation.[79]

- Quadriceps muscle pulling in a direction slightly lateral per the mechanical axis of the limb (Hayat 2020).

2) Medial dislocation or displacement is not as common as lateral displacement. There are several causative factors, including violent force from the lateral patella. Quadriceps atrophy, iatrogenic injury, and congenital conditions such as Down syndrome are the risk factors.
3) Intra-articular dislocation is rear usually occurs during traumatic injury so that the patella is avulsed from the quadriceps tendon and then rotated (Hayat 2020).
4) Inferior dislocation can occur when the violent force ruptures the quadriceps tendon.
5) Superior dislocation can occur when violent force ruptures the patellar tendon.[80]

Chinese Medicine Etiology and Pathology

Qi and Blood Stagnation

A direct blow or cut to the medial, lateral aspect of the knee or to the quadriceps tendon and the patellar tendon causes blockages to the flow of the local Qi and blood in meridians almost immediately after the violence.

Liver and Kidney Deficiency

Ligament injury is considered "Jin Shang 筋傷" or "traumatic injury on tendons." The definition of "tendon" is not just limited to tendons; traditionally, "tendon" in Chinese medicine includes muscle and all of its sounding soft tissues. Therefore, injury to the tendons is the trauma of muscle, ligament, cartilage, or meniscus, intervertebral discs and their outer fibrous rings, blood vessel, nerve, tendon and sheath, bursa, and joint capsule. *Treatise on the Engenderment of the Five Viscera*《素問·五藏生成篇第十》states: "All tendons belong to the joints 諸筋者, 皆屬於節."

In the theory of traditional Chinese medicine (TCM), the liver governs the tendons. *Treatise on the Engenderment of the Five Viscera*, chapter 10《素問·五藏生成篇第十》states:

> The Liver stores the Blood and the Heart moves it. When a person is active, Blood moves through all the meridians to Tendons; when the body is in a state of rest, the Blood returns and stores in the Liver. Why is this so? Because the Liver governs the sea of Blood. 肝藏血，心行之，人動則血運于諸經，人靜則血歸于肝藏。何者？肝主血海故也。...When a person is at the state of rest, the Blood returns to the liver. The liver receives the Blood and the person is able to see. The feet receive the Blood and are able to walk. The palms receive the Blood and are able to grasp. The fingers receive the Blood and are able to hold 人臥血歸於肝，肝受血而能視，足受血而能步，掌受血而能握，指受血而能攝。. (Unschuld and Tessenow 2011)

When injury occurs to tendons, it consumes liver blood. As the trauma becomes chronic, the blood is insufficient. Liver's element is wood, which needs water nourishment from kidney. Deficiency of liver continually demands the sustentation of kidney; kidney then becomes deficient.

Manifestations

- Pain is high intensity, especially when weight bearing until the reduction of the knee joint.
- Tenderness around the patella.
- Instability of the knee and wobbly patella.
- Locking of the knee after the trauma.
- Rapid medial swelling when there is lateral dislocation.
- Hemarthrosis of the knee.
- Visible translation laterally of the kneecap.
- Pain and effusion may not manifest at every episode in habitual displacement.

Physical Examination

- Begin by obtaining the patient's medical history:
  - Injury history such as the direction of the direct violent force and the past history of dislocation or displacement of the knee; indirect injury history such as twisting the leg when the is foot planted.
  - General health conditions that might result in the dislocation of the patella including some congenital diseases such as Down syndrome.
- Inspection:
  - Knee effusion.
  - Surrounding skin for the signs of direct trauma.
  - Hemarthrosis, hematoma, and ecchymosis.
  - Difficulty standing, there is more pain when weight bearing on the affected leg.
  - The patella easily slips out when flexion.
  - Measure the Quadriceps Angle (Q-angle).
- Palpation:
  - Abnormal location of the patella.
  - Tenderness around the patella.
- Range of motion:
  - The patella easily slips out when flexed, reduction when extended.

Orthopedic tests:

- Patella apprehension test:
  1) The patient lies supine with 30 degrees of knee flexion, and the quadriceps muscle relaxes.
  2) A medial force is applied to the patella with the examiner's thumb, and the examiner observes the patient's facial expression.

3) The test is positive if the patient exhibits an apprehensive facial expression or contracts the quadriceps muscle.[81]

- Patella tracking assessment: Positive when the patella moves sideways.
- Patellar hypermobility.[82]

Differential Diagnosis

Patella Fracture

Palpation of closed injury may reveal the crepitus and fracture gap.[83] Patients who suffer from a patella fracture may also present with fracture to the ipsilateral acetabulum and lower extremity.[84]

Quadriceps Tendon and Patellar Tendon Rupture

Quadriceps tendon rupture has a suprapatellar gap, while patellar tendon rupture has an infrapatellar gap.

Treatment

Treatment principle:

Treatment for patella displacement with Chinese medicine is effective for both direct and indirect trauma. It is extremely effective for habitual dislocation or displacement of the patella. It prevents future dislocation by strengthening the tendons and ligaments surrounding the patella.

- Moves Qi and blood, opens meridians, reduces swelling, and stops pain.
- Tonifies liver and kidneys, strengthens tendons and bones.

Herbal Treatments

Direct Trauma to the Knee

Herbal formula:

- Huo Xue Zhi Tong Tang 活血止痛湯.

Formula ingredients:

Dang Gui 當歸 6 g, Su Mu 蘇木 6 g, Luo De Da 落得打 6 g, Chuan Xiong 川芎 2 g, Hong Hua 紅花 1.5 g, Ru Xiang 乳香 3 g, Mo Yao 沒藥 3 g, San Qi 三七 3 g, Chi Shao 赤芍 3 g, Chen Pi 陳皮 3 g, Zi Jing Teng 紫荆藤 9 g, Di Bie Cong 地鱉蟲 9 g.

Ingredient explanations:

- Dang Gui 當歸 moves and tonifies blood, disperses cold, and stops pain due to blood stasis.
- Su Mu 蘇木 moves blood, reduces swelling, stops bleeding.
- Luo De Da 落得打 reduces swelling, clears heat, drains damp.
- Chuan Xiong 川芎 moves blood and Qi, expels wind, and stops pain.

- Hong Hua 紅花 moves blood, releases bruise, opens meridians, stops pain.
- Ru Xiang 乳香 moves blood, stops pain.
- Mo Yao 沒藥 moves blood, reduces swelling, and stops pain.
- San Qi 三七 moves blood, stops bleeding, reduces swelling, stops pain.
- Chi Shao 赤芍 moves blood, reduces swelling, stops pain.
- Chen Pi 陳皮 dries damp, transforms phlegm.
- Zi Jing Pi 紫荆皮 moves blood, reduces swelling, promotes urination.
- Di Bie Cong 地鱉蟲 breaks up stubborn blood stasis.

Formula action:

- Moves Qi and blood, drains damp reduces swelling, stops pain.

Contraindications:

- Contraindicated during pregnancy.

Modifications:

- For chronic cold damp wind Bi syndrome, add Qiang Huo 羌活 for opening meridians so that Zi Jing Pi 紫荆皮 is able to expel and drain the wind, damp, and cold in the joints.
- For swelling, add Ze Lan 澤蘭, Mu Tong 木通, Che Qian Zi 車前子, and Jiang Huang 薑黃.
- For severe pain, add Yan Hu Suo 延胡索, Niu Xi 牛膝, and Mu Xiang 木香.
- For swelling, redness, and hot sensation, add Mu Dan Pi 牡丹皮, Zhi Zi 梔子, and Huang Bai 黃柏.

Topical Application

Herbal formula:

- Xiao Zhong Hua Yu San 消腫化瘀散.

Formula ingredients:

Dang Gui 当归, Chi Shao 赤芍, Sheng Di Huang 生地黄, Yan Hu Suo 延胡索, Xie Jie 血竭, Ru Xiang 乳香, Hong Hua 红花, Da Huang 大黄, Jiang Huang 姜黄, Bie Jia 鳖甲, Qie Zi Gen 茄子根, Hong Qu 红曲, Chi Xiao Dou 赤小豆.

Ingredient explanations:

- Dang Gui 當歸 moves and tonifies blood, disperses cold, and stops pain due to blood stasis.
- Chi Shao 赤芍 moves blood, reduces swelling, stops pain.
- Sheng Di Huang 生地黄 clears heat for heat symptoms due to Yin deficiency, cools the blood, nourishes Yin, and generates fluids and tonifies kidney Yin.

- Yan Hu Suo 延胡索 moves blood, stops pain.
- Xie Jie 血竭 moves blood, stops bleeding, stops pain.
- Ru Xiang 乳香 moves blood, stops pain.
- Hong Hua 紅花 moves blood, releases bruising, opens meridians, stops pain.
- Da Huang 大黃 moves blood and disperses bruising.
- Jiang Huang 薑黃 moves blood, opens meridians, expels wind, reduces swelling.
- Bie Jia 鱉甲 nourishes Yin, anchors Yang, moves blood, and softens hardness and nodules.
- Qie Zi Gen 茄子根 moves blood, stops bleeding, stops pain, clears heat.
- Hong Qu 紅曲 moves blood, tonifies spleen.
- Chi Xiao Dou 赤小豆 clears damp heat, reduces swelling, promotes urination.

Dose:

- All herbs are in equal doses.

Application procedures:

Method 1:

1) Grind the herbs to fine granules.
2) Make a paste by mixing the granules with vinegar.
3) Apply the paste on the quadriceps tendon.

Method 2:

1) Grind the herbs to fine granules.
2) Spread some Vaseline on a piece of gauze.
3) Sprinkle the granules on the Vaseline over about 0.2 cm, then apply the gauze to the torn quadriceps tendon.

Formula action: Moves blood, stops pain.

Indications: Contusion injury, joint dislocation, tendon injury, hemarthrosis in meridians.

Indirect Trauma to the Knee

Herbal formula:

- Shu Jin Huo Xue Pian 舒筋活血片.

Formula ingredients:

Zhi Gou Ji 狗脊 15 g, Sang Ji Sheng 桑寄生 15 g, Zi Ran Tong 自然銅 15 g, Ze Lan 澤蘭 10 g, Hong Hua 紅花 5 g, Wu Jia P i 五加皮 10 g, Shen Jin Cao 伸筋草 10 g, Luo Shi Teng 絡石藤 10 g, Zhi Xiang Fu 香附 10 g, Ji Xue Teng 雞血藤 15 g.

Ingredient explanations:

1. Zhi Gou Ji 狗脊 tonifies liver and kidneys, strengthens tendons and bones, expels wind-dampness, and invigorates blood flow.
2. Sang Ji Sheng 桑寄生 expels wind-dampness, tonifies liver and kidney Yin.
3. Zi Ran Tong 自然銅 dispels blood, stops pain.
4. Ze Lan 澤蘭 moves blood, promotes urination, and reduces swelling.
5. Hong Hua 紅花 moves blood, releases bruising, opens meridians, stops pain.
6. Wu Jia Pi 五加皮 expels wind and damp, nourishes and warms liver and kidneys, strengthens bones and tendons.
7. Shen Jin Cao 伸筋草 expels wind, drains damp, relaxes tendons.
8. Luo Shi Teng 絡石藤 dispels wind-dampness, cools blood, reduces swelling, strengthens tendons and bones.
9. Zhi Xiang Fu 香附 spreads and regulates liver Qi and alleviates pain.
10. Ji Xue Teng 雞血藤 moves and tonifies blood, relaxes tendons.

Formula action:

- Relaxs tendons, moves blood, and disperses bruising.

Contraindications:

- Contraindicated during pregnancy.

Habitual Dislocation

Herbal formula:

- Tao Hong Si Wu Tang 桃紅四物湯.

Formula ingredients:

Dang Gui 當歸 15 g, Shu Di Huang 熟地黃 15 g, Chuan Xiong 川芎 15 g, Bai Shao 白芍 15 g, Tao Ren 桃仁 15 g, Hong Hua 紅花 15 g.

Ingredient explanations:

- Dang Gui 當歸 moves and tonifies blood, disperses cold, and stops pain due to blood stasis.
- Shu Di Huang 熟地黃 nourishes blood, nourishes liver and kidney Yin.
- Chuan Xiong 川芎 moves blood and Qi, expels wind, and stops pain.
- Bai Shao 白芍 nourishes blood, preserves Yin, calms liver Yang and liver wind, softens tendons, and alleviates spasm and pain.
- Tao Ren 桃仁 breaks up stubborn blood stasis.
- Hong Hua 紅花 moves blood, releases bruising, opens meridians, stops pain.

Formula action:

- Tonifies and moves blood, breaks up stubborn blood stasis, regulates blood circulation.

Contraindications:

- Contraindicated during pregnancy.

Topical Application

Formula ingredients:

Gu Sui Bu 骨碎補 20 g, Ji Xing Zi 急性子 20 g, Niu Xi 牛膝 20 g, Tou Gu Cao 透骨草 20 g, Shen Jin Cao 伸筋草 20 g, Chuan Wu 川烏 20 g, Cao Wu 草烏 20 g, Wei Ling Xian 威靈仙 20 g, Hai Tong Pi 海桐皮 20 g, Ze Lan 澤蘭 20 g, Ru Xiang 乳香20 g, Qiang Huo 羌活 20 g, Du Huo 獨活 15 g, Mo Yao 没藥 15 g, Hong Hua 紅花 15 g, Mu Gua 木瓜 15 g, Da Huang 大黄 15 g, Chuan Jiao 川椒 15 g, Sang Ji Sheng 桑寄生 30 g, Ji Xue Teng 雞血藤 30 g, salt 食盐 100 g, wine 白酒 100 ml.

Ingredient explanations:

- Gu Sui Bu 骨碎補 tonifies kidney Yang, strengthens tendons and bones, moves blood.
- Ji Xing Zi 急性子 breaks up stubborn blood stasis, softens hardness, removes toxins.
- Niu Xi 牛膝 expels wind-dampness, tonifies liver and kidneys, directs herbs downward.
- Tou Gu Cao 透骨草 moves blood, breaks up stubborn blood stasis, reduces swelling, clears heat, cools blood, stops pain.
- Shen Jin Cao 伸筋草 expels wind, drains damp, relaxes tendons.
- Zhi Chuan Wu 川烏 expels wind-dampness, warms meridians, stops pain.
- Zhi Cao Wu 草烏 reduces swelling, stops pain.
- Wei Ling Xian 威靈仙 expels wind-dampness, stops pain.
- Hai Tong Pi 海桐皮 expels wind, drains damp, and reduces swelling.
- Ze Lan 澤蘭 moves blood, promotes urination, and reduces swelling.
- Ru Xiang 乳香 moves blood, stops pain.
- Qiang Huo 羌活 expels wind, cold, and damp, unblocks painful obstruction, alleviates pain, and guides Qi to the Tai Yang and Du channels.
- Du Huo 獨活 expels wind-dampness, stops pain.
- Mo Yao 沒藥 moves blood, reduces swelling, and stops pain
- Hong Hua 紅花 moves blood, releases bruising, opens meridians, stops pain.
- Mu Gua 木瓜 promotes Qi and blood circulation and unblocks the channels, releases spasm.
- Da Huang 大黄 moves blood and disperses bruising.
- Chuan Jiao 川椒 warms middle Jiao, disperses cold, dispels damp, and stops pain.
- Sang Ji Sheng 桑寄生 expels wind-dampness, tonifies liver and kidney Yin.

- Ji Xue Teng 雞血藤 moves and tonifies blood, relaxes tendons.
- Salt induces the herbs to go downward to the kidneys, tonifies kidneys, tonifies Yin, cools heat, and harmonizes the dry hot herbs in the formula.
- Wine induces the herbs to go upward to all meridians to invigorate blood and expel wind-dampness, and warms the cold herbs in the formula.

Application procedures:

1) Grind the herbs to powder and store them with the salt and wine in a fabric bag.
2) Steam the bag for ten minutes.
3) Apply the bag on the quadriceps tendon for an hour.
4) Repeat the procedure daily.

Acupuncture Treatments

Direct Trauma to the Knee

Acupuncture point formula:

- GB31 風市 Fengshi, ST35 犊鼻 Dubi, ST36 足三里 Zusanli, GB34 陽陵泉 Yanglingquan, ST32 伏兔 Futu, SP10 血海 Xuehai with electro-acupuncture stimulation at the frequency of 2/100 Hz for 20 minutes per treatment.

Formula action: Moves blood.

Treatment frequency: Daily.

Treatment duration: 30 minutes.

Indirect Trauma to the Knee and Habitual Dislocation

Acupuncture point formula:

LI4 合谷He Gu, LIV8 曲泉 Ququan, ST36 足三里 Zusanli, ST35 犊鼻 Dubi, KD7 復溜 Fuliu, KD3 太溪 Taixi, KD1 涌泉 Yong Quan, GB39 懸鐘 Xuanzhong, SP10 血海 Xuehai, Ren12 中脘 Zhongwan, Ren6 氣海 Qihai with warming needle technique for three to five cones per point.

Formula action: Moves Qi, tonifies liver and kidneys.

Treatment frequency: Daily.

Treatment duration: 30 minutes.

Cupping Treatments

He Ding 鶴頂 (above the knee, in the depression of the midpoint of the upper border of the patella).

Bleeding Treatments

Ashi points 阿是穴 (a tender point that is not included in the classic acupuncture points).

Tui Na Treatments

Reduction procedures:

- The patient lies supine; the therapist stands beside the affected knee.
- The therapist grasps the ipsilateral ankle and places a hand on the affected knee.
- The thumb pushes the lateral aspect of the patella while extending the knee.

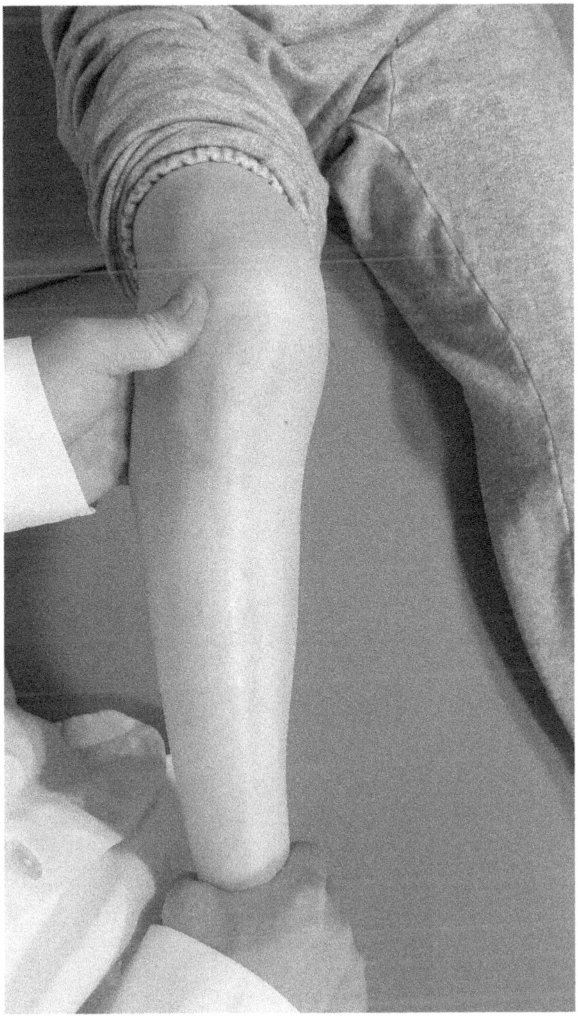

- Immobilize the affected knee for three to four weeks.

Tui Na techniques:

- Rolling, pushing, grasping, pressing, kneading, bending, striking

Tui Na procedures:

1) The patient lies on the left side with left knee and left hip flexed at 90 degrees and the right knee flexed at about 30 degrees. Place a pillow in between knees. The right knee is used as an example for treatment. The therapist stands beside the affected knee.
2) The therapist applies rolling technique on the IT band at about acupuncture points GB31 風市 Fengshi, GB32 中瀆 Zhongdu, GB33 膝陽關 Xiyangguan, GB34 陽陵泉 Yanglingquan for about five minutes.

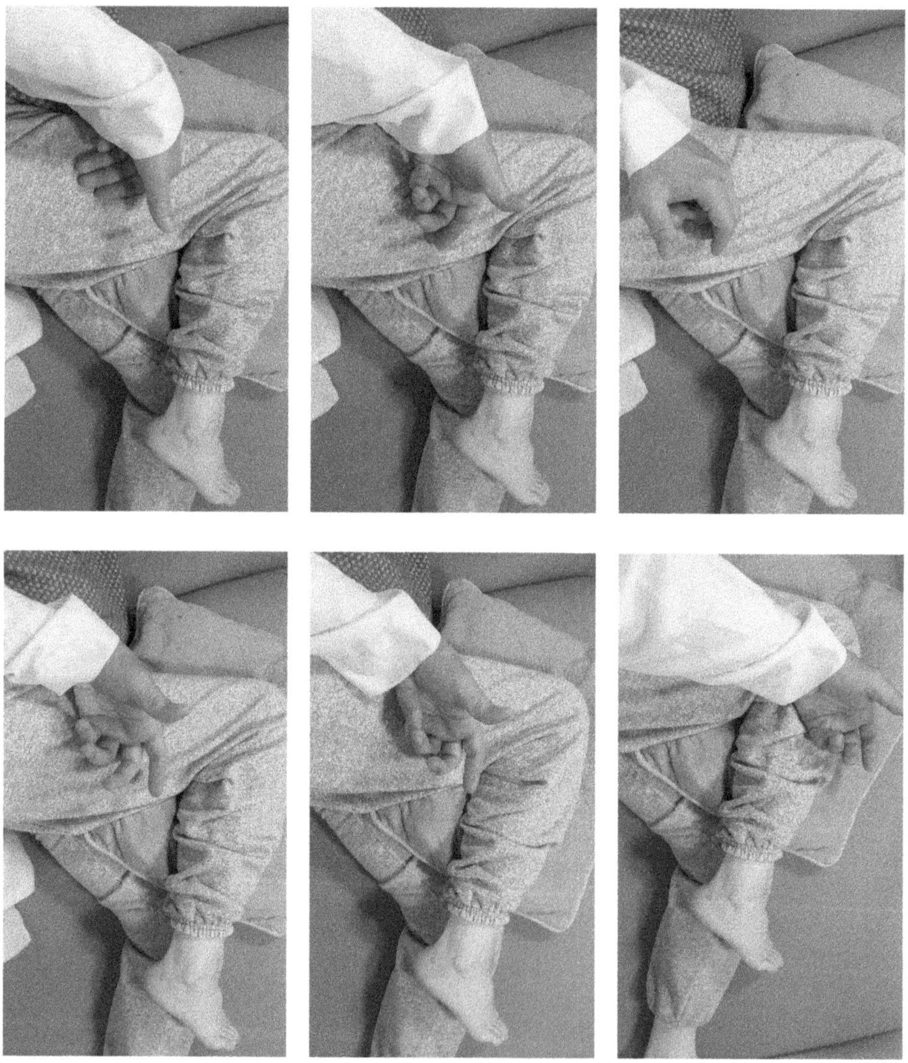

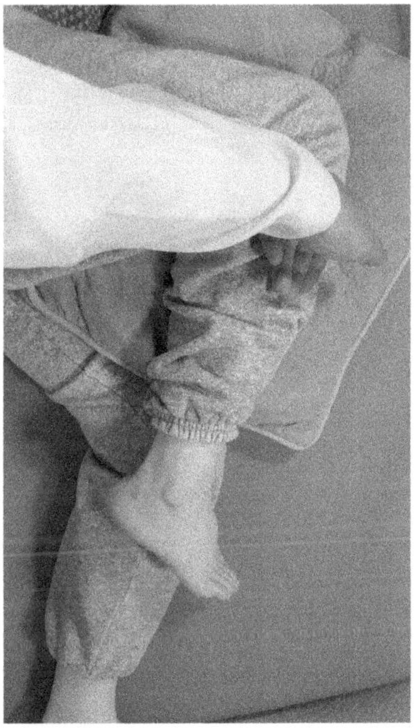

3) The therapist applies pushing techniques on GB31 風市 Fengshi, GB32 中瀆 Zhongdu, GB33 膝陽關 Xiyangguan, GB34 陽陵泉 Yanglingquan for about five minutes or until the skin is warm.

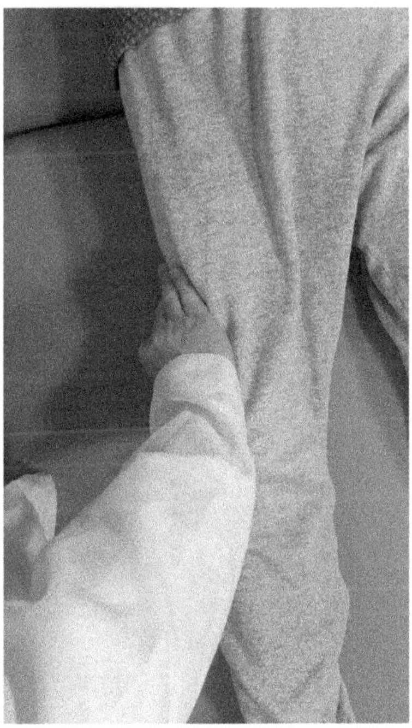

4) The therapist applies grasping techniques on the quadriceps muscle and tendon for about three minutes.

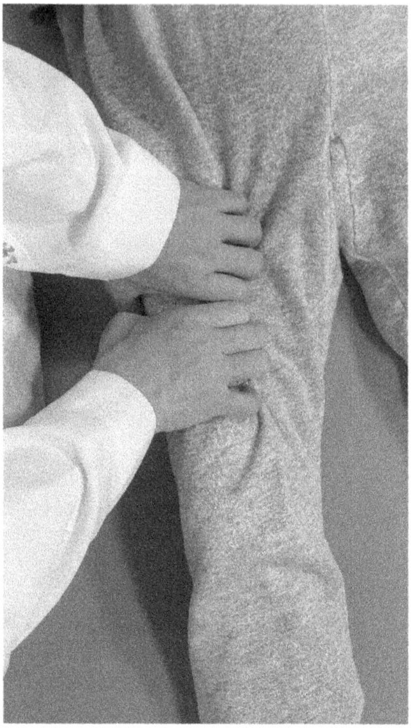

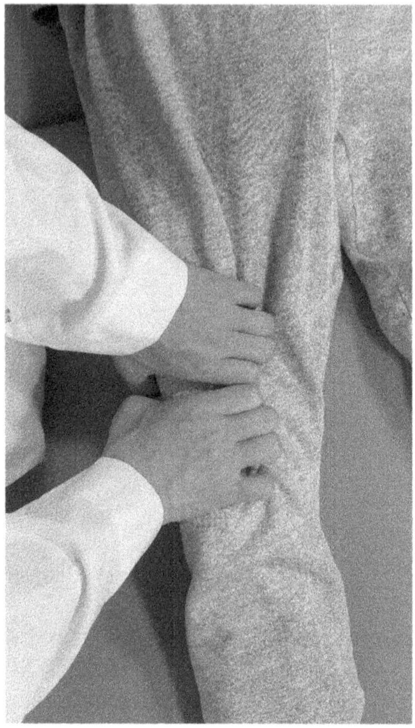

5) The therapist applies pressing techniques with fingertips on UB40 委中 Weizhong and UB39 委陽 Weiyang for about one minute on each point.
6) The therapist applies kneading techniques with fingers on ST35 犊鼻 Dubi, Nei Xi Yan 膝眼 (in the depression on the medial side of the patellar ligament when the kneels flexed), GB34 陽陵泉 Yanglingquan, and ST36 足三里 Zusanli for about one minute on each point.

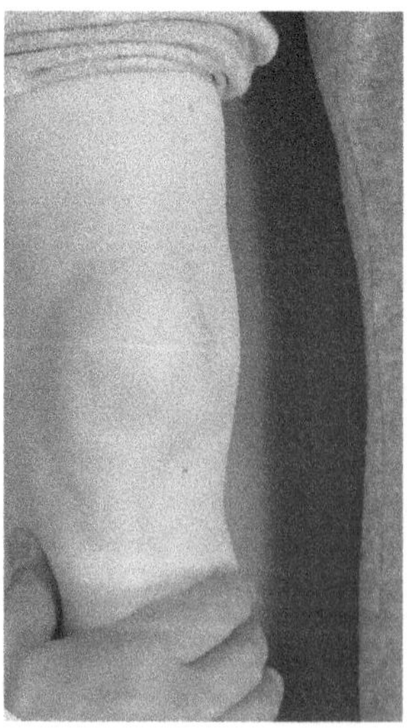

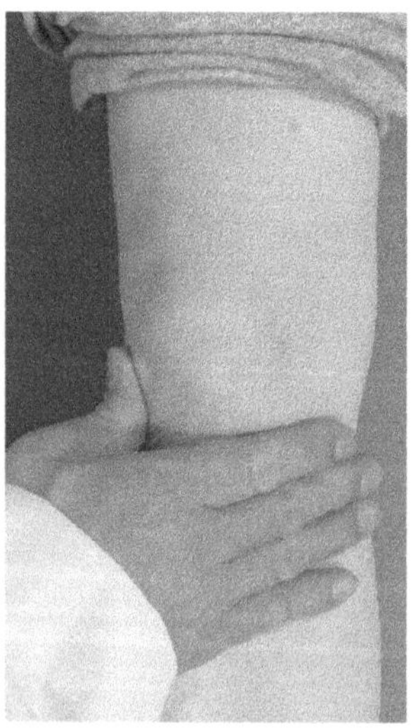

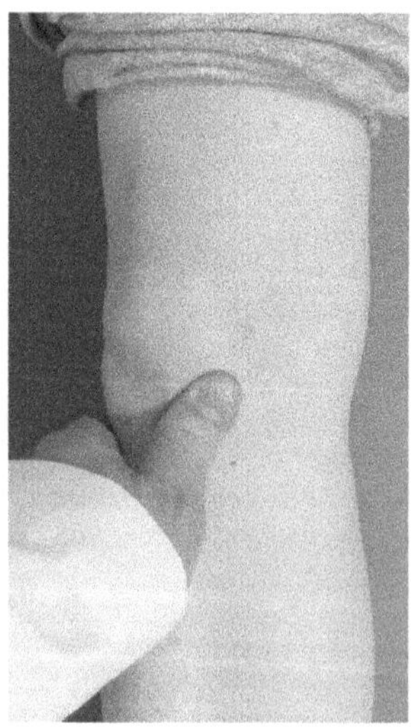

7) Applies pressing techniques with the elbow on GB30 環跳 Huantiao for about three minutes.
8) The patient lies supine with the legs straightened. The therapist stands beside the affected knee.
9) The therapist grasps the ipsilateral ankle with the right hand and places the left hand under the popliteal fossa.
10) Both hands work together to flex the affected knee and hip to 90 degrees then extend the knee. Repeat three to five times. The procedures need to be slow.

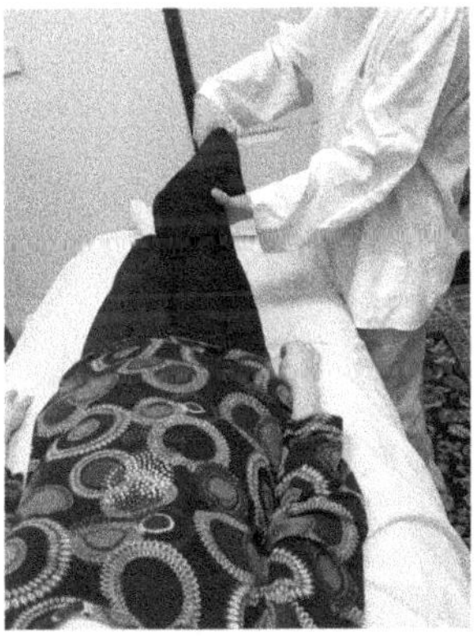
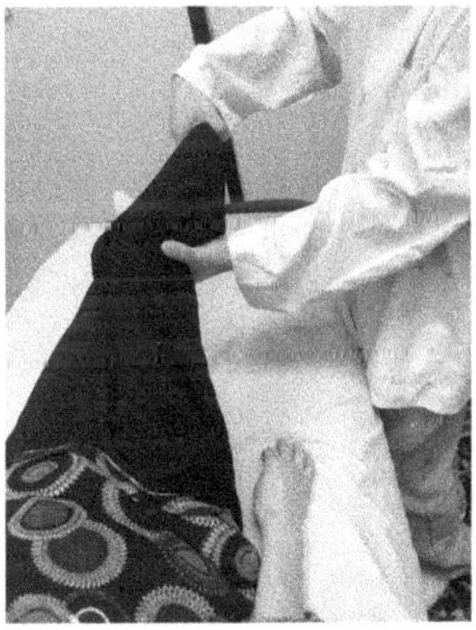
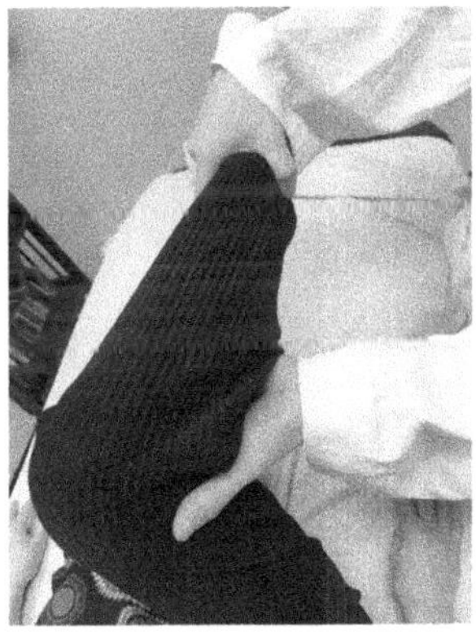

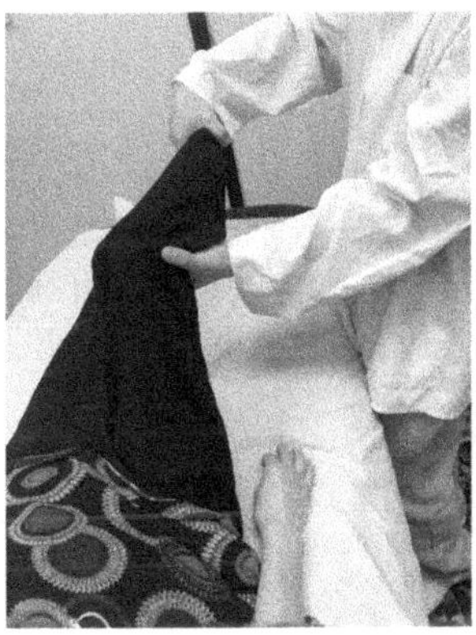
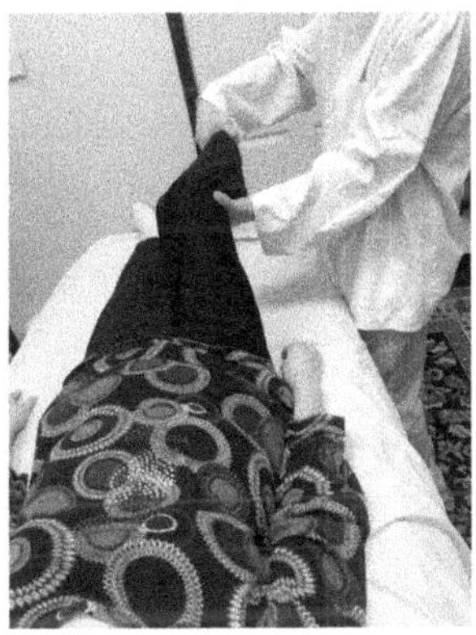

11) The right-hand clenches an empty fist and strikes the following acupuncture points with light force if the case is deficiency that causes habitual dislocation; or heavy force for Qi and blood stagnation. The acupuncture points are GB31 風市 Fengshi, GB34 陽陵泉 Yanglingquan, ST36 足三里 Zusanli, GB39 懸鐘Xuanzhong, SP10 血海 Xuehai, LIV8 曲泉 Ququan.

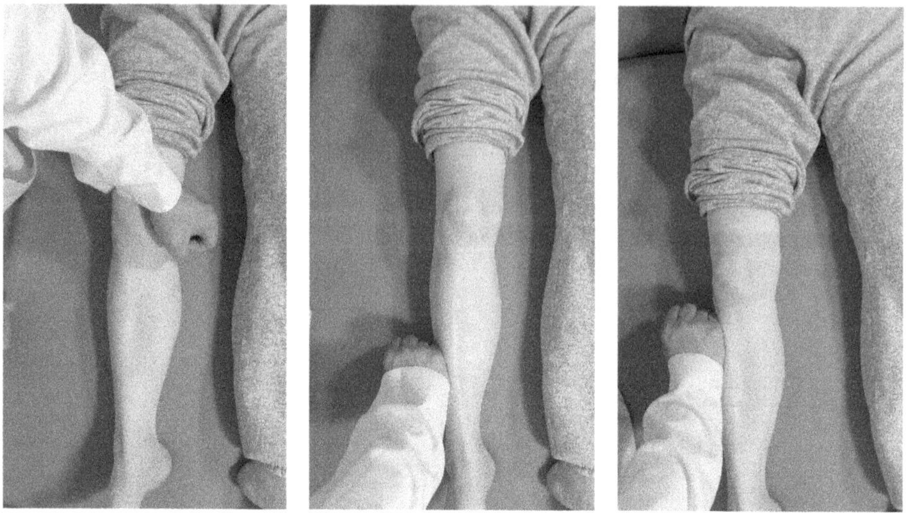

12) Repeat the same procedure as step 11 with a cupping palm. Always start striking techniques with light force.

Patient Advisory

Acute condition: After reduction of the patella, patients should rest the knee.

1) The affected leg should avoid bearing weight for at least four to six weeks.
2) Apply ice packs on the knee for ten to twenty minutes at a time, once every one to two hours for the first two days after the injury. Ice packs should not be put on skin directly; put a thin cloth between the ice and the skin.[85]
3) Apply elastic bandage on the injured knee.
4) Lift the knee higher than the heart when rest. Do this for about three days following your injury.
5) Wearing a knee brace for knee support is helpful for recovery.
6) Because the knee is painful and unstable, patients are recommended to use a crutch to prevent falls.
7) Chinese medicine is able to reduce swelling and stop pain quickly, but it should be applied after ruling out any complications such as fractures, torn ligaments, and tendon and nerve injuries.
8) Take a break from the activities that caused the injury.

Chronic condition:

1) Exercise to strengthen the surrounding muscles and tendons, especially the soft tissues at the medial aspect of the knee and femur:

a) The patient lies supine with the affected leg extended and the contralateral leg flexed.
b) Slowly raise the leg until it is even with the contralateral thigh.
c) Do not raise the leg higher than the contralateral thigh.
d) Hold the affected leg until it is tremoring. Repeat 30–50 times.
e) Repeat on the other side.

2) Before exercise, warm up the knee joints to loosen the tendons and ligaments and promote circulation.
3) Learn the correct postures for sports, activities, and exercises. For example, avoid pivoting the femur in internal rotation and the tibia in external rotation while running at high speed.

Dietary suggestions:

- Foods that benefit blood circulation such as grapes,[86] onions, yams, vinegar, and brown sugar.[87]
- Foods that are rich in fiber such as avocado, beans, bananas, oranges, apples, mangoes, strawberries, and raspberries.
- Foods that are rich in protein such as eggs,[88] almonds, oats, milk, and broccoli.
- Foods that are rich in vitamins and minerals such as soymilk, milk, yogurt, fish, whole grains, mushrooms, nuts, eggs, and cheese (Harvey and Woods 1992).
- Avoid foods that are fatty, deep-fried, smoked, barbecued, raw, cold, or salted.

LUXATION OF THE PROXIMAL TIBIOFIBULAR JOINT 脛腓關節脱位

INTRODUCTION

Luxation or dislocation of the proximal tibiofibular joint is an uncommon injury; it accounts for less than 1% of all knee injuries (Harvey and Woods 1992). Even luxation of the proximal tibiofibular joint is a rare knee injury; it is a commonly misdiagnosed disease that is a potential cause of chronic knee pain and may result in disability.

In 1974, Ogden proposed four types of this dislocation: type I—subluxation; type II—anterolateral dislocation; type III—posteromedial dislocation, and type IV—superior dislocation (Nieuwe Weme et al. 2014).

Among these four types, anterolateral dislocation is the most common injury, accounting for 85% of proximal tibiofibular joint dislocations, posteromedial dislocation occurs in 10%, and superior dislocation in 2% (Nunes 2020).

The knee injury is mostly seen in young patients around 17–30 years old, and it is more often seen in male patients (Horst and LaPrade 2010). The injury relates to sports activities such as soccer, football, ballet dancing, parachuting, snowboarding, the long jump, horse riding, equestrian jumping, and direct trauma (Horan and Quin

2006). Usually patients report that the injury is from a fall with the knee slightly flexed, the ankle inverted and the foot flexed (Horan and Quin 2006).

Anatomy

The lower leg is made up of two bones, the tibia and fibula. The fibula is the slender one and posterolaterally situated to the tibia. It is a non-weight-bearing bone. These two bones are connected at two locations:

- Proximal tibiofibular joint—the fibula articulates with the inferior aspect of the lateral tibial condyle. It is supported by a tense joint capsule; the capsule attaches to the tibia and fibula at the margin of the articular surface and is reinforced anteriorly by the biceps femoris tendon insertion into the fibular capitulum, posteriorly by the popliteus tendon, superiorly by the lateral collateral ligament, and inferiorly by the interosseous membrane.
- Distal tibiofibular joint—the fibula articulates with the fibular notch of the tibia.

These two articulations are connected by the tibiofibular syndesmosis, which includes the interosseous membrane.[89]

Western Medicine Etiology and Pathology

Anterolateral dislocation is the most common injury. The injury is from the external rotation of the leg while the knee is slightly flexed, the ankle inverted, and the foot flexed (Horan and Quin 2006).

Knee flexion relaxes the biceps tendon and the lateral collateral ligament.[90] The tensed muscles when the ankle is inverted and the plantarflexed are fibularis muscles (fibularis longus, fibularis brevis, fibularis tertius), extensor digitorum longus, and extensor hallucis longus. Twisting the leg will transmit the force from the femur to cause an external rotatory torque of the tibia. The torque results the fibular displacing laterally at the same time as the tensed muscles pull the fibula anteriorly.[91]

Chinese Medicine Etiology and Pathology

Qi and Blood Stagnation

A direct blow or cut to the medial, lateral aspect of the knee or to the quadriceps tendon and the patellar tendon causes blockages to the flow of the local Qi and blood in meridians almost immediately after the violence.

Liver and Kidney Deficiency

Ligament injury is considered "Jin Shang 筋傷" or "traumatic injury on tendons." The definition of "tendon" is not just limited to tendons. Traditionally, "tendon" in Chinese medicine includes muscle and all of its sounding soft tissues. Therefore, injury to the tendons is the trauma of muscle, ligament, cartilage, meniscus, intervertebral discs and their outer fibrous rings, blood vessel, nerve, tendon and its sheath, bursa, and joint capsule. *Treatise on the Engenderment of the Five Viscera* 《素問·五藏生成篇第十》states: "All tendons belong to the joints 諸筋者, 皆屬於節" (Unschuld and Tessenow 2011).

In the theory of traditional Chinese medicine (TCM), the liver governs the tendons. *Treatise on the Engenderment of the Five Viscera*, chapter 10 《素問·五藏生成篇第十》states:

> The Liver stores the Blood and the Heart moves it. When a person is active, Blood moves through all the meridians to Tendons; when the body is in a state of rest, the Blood returns and stores in the Liver. Why is this so? Because the Liver governs the sea of Blood. 肝藏血，心行之，人動則血運于諸經, 人靜則血歸于肝藏。何者？肝主血海故也。...When a person is at the state of rest, the Blood returns to the liver. The liver receives the Blood and the person is able to see. The feet receive the Blood and are able to walk. The palms receive the Blood and are able to grasp. The fingers receive the Blood and are able to hold 人臥血歸於肝，肝受血而能視，足受血而能步，掌受血而能握，指受血而能攝。.
>
> **(Unschuld and Tessenow 2011)**

When injury occurs on tendons, it consumes liver blood. As the trauma becomes chronic, the blood is insufficient. Liver's element is wood, which needs water nourishment from kidney. Deficiency of liver continually demands the sustentation of kidney, and kidney becomes deficient.

Manifestations

- Acute dislocation may cause swelling at the lateral aspect of the knee.
- Acute pain at the lateral side of the knee is aggravated by pressure over the fibular capitulum. Ankle movement may exacerbate the pain as well. Patients usually may not feel the knee pain and instability after the reduction.
- There may be transient peroneal nerve palsy on the ipsilateral leg, more common in posteromedial dislocation.
- The knee cannot bear weight.
- The knee is unstable.
- Patients with the type of subluxation may suffer from acute lateral knee pain under unknown factors; the pain may be exacerbated when pressing the fibular capitulum.
- Crepitus.
- Limited knee extension.
- The fibular capitulum has a floating sensation with pressure over the fibular head.
- Visual deformity on the fibular capitulum.
- Knee locking.
- Knee popping.

Physical Examination

- Begin by obtaining the patient's medical history:
  - Injury history such as the direction of the violent force from a motor vehicle accident or direct impact on the lateral knee when riding a horse.
  - History of dislocation of fibula, pain, swelling, and habitual dislocation.
- Inspection:
  - Knee effusion at the lateral side.
  - Surrounding skin for the signs of direct trauma.
  - Hemarthrosis, hematoma, and ecchymosis.
  - Difficulty standing, there is more pain when bearing weight on the affected leg.
  - The fibula has a floating sensation when pressing.
  - Abnormal location of the fibula.
- Palpation:
  - Abnormal location of the fibula.
  - Pain around the patella.
- Range of motion:
  - There is limited knee extension.
- Orthopedic tests:
  - For patients with chronic instability, the examination procedure consists of the following: The patient flexes the knee at 90 degrees; the examiner holds the medial side of the knee with one hand, and the thumb and index fingers of the other hand push back and forth the fibula head. If the symptoms of the patient are induced or aggravated, this suggests instability of the tibiofibular joint.

Differentiasl Diagnosis

Meniscus Tear

- The knee swelling usually takes a few hours to develop, and it can be intermittent.
- Pain is in the knee joint, not in the location of proximal tibiofibular joint.

Lateral Collateral Ligament Injury

- Varus Stress Test is positive.

Treatment

Treatment principle:

Treatment for luxation of the fibular capitulum with Chinese medicine is good for stabilizing the joint, and it is extremely useful for habitual dislocation.

- o Moves Qi and blood, opens meridians, reduces swelling, and stops pain.
- o Tonifies liver and kidneys, strengthens tendons and bones.

Herbal Treatments

Herbal formula:

- Gu Sui Bu Tang 骨碎補湯.

Formula ingredients:

Gu Sui Bu 骨碎补 30 g, Shu Di Huang 熟地黄 25 g, Shan Zhu Yu 山茱萸 15 g, Niu Xi 牛膝 25 g, San Leng 三棱 15 g, E Zhu 莪术 15 g, Wu Jia Pi 五加皮 15 g, Ji Xue Teng 雞血藤 25 g, Huang Qi 黄耆 50 g.

Ingredient explanations:

- Gu Sui Bu 骨碎补 tonifies kidney Yang, strengthens tendons and bones, moves blood especially due to trauma.
- Shu Di Huang 熟地黄 nourishes blood; nourishes liver and kidney Yin.
- Shan Zhu Yu 山茱萸 nourishes liver and kidneys, strengthens kidney Yang.
- Niu Xi 牛膝 expels wind dampness, tonifies liver and kidneys, directs herbs downward.
- San Leng 三棱 moves stubborn blood stasis, regulates Qi, and stops pain.
- E Zhu 莪术 moves blood, regulates Qi, and stops pain.
- Wu Jia Pi 五加皮 expels wind and damp, nourishes and warms liver and kidneys, strengthens bones and tendons.
- Ji Xue Teng 雞血藤 moves and tonifies blood, relaxes tendons.
- Huang Qi 黄耆 tonifies Qi and blood and relieves numbness and pain.

Topical Application

Herbal formula:

- Xiao Yu Gao 消瘀膏.

Formula ingredients:

- Da Huang 大黄 250 g, Zhi Zi 栀子 50 g, Mu Gua 木瓜 100 g, Pu Gong Ying 蒲公英 100 g, Jiang Huang 薑黄 100 g, Huang Bai 黄柏 150 g.

Ingredient explanations:

- Da Huang 大黄 moves blood and disperses bruising.
- Zhi Zi 栀子 clears heat, cools blood, resolves damp, reduces swelling.
- Mu Gua 木瓜 promotes Qi and blood circulation and unblocks the channels, releases spasm.
- Pu Gong Ying 蒲公英 reduces abscesses, clears heat in blood, resolves dampness.

- Jiang Huang 薑黃 moves blood, expels wind, reduces swelling, stops pain.
- Huang Bai 黄柏 clears deficient kidney Yin heat.

Application procedures:

- Grind the herbs to fine powder.
- Make a paste by mixing the powders with honey and water (2:1).
- Apply a certain amount of the paste on the proximal fibular capitulum over about 2 mm. The application region needs to be bigger than the affected area.
- Cover the region with a gauze.
- Change the gauze and the paste daily

Formula action: Cools blood and reduces inflammation and swelling.

Indications: Contusion injury, joint dislocation.

Acupuncture Treatments

Acupuncture point formula:

GB34 陽陵泉 Yanglingquan, LIV2 行間 Xingjian, GB39 懸鐘 Xuanzhong, LIV8 曲泉 Ququan, KD3 太溪 Taixi, KD6 照海 Zhaohai, SP10 血海 Xuehai, LI11 曲池 Quchi, LI10 手三里 Shousanli, LI4 合谷 He Gu, SP9 陰陵泉 Yinlingquan, SP6 三陰交 Sanyinjiao.

Electro-Acupuncture Treatment

GB34 陽陵泉 Yanglingquan ~ LI11 曲池 Quchi 2/100 Hz for 20 minutes.

Formula action: Moves Qi, tonifies liver and kidneys, clears heat, reduces swelling, and stops pain.

Treatment frequency: Daily.

Treatment duration: 30 minutes.

Cupping Treatments

GB34 陽陵泉 Yanglingquan, SP10 血海Xuehai.

Moxibustion Treatments

Warming needling techniques on GB34 陽陵泉 Yanglingquan, GB39 懸鐘 Xuanzhong.

Bleeding Treatments

GB34 陽陵泉 Yanglingquan, SP10 血海 Xuehai.

Tui Na Treatments

1) The patient lies supine with the knee in flexion at about 30 degrees; the therapist stands beside the affected knee.
2) The therapist places a thumb on the medial aspect of tibial tuberosity and places the rest four fingers at the lateral fibula capitulum. The other hand grasps the ipsilateral ankle.

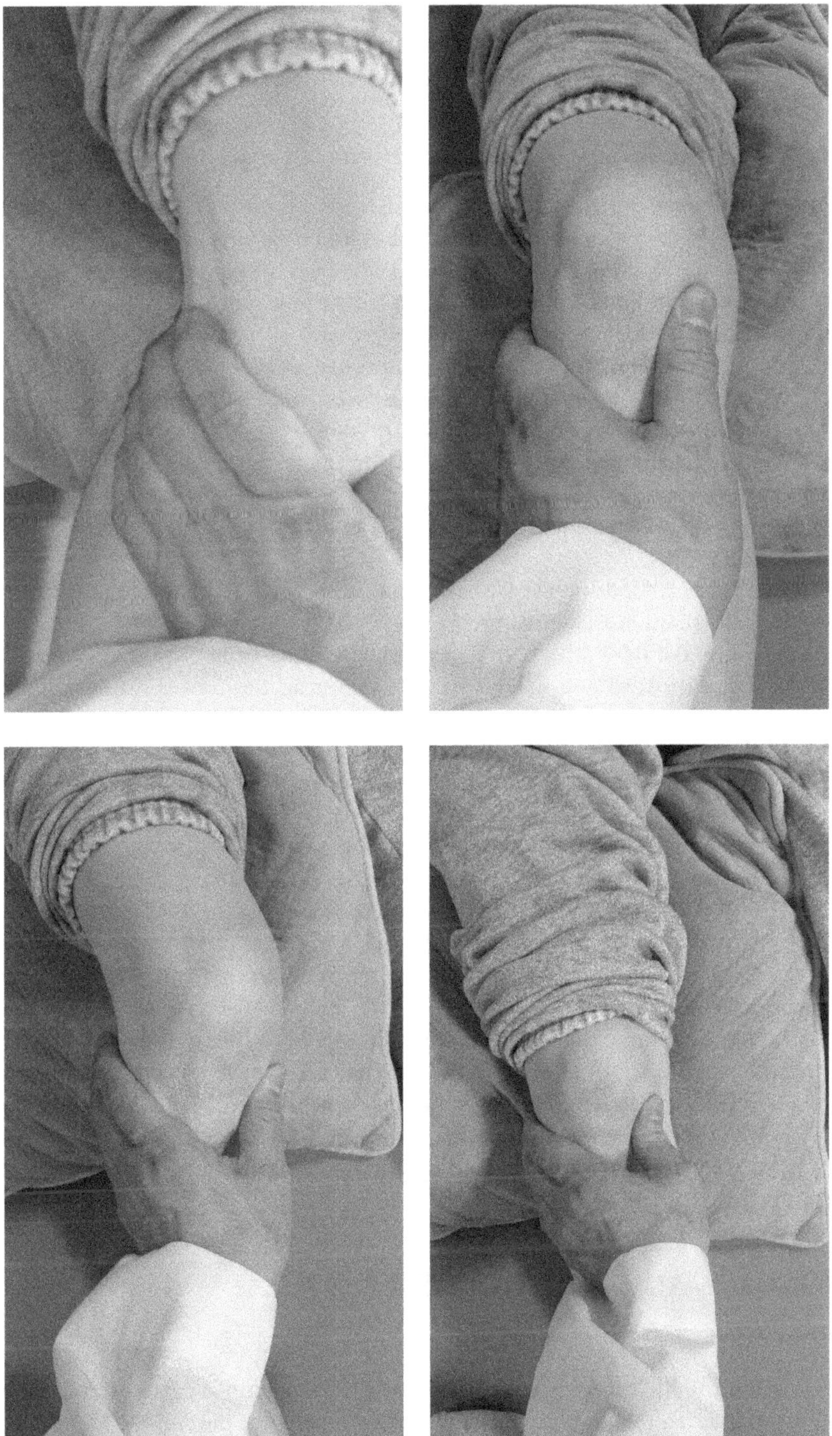

3) The therapist presses the fibula capitulum toward the tibiofibular articulation.
4) Meanwhile, the therapist lifts the ankle and flexes and extends the knee with rotating the tibia in eversion and inversion a few times.

Patient Advisory

Before returning to sports, these are a few self-care reminders to enhance the healing and reduce the risk of sustaining a proximal tibiofibular joint injury:

- Always warm-up and stretch the knee before any sports activity.
- Training the leg muscles prevents knee injury.
- Selecting adequate footwear during sport also prevents knee strain.
- Avoid suddenly increasing or reducing speed while running.
- Avoid sudden change of direction while running.
- Avoid sudden change of sports training frequency and intensity.
- Apply RICE procedures whenever there is a strain to the knee.
- Seek medical service as soon as there is an injury to the proximal tibiofibular joint.
- Exercise will focus on the hamstring muscles, and the biceps femoris muscle is the main muscle to stretch:
  a) Sit on the floor with both legs extended in the front.
  b) Lean forward and extend the arms to grab the toes. Ensure that the torso is straight, the spine is straight, and the neck is straight along with the spine. Also ensure that both knees are not flexed. If the hands are not able to reach the toes, hook a bath towel around the front part of the soles.
  c) The hands hold both ends of the towel and pull the towel to stretch the feet in dorsal flexion until the tension in the hamstring is felt. The torso still needs to lean forward; the towel is just an extension of the hands to stretch the feet. Hold for five seconds and relax.
  d) Practice this exercise for a set of ten repetitions to stretch the bicep femoris muscles.
- Dietary recommendations: Patients should be advised to avoid Fa Wu 發物 diet. Fa Wu foods are the foods that may cause more inflammation and bring old injuries back. Some examples of Fa Wu 發物 foods are shellfish such as crab and shrimp, fish, fried foods, lamb, duck, chocolate, garlic, and green onions. This concept applies to both acute and chronic injuries.

CHONDROMALACIA PATELLAE 髕骨軟骨軟化症

Introduction

This is often called "runner's knee," because the knee pain is often associated with running sports. It's the number one cause of anterior knee pain in the United States with an incidence as high as one in four people (Laprade et al. 1998). The word chondromalacia is derived from the Greek words chrondros, meaning cartilage, and malakia, meaning softening. Chondromalacia patellae means a softening of the cartilage on the posterior surface of the patella which may eventually lead to fibrillation, fissuring, and erosion.

Patients often report that the pain is in the front part of the knee. The pain is caused by an irritation of the patella cartilage that rubs over one side of the knee when running. The smooth cartilage is a natural shock absorber that normally glides effortlessly over the knee during running. With the malalignment issues, the patella cartilage is laterally aligned in the femoral groove and rubbed against one side of the knee joint that irritates the surface of the cartilage, resulting in pain. This is commonly seen in young otherwise healthy athletes. Runner's knee is commonly seen in football players, rowers, tennis players, cyclists, gymnasts, equestrians, volleyball players, skiers, snowboarders, and runners. These sports involve running, jumping, landing, and squatting.

However, this knee pain also occurs in non-sporting persons. The chronic friction between the patellar cartilage and the femoral groove can cause increased deterioration, breakdown of the cartilage, and irritatation to the surface of the cartilage, resulting in anterior knee pain. Chronic friction is often seen in people who often walk up and down stairs.

Anatomy

The knee is comprised of four major bones, the femur, tibia, fibula, and patella. The patella protects the knee and articulates with the femur at the femoral groove. The patella articular cartilage covers the ends of the femur, femoral groove, and underside of the patella. It allows the patella to glide smoothly over the femoral groove during the flexion and extension of the leg.

The quadriceps femoris consists of four muscles: the rectus femoris (RF), vastus lateralis (VL), vastus intermedius (VI), and vastus medialis (VM). The VM has oblique fibers which are referred to as the vastus medialis obliques (VMO). The VMO does not extend the knee but keeps the patella in the center of the femoral groove.

Western Medicine Etiology and Pathology

Chondromalacia patella may result from either acute injury to the patella, or chronic friction between the patella and the femoral groove where the patella articulate cartilage lies and moves during knee flexion and extension.

- Female gender: It often affects women more than men, because women receive greater lateral stress on the patella than men. Women have been shown to generally have wider pelvises. Wider hips provide a larger area for a fetus to develop during pregnancy, and greater structural femoral anteversion and tibiofemoral angle[92] causes an increase in the angle of the knee where the femur and tibia come together. This is called the Quadriceps Angle (Q-angle) (Jones 2013). The normal Q-angle is usually 14 degrees for males and 17 degrees for females (Jones 2013). The wider pelvis and increased Q-angle in females slant the femur inward.[93]

- The possible chronic conditions causing chondromalacia patella may include tight iliotibial band, overuse of knee, malalignment, core instability.
  - Abnormal Q-angle is the most significant causative factor; it may also cause tight iliotibial band and malalignment of the knee joint. The increased Q-angle can increase lateral pull on the patella.
  - Trauma such as patella fracture or dislocation may result in poor blood circulation and core instability.
  - Overuse of the knee joint especially during sports. The pain may not be from an injury history but under certain circumstances an easy sport activity may start the episode, such as riding a bicycle over a short distance compared to the longer distance that the patient often cycles. For example, the patient usually rides a bicycle more than 20 miles, but the knee pain occurs when only riding 10 miles in cold, rainy, and windy weather because the patient just recovered from a flu.

Chinese Medicine Etiology and Pathology

Qi and Blood Stagnation

The knee condition is in the category of Bi syndrome. Injury causes blockages to the flow of the local Qi and blood in meridians, resulting in Qi and blood stagnation.

Qi and Blood Deficiency Phlegm-Damp Accumulation

1) *Qi and blood deficiency*: With spleen Qi deficiency, the Zang-Fu organs lack Qi and blood, tendons, bones, and joints lack Qi and blood nourishment, resulting in insufficiency of Qi and blood in other Zang-Fu organs.
2) *Phlegm-damp accumulation*: With spleen Qi deficiency, the functions of transformation and transportation are poor, damp and phlegm accumulate, tendons are insidious.

Liver and Kidney Deficiency

1) Liver and kidney supply the process of trauma repair. Liver blood is necessary for repairing the injury occurring on any tendon because the liver governs tendons according to traditional Chinese medicine. As the injury becomes chronic, an increase in demand for liver blood for the injury will generally cause liver blood insufficiency, as the body is unable to supply the need soon enough.
2) Kidney is the mother of the liver, according to traditional Chinese medicine. The element of kidney is water, and the element of liver is wood. The growing of wood relies on water; the producing of liver blood and Yin relies on kidney Qi. Prolonged illness leads to consumption of Qi and blood and

deficiency of the liver and kidney. The high demand for liver blood and Yin debilitates kidney Qi. A sufficient amount of Qi and blood is required to maintain the balance of Yin and Yang in the body. The long-term consumption of kidney Qi due to chronic physical trauma results in kidney Yang deficiency.

Bi Syndrome

A person with a Yang-deficienct constitution is susceptible to the effects of wind, cold, and damp invasion and retention. Prolonged retention of pathogenic agents and climatic variations and environmental change result in stagnation of Qi and blood, ecchymosis, subcutaneous and periarticular nodules, swelling, deformity, and limitation of movement of joints. This is called Bi syndrome of wind-cold-dampness type.

Manifestations

General manifestations:

- Anterior knee tenderness on palpation under the medial or lateral border of the patella.
- Dull pain in the knee that worsens after a prolonged period sitting such as sitting while watching a movie or driving long distance.
- Knee pain after or during common daily activities such as kneeling, climbing up or downstairs, fast walking or running, or squatting.
- Knee grinding sensation.
- Crepitation in flexion or extension.[94]
- Minor swelling.[95]

Manifestations in Traditional Chinese Medicine

It is mainly divided into three types.

Qi and Blood Stagnation Type

Anterior knee pain aggravated when going up and downstairs and squatting. The tongue color is purple with bruise marks. The pulse is wiry, tight, and choppy.

Phlegm-Damp and Bi Type

Knee joints are sore and give way; discomfort and dull pain gradually increase, and the pain is not fixed. Pain presents when going up and down the stairs and squatting, local swelling, patellar tap is positive. Lassitude, poor appetite. The tongue is pale and flabby, the coating is pale and greasy, and the pulse is wiry and slippery. This type is due to over-exertion of knee activity, repeated sprains, or accumulated strain involving joint synovial membrane, fat pad, and patellar ligament, resulting in hemarthrosis, effusion, and hypertrophy.

Tongue color is red, or pale red with a greasy, yellow coat. Pulse is rapid and tight or slippery.

Liver and Kidney Deficiency Type

Knee weakness. Pain presents when going up and down the stairs, knees give way, knees "lock," tenderness on infrapatellar fat pad, quadriceps muscle atrophy. The tongue body is pale, coating is thin white, and the pulse is weak.

This type is commonly seen in older patients because kidney Qi is declining, kidney essence is insufficient, and the bones lack nourishment. These factors along with external wind invasion, over-exertion, or trauma lead to Qi and blood stagnation blocking meridians, resulting in weakness of bones and joints. Tongue is pale, and may be swollen with thin, white, and moist coating. Pulse is deep, thready, and/or weak and slow.

Physical Examination

Physical examination of the patellar tendonitis:

- Begin by obtaining the patient's medical history:
  - Injury history to the patella.
  - Career or sports activity that may be factors of the chondromalacia patellae such as sports that involve running, jumping, landing, squatting.
- Inspection:
  - Signs of mild swelling.
  - The patella may not be aligned with the femur.
- Palpation:
  - Applying pressure to one or other margin of the patella when displacing the patella either medially or laterally may elicit pain and crepitus.
  - A grating sensation underneath the patella when the patella is compressed against the femur either vertically or horizontally with the knee in full extension.
- Range of motion:
  - Passive movements are usually full and painless, but repeated extension of the knee from flexion will produce pain.
  - Patellar grind test or Clarke's sign: This test detects the presence of patellofemoral joint disorder. A positive sign on this test is pain in the patellofemoral joint.

Differential Diagnosis

Osteoarthritis of the Knee

- The joint may become difficult to flex and extend. Chondromalacia patella has full range of extension.

Displacement of Patella

- Pain in high intensity especially when weight bearing until the reduction of the knee joint.
- Rapid medial swelling when it is acute lateral dislocation.
- Hemarthrosis of the knee.

Patellofemoral Pain Syndrome

- The knee is painful when flexed completely.

Patellar Tendonitis

- The pain may be more at night, and the patient may feel stiffness in the morning when getting up.
- It may manifest as red, swollen, and warm to touch.
- The affected tendon may appear thickened in comparison to the unaffected side.

Treatment

Treatment principle:

- Moves Qi and blood, tonifies Qi and blood, drains damp, and tonifies liver and kidneys.

Herbal Treatments

Qi and Blood Stagnation Type

Herbal formula:

Tao Hong Si Wu Tang + Si Miao Yong An Tang Jia Jian 桃紅四物湯合四秒勇安湯加減.

Formula ingredients:

- Tao Ren 桃仁 15 g, Hong Hua 紅花 12 g, Dang Gui 當歸 30 g, Chuan Xiong 川芎 10 g, Niu Xi 牛膝 10 g, Huang Qi 黃芪 40 g, Gui Zhi 桂枝 15 g, Jin Ying Hua 銀花 30 g, Xuan Shen 玄参 20 g, Mu Tong 木通 15 g, Chi Shao 赤芍 15 g, Gan Cao 甘草 6 g.

Ingredient explanations:

- Tao Ren 桃仁 breaks up stubborn blood stasis.
- Hong Hua 紅花 moves blood, releases bruising, opens meridians, stops pain.
- Dang Gui 當歸 moves and tonifies blood, disperses cold, and stops pain due to blood stasis.
- Chuan Xiong 川芎 moves blood and Qi, expels wind, and stops pain.
- Niu Xi 牛膝 expels wind-dampness, tonifies liver and kidneys, directs herbs downward.
- Huang Qi 黃芪 strengthens spleen, reduces edema.
- Gui Zhi 桂枝 releases the exterior, warms meridians, and stops pain.
- Jin Ying Hua 銀花 clears heat, disperses wind heat, clears damp heat, cools blood.
- Xuan Shen 玄参 clears heat, cools blood, nourishes Yin.
- Mu Tong 木通 drains heat, promotes urination.
- Chi Shao 赤芍 moves blood, reduces swelling, stops pain.

- Gan Cao 甘草 releases spasms, alleviates pain, harmonizes the other herbs, and guides the herbs to all 12 meridians.

Formula action:

- Cools and moves blood, reduces swelling, stops pain.

Qi and Blood Deficiency Phlegm-Damp Accumulation
Herbal formula:

- Si Miao San Jia Jian 四妙散加減.

Granule formula ingredients:

- Huang Bai 黃柏 10 g, Yi Yi Ren 薏苡仁 20 g, Cang Zhu 蒼朮 10 g, Chuan Niu Xi 川牛膝 9 g, Bai Shao 白芍 6 g, Dang Gui 當歸 6 g, Mu Gua 木瓜 6 g, Bi Xia 萆薢 9 g, Xi Xian Cao 豨薟草 6 g, Gan Cao 甘草 6 g.

Ingredient explanations:

- Huang Bai 黃柏 drains damp heat, especially from the lower Jiao, drains Fire, and relieves Fire toxicity.
  - With Cang Zhu and Niu Xi, for damp heat leg Qi.
  - With Cang Zhu, it eliminates dampness and clears heat for damp heat sores.
- Yi Yi Ren 薏苡仁 promotes urination, leaches out dampness, strengthens the spleen, and clears damp heat.
  - With Huang Bai, Cang Zhu, and Niu Xi, for damp heat pouring downward, leading to atrophy and weakness of the lower extremities.
- Cang Zhu 蒼朮 strongly dries dampness, tonifies the spleen, and clears dampness from the lower Jiao.
  - With Huang Bai, for pain, swelling, and weakness of the lower extremities or skin lesions due to wind-dampness.
- Chuan Niu Xi 川牛膝 moves the blood, expels wind, drains dampness, tonifies the kidneys and liver, and strengthens the tendons and bones.
  - With Huang Bai, Cang Zhu, and Yi Yi Ren, for red, swollen, and painful joints of the lower extremities from damp heat pouring downward.
- Bai Shao 白芍 nourishes blood, preserves Yin, calms liver Yang and liver Wind, softens tendons, and alleviates spasm and pain.
- Dang Gui 當歸 moves and tonifies blood, disperses cold, and stops pain due to blood stasis.
- Mu Gua 木瓜 promotes Qi and blood circulation and unblocks the channels, releases spasm.
- Bi Xia 萆薢 expels wind-dampness, relaxes the sinews, and unblocks the Luo meridians.
- Xi Xian Cao 豨莶草 expels wind-dampness, opens meridians, clears heat.
- Gan Cao 甘草 releases spasms, alleviates pain, harmonizes the other herbs, and guides the herbs to all 12 meridians.

Dosage:

- 70 grams—Take five grams dissolved in warm water, twice a day.

Formula actions:

- Clears heat, drains damp, moves blood.

General indications:

- Atrophy, numbness, and weakness of lower limbs.
- Clears heat and dries dampness in the lower body.

Precautions:

- Use with caution in conditions due to Yin deficiency.
- Not suitable for arthritis pain with chilly sensation.

Caution:

- Not for use during pregnancy.

Liver and Kidney Deficiency

Herbal formula:

- Shen Qi Wan Jia Jian 肾氣丸加減.

Granule formula ingredients:

- Shu Di Huang 熟地黄 24 g, Shan Yao 淮山 12 g, Shan Zhu Yu 山茱萸 10 g, Fu Ling 茯苓 6 g, Dan Pi 丹皮 6 g, Ze Xie 澤瀉 6 g, Zhi Fu Zi 制附子 3 g, Rou Gui 肉桂 3 g.

Ingredient explanations:

- Shu Di Huang 熟地黄 nourishes blood, nourishes liver and kidney Yin.
- Shan Yao 淮山 tonifies kidneys, benefits the Yin and the spleen Qi.
- Shan Zhu Yu 山茱萸 nourishes liver and kidneys, strengthens kidney Yang.
- Fu Ling 茯苓drains damp, tonifies spleen.
- Mu Dan Pi 丹皮 clears deficient heat and cools blood, moves blood, drains pus, and reduces swelling.
- Ze Xie 澤瀉 drains damp, especially damp heat in lower Jiao, clears kidney deficient heat.
- Zhi Fu Zi 制附子 warms Ming Men Fire.
- Rou Gui 肉桂 warms the kidney Yang and Ming Men Fire and assists the generation of Qi and blood.

Dosage:

- 70 grams—Take five grams dissolved in warm water, twice a day.

Formula actions:

- Warms and tonifies kidney Yang, Atrophy disorder (Wei Syndrome) due to liver and kidney deficiency.

General indications:

- Atrophy, numbness, and weakness of lower limbs.

Precautions:

- Use with caution in conditions due to Yin deficiency with dry mouth and throat and a red tongue with no coat.
- Use with caution for those with gastrointestinal weakness.

Caution:

- Not for use during pregnancy.

Acupuncture Treatments

Acupuncture point formula:

- "Five Tiger Catching Sheep" Protocol 五虎擒羊針法.

Procedure:

1) The patient lies supine with the affected knee flexed at 30 degrees.
2) Insert needles to acupuncture points ST35 犊鼻 Dubi, Nei Xi Yan 膝眼 (in the depression on both sides of the patellar ligament when the kneels flexed), He Ding 鶴頂 (above the knee, in the depression of the midpoint of the upper border of the patella), Ashi Points 阿是穴, SP10 血海 Xuehai, and ST34 梁丘 Liangqiu.
3) For chronic condition, add acupuncture points SP9 肩貞Jianzhen and GB34 陽陵泉 Yanglingquan.
4) Attach each end of the needle with a 1" moxibustion cone and burn total two to three cones.

Cupping Treatment

He Ding 鶴頂 (above the knee, in the depression of the midpoint of the upper border of the patella).

Moxibustion Treatment

Warming needling techniques on Nei Xi Yan 膝眼 (in the depression on both sides of the patellar ligament when the kneels flexed), ST35 犊鼻 Dubi.

Bleeding Treatment

Ashi Points 阿是穴 (a tender point that is not included in the classic acupuncture points).

Electro-Acupuncture Treatment

SP10 血海 Xuehai, He Ding 鶴頂 2/100 Hz for 20 minutes daily
Formula action: Moves Qi, tonifies liver and kidneys.
Treatment frequency: Daily.
Treatment duration: 30 minutes.

Tui Na Treatments

Tui Na techniques:

- Grasping, lifting, kneading, pressing, "holding a globe"

Tui Na procedure:

1) The patient lies supine with the affected knee in flexion at 30°; the therapist stands beside the knee.
2) The therapist applies grasping and lifting techniques on the distal quadriceps muscle 30 times.
3) The therapist applies grasping and kneading techniques on the distal quadriceps muscle 30 times.

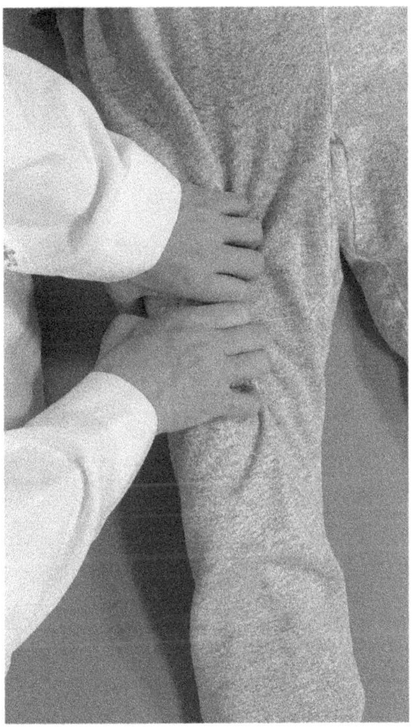

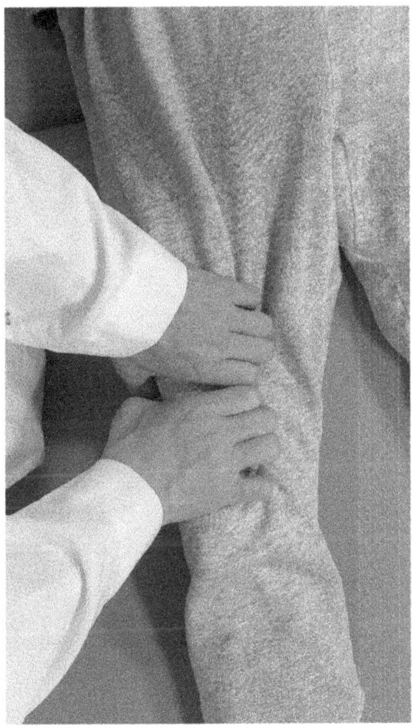

4) The therapist applies kneading and pressing techniques on the distal quadriceps muscle 30 times.
5) The therapist applies "holding a globe" techniques on the affected knee until the skin is warm.
   a) The therapist interlocks the fingers.
   b) Place both palms on the medial and lateral aspects of the affected knee.
   c) Apply rubbing technique with light pressure.

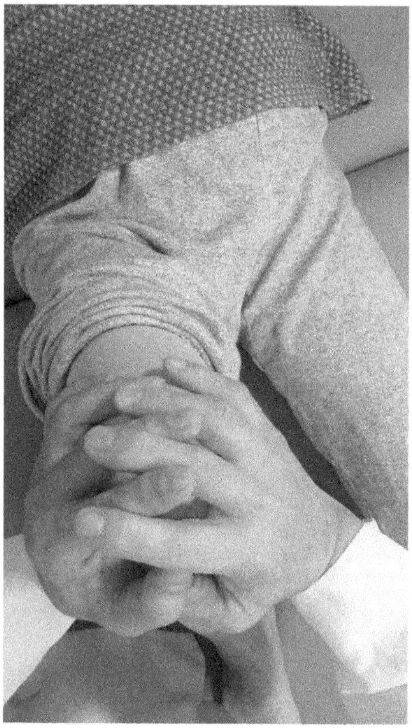

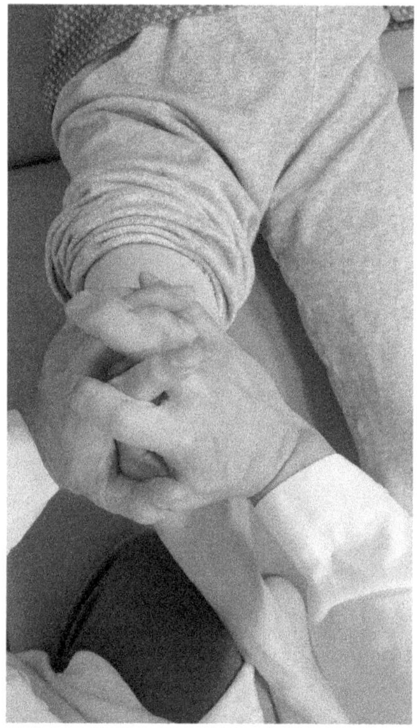

6) After the procedure of "holding a globe," follow by pressing with both palms on both medial and lateral aspects of the knee 30 times.
7) The therapist applies kneading technique on the edges surrounding the patella until warm or red.

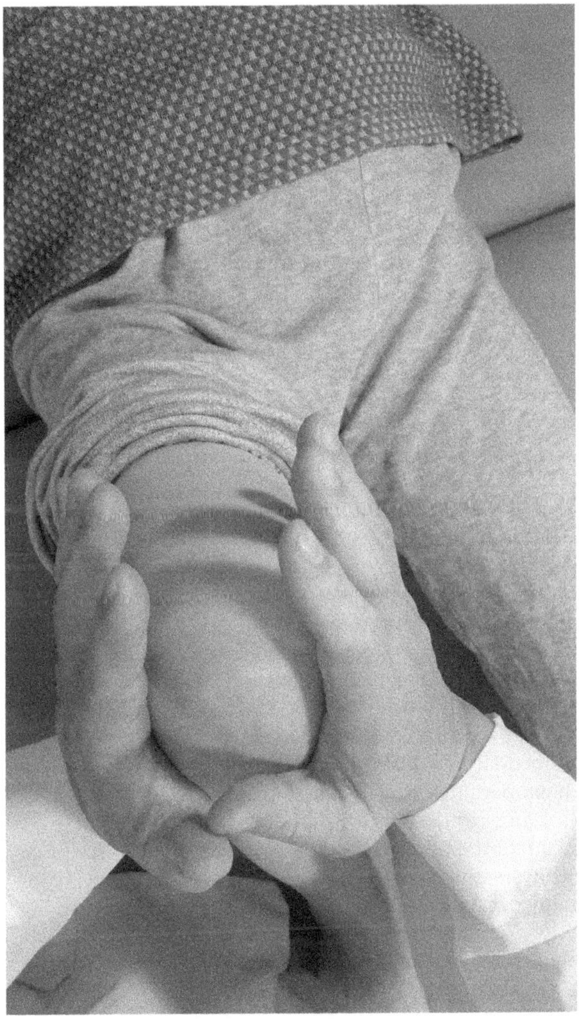

Patient Advisory

- For several weeks when the treatments begin, the patient should avoid activities that cause knee pain such as running, long-distance walking, mountain hiking, excessive stair use, squatting, and kneeling, especially kneeling on a hard surface. The activities can be resumed and intensity can be increased following the medical provider's recommendations

- Non-weight-bearing exercises, such as riding a bicycle and swimming, that can strengthen and increase the flexibility of the quadriceps and hamstring muscle groups may be recommended. The duration and frequency may increase per the recommendation by the medical provider or the athletic trainer; usually it is approximately 30 minutes a day, five times a week
- Lose weight if obese by avoiding food that is raw, cold, or deep-fried, reducing sugar and fat intake, and eating plenty of fresh vegetables, fruits, and whole grains, dairy products and bone broth.
- Keep the knee warm.
- Always warm up and stretch the knee before athletic activities.

NOTES

1. Marques_et_al-2019-Sports_Medicine.pdf
2. https://www.mayoclinic.org/diseases-conditions/acl-injury/symptoms-causes/syc-20350738
3. https://www.cedars-sinai.edu/Patients/Health-Conditions/Anterior-Cruciate-Ligament-ACL-Injuries.aspx?_ga=2.16980156.427522529.1571518092-106724449.1571518092
4. https://www.physio-pedia.com/Anterior_Cruciate_Ligament_(ACL)
5. https://www.physio-pedia.com/Anterior_Cruciate_Ligament_(ACL)
6. https://www.physio-pedia.com/Anterior_Cruciate_Ligament_(ACL)
7. https://orthoinfo.aaos.org/en/diseases--conditions/collateral-ligament-injuries/
8. https://www.ncbi.nlm.nih.gov/pmc/articles/PMC5420825/
9. https://www.physio-pedia.com/Posterior_Cruciate_Ligament_Injury
10. http://www.wheelessonline.com/ortho/anatomy_of_the_posterior_cruciate_ligament
11. https://en.wikipedia.org/wiki/Posterior_cruciate_ligament
12. https://now.aapmr.org/posterior-cruciate-ligament-pcl-injuries/ 已改良
13. https://now.aapmr.org/posterior-cruciate-ligament-pcl-injuries/
14. https://now.aapmr.org/posterior-cruciate-ligament-pcl-injuries/
15. https://www.physio-pedia.com/Posterior_Cruciate_Ligament_Injury
16. http://sterlingmedgroup.com/injuries-view/mcl-and-pcl-injuries/
17. https://medisavvy.com/reverse-lachmans-test/
18. http://www.gdhtcm.com/cbq/caibingqin/jdya/info_185.aspx?itemid=870
19. https://zhuanlan.zhihu.com/p/33648385
20. https://www.medicinenet.com/medial_collateral_ligament_mcl_injury/article.htm#what_tests_diagnose_and_assess_medial_collateral_ligament_mcl_injuries
21. https://physioworks.com.au/injuries-conditions-1/medial-collateral-ligament-sprain
22. https://www.healthline.com/health/hemarthrosis#symptoms
23. https://www.physio-pedia.com/Pes_Anserinus_Bursitis
24. http://www.gdhtcm.com/cbq/caibingqin/jdya/info_185.aspx?itemid=870
25. https://www.healthline.com/health/lateral-collateral-ligament-lcl-injury#outlook
26. https://www.healthline.com/health/hemarthrosis#symptoms
27. https://www.ucsfhealth.org/conditions/lcl_tear/
28. http://www.gdhtcm.com/cbq/caibingqin/jdya/info_185.aspx?itemid=870
29. https://orthoinfo.aaos.org/en/diseases--conditions/meniscus-tears/
30. https://www.medicalnewstoday.com/articles/299204.php
31. https://www.ncbi.nlm.nih.gov/pubmed/28613721
32. https://www.ncbi.nlm.nih.gov/pubmed/28613721
33. https://www.medicalnewstoday.com/articles/299204.php

34. https://www.healthline.com/health/meniscus-tears#outlook
35. https://www.medicinenet.com/pain_management_knee_pain/article.htm
36. https://www.ncbi.nlm.nih.gov/pmc/articles/PMC2384263/
37. https://www.chiroweb.com/hg/10/14/27.html
38. https://www.chiroweb.com/hg/10/14/27.html
39. https://www.physio-pedia.com/Steinman_Test
40. https://www.chiroweb.com/hg/10/14/27.html
41. https://www.verywellhealth.com/eges-test-detecting-a-meniscus-tear-2549648
42. https://baike.baidu.com/item/%E5%8D%8A%E6%9C%88%E6%9D%BF%E6%97%8B%E8%BD%AC%E8%AF%95%E9%AA%8C
43. https://www.mdedge.com/surgery/article/198043/knee/snapping-biceps-femoris-tendon/page/0/1
44. https://www.mayoclinic.org/diseases-conditions/patellar-tendinitis/symptoms-causes/syc-20376113
45. http://blog.sina.com.cn/s/blog_67b17fd20102v1ys.html
46. https://www.jd.com/phb/zhishi/90c73752eaf351e2.html
47. Whole eggs are high in protein, but egg whites are almost pure protein.
48. https://www.jointpreservationinstitute.com/patella-tendon-rupture-amir-jamali-walnut-creek-sacramento-california.html
49. https://www.kenhub.com/en/library/anatomy/patellar-tendon
50. https://physioworks.com.au/injuries-conditions-1/patella-tendonitis-tendinopathy
51. https://www.jointpreservationinstitute.com/patella-tendon-rupture-amir-jamali-walnut-creek-sacramento-california.html
52. https://orthopedicspecialistsofseattle.com/physicians-staff/dr-grant-h-garcia/patellar-tendon-rupture/
53. https://orthoinfo.aaos.org/en/diseases--conditions/patellar-tendon-tear/
54. https://www.reboundmd.com/news/patellar-tendon-tears-athletes
55. https://www.verywellhealth.com/treatment-of-a-patellar-tendon-tear-2549591
56. https://www.verywellhealth.com/treatment-of-a-patellar-tendon-tear-2549591
57. https://orthoinfo.aaos.org/en/diseases--conditions/patellar-tendon-tear/
58. https://intermountainhealthcare.org/services/orthopedics/services/knee/patellar-tendon-rupture/
59. https://intermountainhealthcare.org/services/orthopedics/services/knee/patellar-tendon-rupture/
60. https://orthopedicspecialistsofseattle.com/physicians-staff/dr-grant-h-garcia/patellar-tendon-rupture/
61. https://www.ncbi.nlm.nih.gov/books/NBK513330/
62. https://www.ncbi.nlm.nih.gov/books/NBK513330/
63. https://www.ncbi.nlm.nih.gov/books/NBK513334/
64. https://www.ncbi.nlm.nih.gov/books/NBK482389/
65. http://dlv.org.rs/wp-content/uploads/2015/05/MedicinskiPregled3-4-2013comp.pdf#page=10
66. https://www.clinicalpainadvisor.com/home/decision-support-in-medicine/anesthesiology/repair-of-ruptured-quadriceps/
67. https://www.verywellhealth.com/treatment-of-a-patellar-tendon-tear-2549591
68. https://intermountainhealthcare.org/services/orthopedics/services/knee/patellar-tendon-rupture/
69. https://www.physio-pedia.com/Quadriceps_tendon_tear
70. https://www.youtube.com/watch?v=KEMj1oYGFEg
71. https://www.physio-pedia.com/Quadriceps_tendon_tear
72. https://www.ncbi.nlm.nih.gov/books/NBK513330/

73. https://www.ncbi.nlm.nih.gov/books/NBK513330/
74. https://teachmeanatomy.info/lower-limb/bones/patella/
75. https://health.uconn.edu/orthopedics-sports-medicine/conditions-and-treatments/where-does-it-hurt/knee/patellar-dislocation/
76. https://radiopaedia.org/articles/lateral-patellar-dislocation?lang=us
77. https://www.hss.edu/conditions_patellofemoral-kneecap-instability.asp
78. https://www.kneesurgeon.com.au/patella-trochlear-dysplasia.html
79. https://www.physio-pedia.com/Patellar_dislocation
80. https://baike.baidu.com/item/急性髌骨脱位
81. https://www.youtube.com/watch?v=xXmjYVDkmVg
82. https://www.physio-pedia.com/Patellar_dislocation
83. https://www.ncbi.nlm.nih.gov/books/NBK513330/
84. https://www.ncbi.nlm.nih.gov/books/NBK513330/
85. https://www.uofmhealth.org/health-library/abn2411
86. https://www.jd.com/phb/zhishi/90c73752eaf351e2.html
87. https://www.health.harvard.edu/staying-healthy/the-best-foods-for-vitamins-and-minerals Whole eggs are high in protein, but egg whites are almost pure protein.
88. https://www.physio-pedia.com/Fibula
89. https://drrobertlaprademd.com/lcl-fcl-injury-fibular-collateral-ligament-tear-colorado/
90. https://www.researchgate.net/publication/5370041_Dislocation_of_the_fibular_head_in_an_unusual_sports_injury_A_case_report
91. https://www.ncbi.nlm.nih.gov/pmc/articles/PMC2881465/
92. https://books.google.com/books?id=eYZEBAAAQBAJ&pg=PA459&lpg=PA459&dq=The+wider+pelvis+and+an+increased+Q-angle+in+females+slant+the+femur+inwardly&source=bl&ots=jq3Xi0wHbz&sig=ACfU3U3SH3u7X_EI1rxYeG5ACBUd97P0mw&hl=en&sa=X&ved=2ahUKEwi31aGx6uvmAhVBip4KHQkgCHIQ6AEwDnoECAkQAQ#v=onepage&q=The%20wider%20pelvis%20and%20an%20increased%20Q-angle%20in%20females%20slant%20the%20femur%20inwardly&f=false page 459.
93. https://www.physio-pedia.com/Chondromalacia_Patellae#cite_note-Shultz-32
94. https://www.physio-pedia.com/Chondromalacia_Patellae#cite_note-Shultz-32

6 Leg Pain

HEMIPLEGIA 半身不遂

INTRODUCTION

The definition of hemiplegia is loss of strength in the arm, leg, and face on one side of the body caused by a brain injury.

If this condition injury occurs before, during, or within two years after birth, it is called congenital hemiplegia.

If this brain condition occurs later in life as a result of a stroke; head injury; injury to the spinal cord from violence, such as sports trauma, motor vehicle accidents, gunshot wounds, or knife wounds; brain infection; or illness such as a tumor, it is called acquired hemiplegia.[1]

Fifteen million people worldwide suffer from stroke annually, five million die, and another five million are left permanently disabled (Zhang et al. 2009).

Traditional Chinese medicine has been the main therapy for hemiplegia in Asia. Acupuncture has been shown to promote cell proliferation in tissue damaged by ischemia (Chen et al. 2015).

Clinical trials have shown that acupuncture can enhance balance, reduce spasticity, and increase muscle strength (Liu et al. 2009; Zhao et al. 2009; Yan and Hui-Chan 2009).

ANATOMY

The pyramidal tracts are comprised of the corticospinal tract and the corticobulbar tract. Because their nerve fibers are mainly derived from pyramidal cells in the cerebral cortex, they are called pyramidal tracts. After leaving the cerebral cortex, the pyramidal tract moves through the inner capsule and the cerebra peduncle to the medulla. Most of the nerve fibers cross over at the caudal end of the medulla to the contralateral side of the body, which is referred to as decussation of the pyramids (Rea 2015), enter the spinal cord (corticospinal) or the brain stem (corticobulbar), and control the motor function of the body.

The corticospinal tract is the largest descending tract present in humans[2] and is comprised of a lateral (85%) and an anterior (ventral) tract (15%) (Rea 2015). The lateral corticospinal tract sends fibers predominantly to the muscles of the extremities to control the voluntary movement of the limbs, and the cortical innervation is contralateral; in other words, the left motor cortex controls the right extremities. When a stimulus is engaged, the cell body of the lateral corticospinal tract will send an impulse through the tract that travels to the anterior horn of the spinal cord, proceeds through the spinal nerve root, plexus, and peripheral nerve, and is then transmitted via the lower motor neurons into the muscle fibers, resulting in contraction of the limb muscles. The anterior corticospinal tract sends fibers mainly to the trunk or axial muscles. The control is both ipsilateral and contralateral. Therefore, the trunk muscles are generally bilaterally cortically innervated.[3]

Because of the pyramidal decussation of the lateral corticospinal tract in the caudal medulla, injury to the pyramidal tract above the medulla generally causes contralateral hemiparesis; for example, following injury to the left cerebral cortex, the hemiplegia symptoms will present on the left side of the body.[4] Injuries to the lower medulla, spinal cord, and peripheral nerves result in ipsilateral hemiparesis; for example, following injury to the left anterior horn cells, clinical findings such as decreased deep tendon reflexes and muscle wasting will present on the left side of the body.

Western Medicine Etiology and Pathology

Generally speaking, the affected lower limb actually is healthy even though the symptoms and signs of hemiplegia manifest on the leg. The main causative factor resides in the brain or the nerves, which are unable to produce, send, and interpret signals due to traumatic injuries or health conditions.

- Spinal cord injury. Hemiplegia can result from an incomplete spinal cord injury, such as motor vehicle accidents, falls, gunshot wounds, and knife wounds that cause injury to the nerves.
- Cardiovascular problems such as aneurysms and hemorrhages in the brain.
- Brain infections, such as encephalitis from viral infection and meningitis from viral or bacterial infection, or infections elsewhere spreading to the brain, such as sepsis and abscesses in the neck.
- Brain injuries, such as motor vehicle accidents and falls that directly impact the head.
- Brain tumor.
- Mini-stroke, such as ischemic stroke from thrombus or embolus or cerebral hypoperfusion.
- Hemorrhagic stroke or cerebral vascular accident.
- Demyelination of the brain, such as multiple sclerosis.
- Reactions to surgery, medication, or anesthesia.
- Loss of oxygen supply to the brain, such as choking or anaphylactic shock.

Chinese Medicine Etiology and Pathology

Stroke in Chinese medicine is called wind-stroke, because the main factor is wind, both internal wind and external wind. When wind and phlegm block the flow of Qi and blood or disturb the mind, the meridians may be afflicted, resulting in stroke and hemiplegia.

It is caused mainly by the disorders of heart, liver, spleen, and kidney; these Zang-Fu 脏腑 produce wind, fire, and phlegm.

- *Improper diet*: This impairs the spleen's functions of transformation and transportation. Such impairment causes the generation of internal turbid phlegm, which mists the heart orifice or obstructs the flow of the channels. Corresponding manifestations include loss of consciousness or hemiplegia and aphasia. Heavy alcohol drinking causes a high risk of stroke. Smoking cigarettes causes the Yin body fluid to be dried by heat. Foods that are sweet and fatty may impair the spleen, such that it loses the functions of transportation and transformation, and then, phlegm-damp accumulates, heat then rises; liver wind stirs upward with the phlegm-damp to block the orifice.

- *Emotional stress*: Excessive emotional stimulation, such as anger, worry, sadness, fear, and stress, affects both heart and liver. When disturbed by intense emotional stress, heart fire explosively flares upward; meanwhile, uprising liver Yang generates liver wind, and then, these two Zang-Fu mutually aggravate each other. Heart fire and liver wind will further cause Qi and blood to ascend and rush upward to the brain; sudden loss of consciousness is then the result.
- *Prolonged exhaustion*: Aging, chronic illness, excessive exertion, and a weak constitution may cause liver and kidney Yin deficiency accompanied by uprising of liver Yang. In these circumstances, Qi and blood rise upward following the uprising of liver Yang. Moreover, phlegm may follow and ascend upward to mist the orifices or obstruct the flow in the meridians.
- *Sudden change of weather*:
  a) Researchers have found that colder temperatures are associated with stroke and death.[5] Chinese medicine holds that the movement of the flow of blood is related to Yang energy. Yang deficiency results in slow or sluggish blood flow.
  b) Another season that brings a high risk of stroke is the early spring. Even as the temperature gets warmer, it is the season of wind. Spring is the season of liver, and liver produces wind.
- *Blood stasis*:
  a) Qi stagnation results in blood stasis. "Qi is the commander of Blood." Emotional stress and anger cause Qi stagnation. Blood moves when Qi moves, and sluggish Qi movement causes blood stasis.
  b) Qi deficiency results in blood stasis. "Qi is the commander of blood." When Qi is deficient, Qi is too weak to move blood.
  c) Cold invasion results in blood stasis. Cold causes sluggish blood movement.
  d) Heat dries Yin, resulting in blood stasis. Heat dries body Yin fluid, and Yin deficiency may be a result of thick body fluids. Yin deficiency and dryness can affect the blood flow; it is difficult for blood to flow when it is very thick.[6]

Manifestations

Manifestations of stroke in traditional Chinese medicine are divided into two types:

1) The mild type is called Meridian Stroke 中經絡. Wind attacks the meridians; the symptoms and signs pertain to the meridians.
2) The severe type is called Zang-Fu Stroke 中臟腑. Wind attacks both the meridians and Zang-Fu organs; the symptoms and signs pertain to the organs.

1) Meridian Stroke 中經絡

a) Wind invading meridians 風痰入絡證
   - Numbness of the lower limb
   - Sudden onset of mouth and eye deviation
   - Slurred speech
   - Copious saliva
   - Unilateral paralysis
   - May have aversion to wind-cold
   - Fever
   - Pain or stiff joints
   - Tongue: Thin white/yellow coating
   - Pulse: Floating tight or rapid pulse

b) Hyperactive liver Yang 肝陽暴亢
   - Sudden hemiplegia, or with numbness and tingling
   - Or sudden collapse, loss of consciousness
   - Mouth deviation
   - Frequent headache, dizziness
   - Irritability and restlessness
   - Bitter taste
   - Thirst
   - Anger
   - Red face and eyes
   - Coarse breathing
   - Dry stools, dark urine
   - Tongue: Red tongue, yellow coating
   - Pulse: Wiry rapid pulse

c) Phlegm heat and excess Fu 痰热腑實
   - Sudden hemiplegia
   - Mouth deviation
   - Speech difficulty
   - Abdominal distention
   - Constipation
   - Overweight
   - Sticky excess saliva
   - Dizziness
   - Tongue: Red tongue with yellow greasy or dry coat
   - Pulse: Wiry slippery pulse

d) Qi Xu with blood stasis 氣虛血瘀
   - Hemiplegia
   - Weak and flaccid limbs
   - Slurred speech
   - Deviation of mouth and eye
   - Edema on affected side
   - Frequent urination or dribbling urination
   - Pale lustrous less complexion
   - Tongue: Deviated purple tongue with thin white coating

- Pulse: Thin moderate pulse

e) Liver Yang rising 肝陽上亢
- Dizziness
- Headache, distention
- Tinnitus
- Sudden onset of mouth and eye deviation
- Stiff tongue with slurred speech
- Heavy stiff limbs, lumbar soreness
- Hemiplegia
- Red face, red eyes
- Irritability, restlessness
- Insomnia, five-center heat
- Tongue: Red or deep red tongue, thick coating
- Pulse: Thin or wiry slippery pulse

2) Zang-Fu Stroke 中臟腑

The severe stroke is divided into (a) excess type and (b) deficiency type.

a) Excess Type 閉證

1) Phlegm heat obstructing clear orifices 痰熱內閉清竅
- Sudden loss of consciousness
- Tight jaw
- Eyes open, staring eyes, eye deviation
- Clenched fists
- Tight limbs
- Closed mouth
- Hemiplegia
- Red face
- Hot body temperature
- Rough breathing
- Constipation, no loose bowel movements
- No incontinence in sweating and urination
- Coarse breathing
- Irritability and restlessness
- Tongue: Red tongue, yellow greasy coating
- Pulse: Wiry, slippery, rapid pulse

2) Phlegm damp misting the mind 痰濁蒙竅證
- Sudden collapse with convulsion
- Loss of consciousness
- Clenched teeth
- Clenched fists
- Lockjaw
- Eye closed
- No incontinence of bowel movement
- No sweat
- Rattling sound of phlegm in the throat
- Profuse sputum
- Sudden hemiplegia

- Stiffness of limbs
- Cold limbs, cooler body temperature
- Pale complexion
- Tight face
- Dark lips
- Rapid breathing or wheezing
- No irritability
- Tongue: Pale and dimmer tongue body; white, sticky, greasy, coating
- Pulse: Deep, slippery, and slow pulse

b) Deficiency Type 脫證

1) Yang collapse condition 陽脫

- Sudden collapse
- Loss of consciousness
- Coma
- Hands and mouth are open
- Eyes closed
- Pale face
- Hemiplegia
- Flaccid limbs, cold limbs
- Weak breathing
- Incontinence of body fluid, such as copious sweat, cold sweat, tears, nasal discharge, saliva
- Incontinence of stools and urine
- Tongue: Pale, flabby with teeth marks, moist coating
- Pulse: Slow, deep, weak, thready

2) Yin collapse condition 陰脫

- Sudden collapse with loss of consciousness
- Slurred speech or aphonia
- Weak and flaccid tongue
- Paralysis of lower limbs, stiffness or deformity
- Palpitation
- Insomnia
- Profuse, sticky, warm sweating
- Weak or sore back or knee
- Poor memory
- Red cheeks
- Warm limbs
- Incontinence of urination and bowel movement
- Rough breath
- Tinnitus
- Tongue: Dry, scarlet, red tongue, thin body, peeled coating
- Pulse: Thready, weak, racing, deep pulse

Convalescence Stage

There are three patterns for the convalescence stage.

1) Liver and kidney Yin deficiency 肝腎陰虛

- Headache
- Dizziness
- Twitching of muscles and skin
- Red face
- Trembling or stiff tongue
- Hemiplegia
- Flaccidity
- Weakness of lower extremities
- Tinnitus
- Dysphagia
- Aphasia
- Low back pain
- Five-center heat
- Low-grade fever
- Constipation
- Tongue: scarlet, small, and thin; peeled with cracks
- Pulse: deep, wiry, thin, rapid

2) Damp phlegm accumulation 痰濕蒙塞
- Heavy sensation of body
- Pale face
- Chest distension
- Nausea
- Vomiting with mucus
- Poor appetite
- Snoring with rattling sound of phlegm in throat
- Loose stool
- Tongue: Pale, flabby with teeth marks; thick and greasy coating
- Pulse: Slippery, weak (deep and weak on right side; if damp phlegm is more excessive, left side will be more wiry and slippery)

3) Qi and blood deficiency 氣血虛
- Exhaustion
- Pale face
- Poor spirit
- Poor memory, especially short term
- Forgetfulness
- Bad concentration
- Insomnia with dreams disturbing sleep
- Nightmares
- Numbness or flaccidity of one side
- Local weakness
- Twitching of muscles
- Skin and eyelids
- Hemiplegia
- Aphasia
- Palpitations
- Poor appetite

- Loose stool or diarrhea
- Alopecia
- Tongue: Pale with thin white coating
- Pulse: Thin and weak, deep, thin, thready

Physical Examination

More than 700,000 people suffer a stroke each year in the United States, and the majority of stroke survivors suffer from hemiplegia. Physical examination of hemiplegia must include a detailed history of stroke.

- Begin by obtaining the patient's medical history:
    - The history of stroke and other medical history may be obtained from the patient or from the family members in the case of an unconscious or non-communicative patient.
    - History of stroke includes the exact time, symptoms of prestroke such as severe headache that is different from the past (usually patients say it was the worst one in their lives), and history of TIA (transient ischemic attack).
    - History of heart disease.
    - History of hypertension.
    - History of head trauma.
    - Pertinent family history.
- Inspection:
    - Vital signs (heart rate, respiratory rate, blood pressure, and body temperature).
    - Inspect movement coordination, difficulty in walking, balance issues, and difficulty in grasping objects.
    - Soft tissue presentations such as muscle atrophy, edema, and skin temperature.
    - Gait, such as Romberg's test.
- Range of motion:
    - Joint flexibility includes passive range of motion, with or without pain and contracture, and active range of motion, which may be limited or may have disappeared. Measure both ranges with a goniometer.
    - Joint hypermobility or hypomobility.

Differential Diagnosis

Brain Tumor

Brain tumors may also cause hemiplegia. Unlike the rapid onset of stroke, the presentation of limb paralysis due to a brain tumor is gradual.

Treatment

Meridian Stroke 中經絡

Treatment principle:

a) Wind invading meridians 風痰入絡證
- Expels wind.

- Nourishes blood.
- Unblocks meridians.

b) Hyperactive liver Yang 肝陽暴亢
- Clears liver Fire.
- Anchors Yang.
- Extinguishes wind.

c) Phlegm heat and excess Fu 痰熱腑實證
- Unblocks Fu organs.
- Drains heat.
- Transforms phlegm

d) Qi Xu with blood stasis 氣虛血瘀
- Tonifies Qi.
- Moves blood.
- Unblocks meridians.

e) Liver Yang rising 肝陽上亢
- Nourishes Yin.
- Calms liver.
- Subdues Yang.
- Extinguishes wind.

Zang-Fu Stroke 中臟腑

The severe stroke is divided into (a) excess type and (b) deficiency type.

a) Excess Type 閉證
1) Phlegm heat obstructing clear orifices 痰熱內閉清竅
   - Sedates liver Yang.
   - Clears heat.
   - Removes inner wind.
   - Causes Qi and blood to descend.
   - Transforms phlegm.
   - Opens orifices.
2) Phlegm damp misting the mind 痰濁蒙竅證
   - Transforms phlegm.
   - Extinguishes inner wind.
   - Opens orifices.

b) Deficiency Type 脫證
1) Yang collapse condition 陽脫
   - Tonifies Qi.
   - Rescues Yang.
   - Induces resuscitation.
2) Yin collapse condition 陰脫
   - Nourishes Yin.
   - Tonifies Yang.
   - Benefits the orifices.

Convalescence

There are three patterns for the convalescence stage.

1) Liver and kidney Yin deficiency 肝腎陰虛
    - Nourishes liver and kidney Yin.
    - Expels inner wind.
    - Prevents liver Yang rising.
2) Damp phlegm accumulation 痰濕蒙塞
    - Tonifies spleen Qi.
    - Drains damp phlegm.
    - Expels inner wind.
3) Qi and blood deficiency 氣血虛
    - Tonifies Qi.
    - Nourishes blood.

Herbal Treatments

Wind Invading Meridians 風痰入絡證

Herbal formula:

- Da Qin Jiao Tang 大秦艽湯.

Formula ingredients:
Qin Jiao 秦艽 12 g, Gan Cao 甘草 6 g, Chuan Xiong 川芎 6 g, Dang Gui 當歸 9 g, Bai Shao 白芍 9 g, Xi Xin 细辛 3 g, Qiang Huo 羌活 6 g, Fang Feng 防風 6 g, Huang Qin 黄芩 6 g, Shi Gao 9 石膏 9 g, Bai Zhu 白术 6 g, Bai Zhi 白芷 9 g, Sheng Di Huang 生地黄 6 g, Shu Di Huang 熟地黄 6 g, Fu Ling 茯苓 6 g, Du Huo 獨活 6 g.

Ingredient explanation:

- Qin Jiao 秦艽 expels wind damp and relaxes the sinews.
- Gan Cao 甘草 releases spasms, alleviates pain, harmonizes the other herbs, and guides the herbs to all 12 meridians.
- Chuan Xiong 川芎 moves blood and Qi, expels wind, and stops pain.
- Dang Gui 當歸 moves and tonifies blood, disperses cold, and stops pain due to blood stasis.
- Bai Shao 白芍 nourishes blood, preserves Yin, calms liver Yang and liver Wind, softens tendons, and alleviates spasm and pain.
- Xi Xin 细辛 expels wind and internal cold and stops pain.
- Qiang Huo 羌活 expels wind-cold-dampness, unblocks painful obstruction, alleviates or stops pain, and guides Qi to the Tai Yang and Du channels.
- Fang Feng 防風 releases the exterior, expels external wind, expels wind-dampness.
- Huang Qin 黄芩 clears heat, dries damp, cools blood, stops bleeding, calms liver Yang.
- Shi Gao 石膏 clears excess heat.

- Bai Zhu 白术 tonifies spleen, dries damp, promotes water metabolism.
- Bai Zhi 白芷 expels wind, drains dampness, dispels cold, and stops pain.
- Sheng Di Huang 生地黄 clears heat for heat symptoms due to Yin deficiency, cools the blood, nourishes Yin and generates fluids, tonifies kidney Yin.
- Shu Di Huang 熟地黄 nourishes blood; nourishes liver and kidney Yin.
- Fu Ling 茯苓 drains damp, tonifies spleen.
- Du Huo 獨活 expels wind damp, stops pain.

Hyperactive Liver Yang 肝陽暴亢

Herbal formula:

- Tian Ma Gou Teng Yin 天麻钩藤饮.

Formula ingredients:

Tian Ma 天麻 9 g, Gou Teng 鉤藤 12 g, Shi Jue Ming 石决明 18 g, Zhi Zi 栀子 9 g, Huang Qin 黄芩 9 g, Yi Mu Cao 益母草 9 g, Chuan Niu Xi 川牛膝 9 g, Du Zhong 杜仲 9 g, Sang Ji Sheng 桑寄生 12 g, Ye Jiao Teng 夜交藤 9 g, Fu Shen 茯神 9 g.

Ingredient explanation:

- Tian Ma 天麻 expels wind, calms liver, stops spasm and tremors.
- Gou Teng 鉤藤 expels wind, releases spasm, calms liver Yang.
- Shi Jue Ming 石决明 drains liver fire, expels wind.
- Zhi Zi 栀子 clears heat, cools blood, resolves damp, reduces swelling.
- Huang Qin 黄芩clears heat, dries damp, cools blood, stops bleeding, calms liver Yang.
- Yi Mu Cao 益母草 moves blood, promotes urination, reduces swelling, clears heat.
- Chuan Niu Xi 川牛膝 moves the blood, expels wind, drains dampness, tonifies the kidneys and liver, and strengthens the tendons and bones.
- Du Zhong 杜仲 tonifies the kidneys and liver, strengthens bones and tendons, and lowers blood pressure.
- Sang Ji Sheng 桑寄生 expels wind damp, tonifies liver and kidney Yin.
- Ye Jiao Teng 夜交藤 nourishes heart Yin and blood, calms Shen, expels wind and damp.
- Fu Shen 茯神 calms Shen.

Modifications:

- For phlegm, add Dan Nan Xing 胆南星 and Shi Chang Pu 石菖蒲.
- For restlessness and dry mouth, add Sheng Di Huang 生地 and Shan Zhu Yu 山茱萸.

Phlegm Heat and Excess Fu 痰熱腑實證

Herbal formula:

- Xing Lou Cheng Qi Tang 星蔞承氣湯.

Formula ingredients:

- Dan Nan Xing 胆南星 9 g, Gua Lou 瓜蔞 15 g, Da Huang 大黄 15 g, Mang Xiao 芒硝 15 g.

Ingredient explanation:

- Dan Nan Xing 胆南星 clears heat, dissolves phlegm, calms wind, and stops convulsions.
- Gua Lou 瓜蔞 clears lung heat, transforms hot phlegm.
- Da Huang 大黄 moves blood and disperses bruising.
- Mang Xiao 芒硝 clears excess heat, reduces swelling.

Modifications:

- For clearing heat, add Huang Qin 黄芩 and Zhi Zi 栀子.
- For Yin Xu, add Sheng Di 生地 and Mai Dong 麦冬.
- For dizziness, vertigo, and headache, add Tian Ma 天麻, Gou Teng 鉤藤, and Ju Hua 菊花.
- For blood stasis, add Dan Shen 丹参 and Chi Shao 赤芍.

Qi Deficiency with Blood Stasis 氣虛血瘀

Herbal formula:

- Bu Yang Huan Wu Tang 補陽還五湯.

Formula ingredients:

Huang Qi 黄耆 15 g, Dang Gui 當歸 9 g, Chuan Xiong 川芎 6 g, Chi Shao 赤芍 9 g, Tao Ren 桃仁 6 g, Hong Hua 紅花 6 g, Di Long 地龍 6 g.

Ingredient explanations:

- Huang Qi 黄耆 tonifies Qi and blood and relieves numbness and pain.
- Dang Gui 當歸 moves and tonifies blood, disperses cold, and stops pain due to blood stasis.
- Chuan Xiong 川芎 moves blood and Qi, expels wind, and stops pain.
- Chi Shao 赤芍 moves blood, reduces swelling, stops pain.
- Tao Ren 桃仁 breaks up stubborn blood stasis.
- Hong Hua 紅花 moves blood, releases bruising, opens meridians, stops pain.
- Di Long 地龍 drains heat, stops spasms and convulsions, and unblocks and promotes movement in the channels and collaterals.

Modifications:

- For mouth and eye deviations, add Bai Fu Zi 白附子, Quan Xie 全蠍, and Jiang Can 殭蠶.

- For edema on the affected side, add Fu Ling 茯苓 and Ze Xie 澤瀉.
- For atrophy of upper limb, add Gui Zhi 桂枝 and Sang Zhi 桑枝.

Liver Yang Rising 肝陽上亢

Herbal formula:

- Zhen Gan Xi Feng Tang 鎮肝熄風湯.

Formula ingredients:

Huai Niu Xi 淮牛膝 30 g, Dai Zhe Shi 代赭石 30 g, Long Gu 龍骨 15 g, Mu Li 牡蠣 15 g, Gui Ban 龜板 15 g, Xuan Shen 玄参 15 g, Tian Men Dong 天門冬 15 g, Bai Shao 白芍 15 g, Yin Chen Hao 茵陳蒿 6 g, Chuan Lian Zi 川楝子 6 g, Mai Ya 麥芽 6 g, Gan Cao 甘草 5 g.

Ingredient explanation:

- Huai Niu Xi 淮牛膝 tonifies liver and kidneys, directs herbs downward.
- Dai Zhe Shi 代赭石 calms liver Yang and Fire, cools blood.
- Long Gu 龍骨 calms Shen, calms liver Yang.
- Mu Li 牡蠣 calms Shen, calms liver Yang, softens hardness and nodules.
- Gui Ban 龜板 nourishes Yin and Yang, tonifies kidney and strengthens bone, nourishes blood and Yin.
- Xuan Shen 玄参 clears heat, cools blood, nourishes Yin, softens hardness and nodules, drains Fire.
- Tian Men Dong 天門冬 nourishes kidney Yin, clears heat.
- Bai Shao 白芍 nourishes blood, preserves Yin, calms liver Yang and liver Wind, softens tendons, and alleviates spasm and pain.
- Yin Chen Hao 茵陳蒿 clears liver damp heat.
- Chuan Lian Zi 川楝子 moves liver Qi, removes stagnation, stops pain.
- Mai Ya 麥芽 regulates Liver Qi, clears liver heat, reduces food stagnation.
- Gan Cao 甘草 releases spasms, alleviates pain, harmonizes the other herbs, and guides the herbs to all 12 meridians.

Modifications:

- With more Yin deficiency signs, add Huang Bai 黄柏, Zhi Mu 知母, Di Gu Pi 地骨皮, Nu Zhen Zi 女貞子, Sheng Di Huang 生地, and Shan Zhu Yu 山茱萸.
- With severe restlessness, add Zhi Zi 梔子 and Huang Qin 黄芩.
- With slurred speech, add Dan Nan Xing 胆南星 and Shi Chang Pu 石菖蒲.
- With insomnia and restlessness, add Ye Jiao Teng 夜交藤.

Zang-Fu Stroke 中臟腑

a) Excess Type 閉證

1) Phlegm heat obstructing clear orifices 痰熱內閉清竅

Herbal formula:

- Ling Jiao Gou Teng Tang 羚角鉤藤湯.

Formula ingredients:

- Ling Yang Jiao 羚羊角 4 g, Gou Teng 鉤藤 9 g, Sheng Di Huang 生地黄 15 g, Zhu Ru 竹茹 15 g, Sang Ye 桑葉 6 g, Chuan Bei Mu 川貝母 12 g, Ju Hua 菊花 9 g, Bai Shao 白芍 9 g, Fu Ling 茯苓 9 g, Gan Cao 甘草 3 g.

Ingredient explanation:

- Ling Yang Jiao 羚羊角 calms liver, relieves convulsions, alleviates pain.
- Gou Teng 鉤藤 expels wind, releases spasm, calms liver Yang.
- Sheng Di Huang 生地黄 clears heat for heat symptoms due to Yin deficiency, cools the blood, nourishes Yin and generates fluids, tonifies kidney Yin.
- Zhu Ru 竹茹 clears heat, transforms hot phlegm, cools blood.
- Sang Ye 桑葉 calms Liver, clears Heat, cools Blood.
- Chuan Bei Mu 川貝母 clears heat, transforms phlegm, moistens lungs.
- Ju Hua 菊花 clears Heat, expels Wind, calms Liver Yang and Wind.
- Bai Shao 白芍 nourishes blood, preserves Yin, calms liver Yang and liver Wind, softens tendons, and alleviates spasm and pain.
- Fu Ling 茯苓 drains damp, tonifies spleen.
- Gan Cao 甘草 releases spasms, alleviates pain, harmonizes the other herbs, and guides the herbs to all 12 meridians.

Modifications:

- In the acute situation, add Zhi Bao Dan 至寶丹 or An Gong Niu Huang Wan 安宮牛黄丸.
- With constipation, add Da Huang 大黄 or Da Cheng Qi Tang 大承氣湯.

2) Phlegm damp misting the mind 痰濁蒙竅證

Herbal formula:

- Di Tan Tang 滌痰湯.

Formula ingredients:

- Ban Xia 半夏 9 g, Chen Pi 陳皮 6 g, Fu Ling 茯苓 6 g, Zhi Shi 枳實 6 g, Zhu Ru 竹茹 6 g, Dan Nan Xing 胆南星 9 g, Shi Chang Pu 石菖蒲 6 g, Ren Shen 人参 3 g, Gan Cao 甘草 3 g, Sheng Jiang 生姜 3 pc.

Ingredient explanation:

- Ban Xia 半夏 transforms phlegm.

- Chen Pi 陳皮 regulates middle Jiao Qi, dries dampness, and transforms phlegm.
- Fu Ling 茯苓 drains damp, tonifies spleen.
- Zhi Shi 枳實 breaks up stagnation, transforms phlegm.
- Zhu Ru 竹茹 clears heat, transforms hot phlegm, coolsbBlood.
- Dan Nan Xing 胆南星 clears heat, dissolves phlegm, calms wind, and stops convulsions.
- Shi Chang Pu 石菖蒲 opens orifices, drains phlegm, calms Shen, expels wind.
- Ren Shen 人参 tonifies Yuan Qi.
- Gan Cao 甘草 releases spasms, alleviates pain, harmonizes the other herbs, and guides the herbs to all 12 meridians.
- Sheng Jiang 生姜 releases the exterior, induces perspiration, disperses cold, warms the middle Jiao, and reduces the toxicity of the other herbs.

Modifications:

- For acute stage, add Su He Xiang Wan 蘇合香丸.
- With blood stasis, add Tao Ren 桃仁, Hong Hua 紅花, and Dan Shen 丹参.
- For calming wind, add Tian Ma 天麻 and Gou Teng 鉤藤.
- For cold limbs, add Fu Zi 附子, Gui Zhi 桂枝, and Xi Xin 细辛.

b) Deficiency Type 脫證
  1) Yang collapse condition 陽脫

Herbal formula:

- Shen Fu Tang 参附湯.

Formula ingredients:

- Ren Shen 人参 12 g, Zhi Fu Zi 附子 9 g.

Ingredient explanation:

- Ren Shen 人参 tonifies Yuan Qi.
- Zhi Fu Zi 附子 warms Ming Men Fire.

Modifications:

- For tonifying Yin and generating fluids, add Sheng Mai San 生脉散.
- With copious sweating, add Shan Zhu Yu 山茱萸, Huang Qi 黄芪, Duan Long Gu 煅龍骨, Duan Mu Li 煅牡蠣, and Wu Wei Zi 五味子.
- With stasis, add Dan Shen 丹参 and Chi Shao 赤芍.

2) Yin collapse condition 陰脫

Herbal formula:

- Di Huang Yin Zi 地黃飲子.

Formula ingredients:

- Shu Di Huang 熟地黄 9 g, Shan Zhu Yu 山茱萸 12 g, Rou Cong Rong 肉苁蓉 12 g, Ba Ji Tian 巴戟天 9 g, Zhi Fu Zi 附子 6 g, Rou Gui 肉桂 3 g, Shi Hu 石斛 9 g, Mai Men Dong 麥門冬 9 g, Shi Chang Pu 石菖蒲 6 g, Yuan Zhi 遠志 6 g, Fu Ling 茯苓 9 g, Wu Wei Zi 五味子 6 g, Bo He 薄荷 6 g.

Ingredient explanation:

- Shu Di Huang 熟地黄 nourishes blood; nourishes liver and kidney Yin.
- Shan Zhu Yu 山茱萸 nourishes the liver and kidneys, strengthens kidney Yang.
- Rou Cong Rong 肉苁蓉 tonifies the kidney Yang, benefits essence and marrow.
- Ba Ji Tian 巴戟天 tonifies kidneys, strengthens Yang, expels wind cold damp, strengthens tendons and bones.
- Zhi Fu Zi 附子 warms Ming Men Fire.
- Rou Gui 肉桂 warms the kidneys Yang and Ming Men Fire and assists the generation of Qi and blood.
- Shi Hu 石斛 nourishes Yin, clears heat, generates fluids.
- Mai Men Dong 麥門冬 moistens lungs, nourishes Yin.
- Shi Chang Pu 石菖蒲 opens orifices, drains phlegm, calms Shen, expels wind.
- Yuan Zhi 遠 expels phlegm, calms Shen, reduces abscesses.
- Fu Ling 茯苓 drains damp, tonifies spleen.
- Wu Wei Zi 五味子 tonifies kidney and lung, clears heat by generating fluids.
- Bo He 薄荷 clears wind heat, relieves liver Qi stagnation.

Modifications:

- For Yin deficiency, Zuo Gui Wan 左歸丸 may also be prescribed.

Cooking instructions:

1) Place the herbs into an herb pot. Traditionally, a clay cooker is used. Gently rinse the herbs once with a strainer.
2) Add water into the pot until it is just enough to cover the herbs, and then add one more cup of water.
3) Presoak the herbs for 15 minutes.

4) Turn on the stove to high heat and bring the herbs to a boil. Then reduce the heat to medium low to simmer but with slight bubbling. Turn off the heat when the decoction is reduced to one cup.
5) Drain the decoction out into a container.
6) Repeat steps 2–5 one more time and drain each decoction into the same container.
7) Divide these two cup decoctions into two separated cups.

Dose:

- Drink the decoction one cup each time, twice a day.

Convalescence:

There are three patterns for the convalescence stage.

1) Liver and kidney Yin deficiency 肝腎陰虛
   Herbal formula: Qi Ju Di Huang Wan 杞菊地黃丸.
2) Damp phlegm accumulation 痰濕蒙塞
   Herbal formula: Ban Xia Bai Zhu Tian Ma Wan 半夏白朮天麻丸.
3) Qi and blood deficiency 氣血虛
   Herbal Formula: Gui Pi Wan 歸脾丸.

Suggestion for taking these Wan (pills):

The dose depends on the brand name. Read the instruction carefully and follow the suggestion. Usually, taking the pills with warm water is suggested.

Acupuncture Treatments

The treatments of the excess type and deficiency type are opposite; the excessive type uses reducing treatments, while the deficiency type uses tonifying treatments.

Even though our focus is on the treatment for the lower limb hemiplegia, the patient's general health constitution still needs to be included in the treatment plans for acupuncture point selection; improvement of the general health condition improves the prognosis of the paralysis of the limbs.

Meridian Stroke

Acupuncture point formula:

GB30 環跳 Huantiao, GB34 陽陵泉 Yanglingquan, SP9 陰陵泉 Yinlingquan, GB31 風市 Fengshi.

Modifications:

- For wind invading meridians, add ST40 豐隆 Fenglong and LI4 合谷 Hegu.
- For hyperactive liver Yang, add LIV3 太衝 Taichong and KD3 太溪 Taixi.
- Phlegm heat and excess Fu, add LI11 曲池 Quchi, ST44 內庭 Neiting, and ST40 豐隆 Fenglong.
- For Qi deficiency with blood stasis, add ST36 足三里 Zusanli and Ren6 氣海 Qihai.

- For liver Yang rising, add KD3 太溪 Taixi and LI11 風池 Fengchi.
- For drop-foot, add GB40 丘墟 Quxu through KD6 照海 Zhaohai.
- For constipation, add ST28 水道 Shuidao, ST29 歸來 Guilai, ST40 豐隆 Fenglong, and SJ6 支溝 Zhigou.
- For urinary incontinence, add Ren3 中極 Zhongji, Ren2 曲骨 Qugu, and Ren4 關元 Guanyuan.

Zang-Fu Stroke 中腑脏

The severe stroke is divided into (a) excess type and (b) deficiency type.

a) Excess Type 閉證

1) Phlegm heat obstructing clear orifices 痰熱內閉清竅

Acupuncture point formula:

Du20 百會 Baihui, Du26 人中 Renzhong, Ht1 極泉 Jiquan, PC8 勞宮 Laogong, KD1 涌泉 Yong Quan, GB20 風池 Fengchi, Liv2 行間 Xingjian, Liv3 太沖 Taichong, SJ5 外關 Waiguan, Du15 啞門 Yamen, Du16 風府 Fengfu.

2) Phlegm damp misting the mind 痰濁蒙竅證痰湿蒙塞心神 (阴闭)

Acupuncture point formula:

Du16 風府 Fengfu, Du20 百會 Baihui, Du26 人中 Renzhong, Ht1 極泉 Jiquan, PC8 勞宮 Laogong, KD1 涌泉 Yong Quan, Ren12 中脘 Zhongwan, Ren17 膻中 Danzhong, ST40 豐隆 Fenglong, SP9 水分 Shuifen, PC6 內關 Neiguan, GB20 風池 Fengchi, GB34 陽陵泉 Yanglingquan, LI4 合谷 He Gu, LIV3 太沖 Taichong.

b) Deficiency Type 脫證

1) Yang collapse condition 陽脱

Acupuncture point formula:

Ren4 關元 Guanyuan with moxa, Ren6 氣海 Qihai with moxa, Du20 百會 Baihui with moxa, Du4 命門 Mingmen with moxa, UB35 會陽 Huiyang, Ren8 神闕 Shenque with moxa and salt.

2) Yin collapse condition 陰脫

Acupuncture point formula:

Ren1 會陰 Huiyin (with strong stimulation), KD1 涌泉 Yong Quan, KD3 太溪 Taixi, SP6 三陰交 Sanyinjiao, DU20 百會 Baihui, Ren4 關元 Guanyuan, Ren6 氣海 Qihai, Ren8 神闕 Shenque moxa with salt.

Convalescence Stage

There are three patterns for the convalescence stage:

1) Liver and kidney Yin deficiency 肝腎陰虛

Acupuncture point formula:

Kid1 涌泉 Yong Quan, Kid3 太溪 Taixi, Kid6 照海 Zhaohai, SP6 三陰交 Sanyinjiao, Liv3 太沖 Taichong, Liv8 曲泉 Ququan, GB39 懸鐘 Xuanzhong, Du15 啞門 Yamen, Du16 風府 Fengfu, UB23 腎俞 Shenshu, UB52 志室 Zhishi, UB43 膏肓俞 Gaohuangshu.

2) Damp phlegm accumulation 痰濕蒙塞

Acupuncture point formula:

GB8 率谷 Shuaigu, GB20 風池 Fengchi, Yintang 印堂 (on the forehead, at the midpoint between the two medial ends of the eyebrow), Ren17 膻中 Danzhong, Ren22 天突 Tiantu, ST36 足三里 Zusanli, SP6 三陰交 Sanyinjiao, UB20 脾俞 Pishu, UB21 胃俞 Weishu, SP9 陰陵泉 Yinlingquan, Ren12 中脘 Zhongwan, Ren17 膻中 Danzhong, ST40 豐隆 Fenglong, Wind-GB20 風池 Fengchi, DU16 風府 Fengfu, GB34 陽陵泉 Yanglingquan.

3) Qi and blood deficiency 氣血虛

Acupuncture point formula:

Si Shen Cong 四神聰 (at the vertex of the head, a group of four points, 1 cun respectively anterior, posterior and lateral Du-20 百會 Baihui), ST36 足三里 Zusanli, SP6 三陰交 Sanyinjiao, UB15 心俞 Xinshu, UB17 膈俞 Geshu, UB20 脾俞 Pishu, UB21 胃俞 Weishu, SP10 血海 Xuehai, Ren14 巨闕 Juque, Ren15 鳩尾 Jiuwei, GB20 風池 Fengchi, GB34 陽陵泉 Yanglingquan, Du16 風府 Fengfu.

Scalp acupuncture with electroacupuncture stimulation.
Motor area, foot sensory area.

Procedures:

- The patient sits in a chair; the therapist stands beside and facing the patient.
- Insert three needles on each area on the contralateral side.
  a) Each line is equally divided into three portions.
  b) Insert a needle in the middle of the upper portion at an angle of approximately 10–15 degrees to a depth of 0.5–0.8 cun.
  c) Insert the remaining two needles in the middle and the lower portion using the same method as step (b).
- Connect two needles with a cord of the electro machine. Set the continuous frequency at 2/100 Hz for 20 minutes.

Treatment frequency: 5 consecutive days of daily treatment with 2 days of rest afterward.
Treatment duration: 30 minutes.

Tui Na Treatments

Treatment method:

1) The patient lies in a prone position. The Tui Na therapist stands beside the patient.

a) Applies pressing technique on both urinary bladder meridians two or three times.
b) Applies rolling technique from both urinary bladder meridians to the ankle region.
c) Applies stretching and rolling on the lower back and hip.

2) The patient lies on the unaffected side. The Tui Na therapist stands beside the patient.

   Applies rolling technique from the ipsilateral hip region to the ankle for 2 minutes.

3) The patient lies in the supine position. The Tui Na therapist stands beside the patient.
   a) Applies rolling technique from the inguinal region to the feet for 2 minutes.
   b) Applies grasping technique on the affected leg, focusing on the thigh and the knee.
   c) Applies rotating techniques on the ipsilateral hip, knee, and ankle.
      i) The therapist places a hand on the patella and grasps the ankle with the other hand.
      ii) Flexes the knee joint toward the chest.

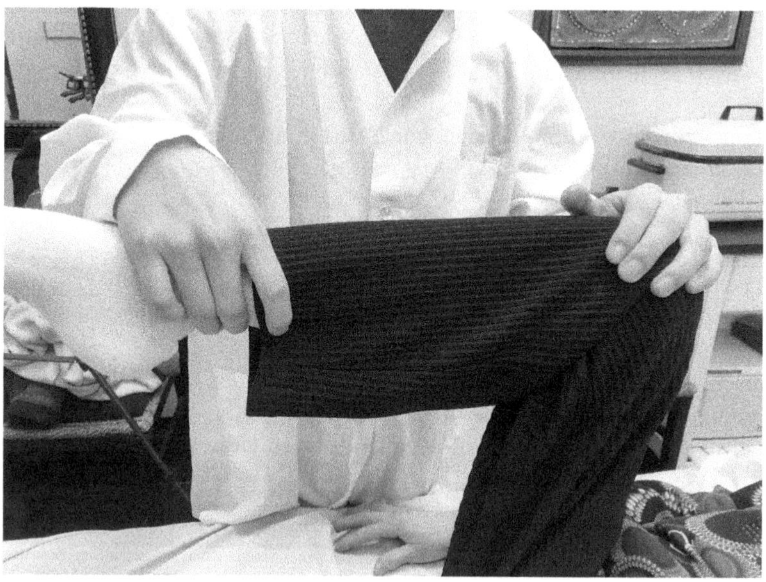

      iii) Rotates the hip by moving the knee in circular motions for 5–10 times in each direction.
      iv) Flexes the hip at 90 degrees, then flexes the knee and moves the tibia in circular motions for 5–10 minutes in each direction.
      v) Extends the leg, lifts the leg by one hand and grasps the toes with the other hand.

vi) The hand grasping the toes moves the foot in circular motions for 5–10 times in each direction.

d) Applies kneading techniques on the GB31 風市 Fengshi, Nei Xi Yan 膝眼 (in the depression on both sides of the patellar ligament when the knee is flexed), GB34 陽陵泉 Yanglingquan, ST41 解谿 Jiexi.
e) Applies kneading techniques on the lower limbs for 3 minutes.
f) Applies striking techniques with the palm and the back of the hand on the gall bladder meridian on the ipsilateral leg about 10–15 times.

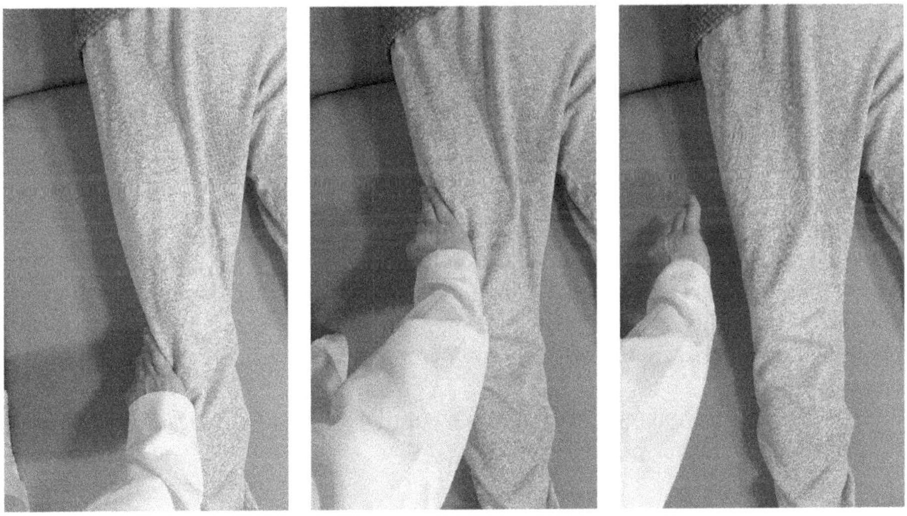

Patient Advisory

Dietary recommendations:

- More liquid food for dysphagia patients.
- The temperature of food should not be cold or too hot.
- Limit oily food and eat more easily digested food for patients with constipation. Use healthy fats such as olive, canola, and other vegetable oils or plant-based foods; stanols are plant-based chemical compounds found in fruits, vegetables, nuts, and seeds and are added to everyday foods, such as margarine spreads, to help people reduce their cholesterol levels. And look for trans-free margarine. Try wholegrain cereals, porridge, brown rice, wholewheat bread, and pasta and grains such as couscous.
- It is recommended to have at least one vegetarian food a week.
- Avoid animal fat.
- Appropriate amount of protein.
- More variety of fresh vegetables and fruits.

- More dietary fiber in the diet.
- Limit salt intake.
- Avoid alcohol, coffee, and strong tea.
- Consider adding blood-nourishing food into the regular diet.
- Read food labels and avoid foods high in sugar, sodium, saturated fat, and trans fat.

Lifestyle:

1) Quit smoking. Smoking leads to furring and narrowing of the arteries (atherosclerosis). It also makes the blood sticky and more likely to clot.
2) Safe exercise, and exercise regularly. Hemiplegia patients may start with range of motion exercises for the affected joints by sitting on a chair or lying on a bed.
3) Manage blood sugar if the patient has diabetes.
4) Keep blood pressure low.

Exercise:

It is generally recommended to start physical training early. For patients with severe paralysis, passive exercise methods such as Tui Na therapy are used to enhance efficacy and recovery. For patients with incomplete paralysis, active movement is mainly adopted. Exercises mainly emphasize training of the hip, knee, and ankle joints.

Exercise can prevent muscle atrophy. Generally, it is 5–10 minutes for active exercise and 30 minutes for passive exercise.

INJURY OF GASTROCNEMIUS MUSCLE 腓腸肌損傷

Introduction

This injury commonly occurs in tennis athletes.[7] It was first described in 1883 in association with tennis; therefore, it is sometimes termed "tennis leg."

Injury of the gastrocnemius muscle or the calf muscle is not limited to the athlete; it is also commonly seen in middle-aged or older patients. At this age, with muscle aging and a general loss of flexibility, the muscle may become predisposed to injury.

The injury also may occur due to certain external factors, from just simply walking to strenuous exercise, such as increasing speed or changing direction while running.

Anatomy

The calf muscle is made up of three muscles:

1) The largest one is called the gastrocnemius. It is a powerful superficial muscle located at the back of the tibia. It has two parts, also called heads.

a) The medial head originates from the posterior surface of the femur superior to the medial condyle and posterior to the insertion of the adductor magnus muscle.
b) The lateral head originates from the lateral epicondyle of the femur.

2) A smaller one is called the soleus; it lies underneath the gastrocnemius muscle.
3) A tiny one is called the plantaris; it originates from the lateral supracondylar line of the femur, just superior and medial to the lateral head of the gastrocnemius muscle, as well as from the oblique popliteal ligament in the posterior aspect of the knee.[8]

It is considered to be at high risk of injury because of its position spanning two joints: the knee and the ankle. Both the medial head and the lateral head of the gastrocnemius muscle run side by side, united into one band of tissue; then, this tough connective tissue and the aponeurosis of the soleus form the Achilles tendon at the lower end of the calf. The Achilles tendon inserts into the calcaneus to lift up the heel to allow forward movement for walking, running, and jumping, while the gastrocnemius muscles assist in flexing the knee and pull the Achilles tendon to allow plantar flexion of the foot.[9]

Western Medicine Etiology and Pathology

A gastrocnemius muscle injury most often occurs when the knee is in maximal extension and simultaneously, the ankle is in fully forced dorsiflexion, which maximally stretches the gastrocnemius muscles,[10] such as in sports that require quick acceleration from a stationary position and a quick halt in movement; an example of this is when a tennis player runs and the foot lands with the ankle suddenly dropping into dorsiflexion. Such sudden movement can overstress the gastrocnemius muscle. The injury can occur from a force stretching the muscle beyond its normal limits. The injury can happen suddenly from a brief overpulling, or the muscle can be predisposed to tearing by long-term overuse.

This injury is not just a strain but a tear. The most common place to incur this injury is at the medial head near the myotendinous junction or roughly halfway between the knee and the heel.[11]

The strain or tear can be mild or severe depending on the type of injury and the predisposition of the gastrocnemius muscle. The severity is generally classified into three grades.

Grade 1

The muscle tear is a result of mild overstrain with some small micro-tears in the muscle fibers. The patient may feel a sharp pain at the time of injury and with activity, but patients usually are still able to continue the activity. The muscles may be in mild spasm and swelling but with no or minimal loss of strength and range of motion.[12]

Grade 2

The muscle tear is a partial tearing of the muscle fibers. The patient is unable to walk at the moment of injury. There is weakness during ankle dorsiflexion and plantar flexion with edema or hemorrhage at the myotendinous junction.[13]

Grade 3

This is the most severe tear; it is a complete tearing or rupture of the muscle fibers. There may be a palpable defect on the affected calf.[14] There is extensive hemorrhage or edema, and the pain is severe.[15] The muscle functions are completely lost.[16]

Chinese Medicine Etiology and Pathology

Acute Stage

Acute traumatic injury to the ligaments leads to blood stasis and swelling. When meridians are blocked by blood stasis, there is a presentation of pain.

Chronic Stage

Injury at the myotendinous junction is considered "Jin Shang 筋傷," meaning "traumatic injury to tendons." Per the theory of traditional Chinese medicine (TCM), the liver governs the tendons. "Traumatic injury to tendons" consumes liver energy; mostly, the liver's Yin and blood are consumed. As the injury becomes chronic, the liver becomes deficient, and the kidney is then deficient, because the liver "Wood" and the kidney "Water" always affect each other. Without the support of liver Yin and blood, the repair is ineffective, and the gastrocnemius muscle fibers are predisposed and are prone to injury when the same activity and the same injury occur on the same fibers repeatedly.

Manifestations

The symptoms depend on the severity of the injury.

- In mild cases, a sudden sharp pain and pulling sensation occur at the halfway point of the lower leg in mild strain; the patient may still be able to walk.
- In severe cases, there is a sudden sharp pain; it may not be just a strain but a tear. It affects mobility; the patient may unable to walk.
- Some patients may have an audible pop in the Grade 3 injury.
- Some patients may report the feeling of being struck or kicked in the calf.
- The leg pain and disability can last for months to years, depending on the severity and the effectiveness of the treatments.

Physical Examination

Begin by obtaining the patient's medical history:

- How the injury occurred; for example, it occurs while walking, running, or jumping. The commonly seen situation is that the person changes direction or suddenly increases speed while running.

Inspection:

- Inspect movement coordination and difficulty in walking.
- Inspect whether or not the patient is able to put weight on the affected foot.
- Inability to stand up on the ball of the foot.
- Even if the patient is able to put some weight on the affected foot, it is with an antalgic gait.
- There is swelling, redness, and bruising in the calf area.

Range of motion:

- Pain on resisted plantar flexion of contracting the muscles against resistance.

Differential Diagnosis

Deep Venous Thrombosis

The pain may manifest at the posterior tibia. Patients often have other presentations such as sudden shortness of breath, chest pain that worsens when taking a deep breath or cough, lightheadedness or dizziness or fainting, and coughing up blood.

Plantaris Strain

The location of pain is at the lateral and posterior aspect of the knee. The pain is exacerbated in knee flexion.

Soleus Muscle Strain

The pain is often lateral. The pain location is often found at the heel when contracting the muscle against resistance with knee flexion and especially pressing on the Achilles tendon.[17]

Treatment

Treatment principle:

Acute stage: Open meridians, move Qi and blood, disperse bruising, stop pain.

Chronic stage: Tonify liver Yin and blood, warm meridians, open meridians, move Qi and blood, stop pain.

Herbal Treatments

Herbal formula:

- Tao Hong Si Wu Tang 桃紅四物湯.

Formula ingredients:

Tao Ren 桃仁 25 pieces, Hong Hua 紅花 2.4 g, Chuan Xiong 川芎 3 g, Dang Gui 當歸 3 g, Chi Shao 赤芍 3 g, Yan Hu Suo 延胡索 3 g, Xiang Fu 香附 3 g, Mu Dan Pi 牡丹皮 3 g, Sheng Di Huang 生地黃 2.4 g.

Cooking instructions:

1) Place the herbs into an herb pot. Traditionally, a clay cooker is used. Gently rinse the herbs once with a strainer. You may notice some red-colored flowers in the strainer; place them back in the herb pot.
2) Add water into the pot until it is just enough to cover the herbs, and then add one more cup of water.
3) Presoak the herbs for 15 minutes.
4) Turn on the stove to high heat and bring the herbs to a boil. Then reduce the heat to medium low to simmer but with slight bubbling. Turn off the heat when the decoction is reduced to one cup.
5) Drain the decoction out into a container.
6) Repeat steps 2–5 twice and drain each decoction into the same container.
7) Divide these three cup decoctions into three separate cups.

Dose:

- Drink the decoction one cup each time, three times a day.

Herbal Heating Pad

Herbal ingredients:

Shen Jin Cao 伸筋草 10 g, Hai Tong Pi 海桐皮 10 g, Qin Jiao 秦艽 10 g, Du Huo 獨活 10 g, Feng Feng 防風 10 g, Mu Gua 木瓜 10 g, Ru Xiang 乳香 6 g, Mo Yao 沒藥 6 g, Hong Hua 紅花 6 g, Yan Hu Suo 延胡索 20 g, Huai Niu Xi 懷牛膝 10 g, Gui Zhi 桂枝 10 g, Yi Yi Ren 薏苡仁 20 g, Ai Ye 艾葉 10 g.

Instructions:

1) Place the herbs in a sealable fabric bag.
2) Steam the bag until hot; let it sit until the temperature is reduced to be appropriate for the skin.
3) Place the bag directly on the popliteal fossa until the bag is no longer warm.
4) Repeat steps 2 and 3 twice, three times a day.

Acupuncture Treatments

Acupuncture point formula:

- GB34 陽陵泉 Yanglingquan, ST36 足三里 Zusanli, UB57 承山 Chengshan, Ashi points 阿是穴 on the calf.

Instructions:

1) Insert a 1 cun needle perpendicularly on GB34 陽陵泉 Yanglingquan and manipulate the needle with sedating techniques.

2) Insert a 1.5 cun needle perpendicularly on UB57 承山 Chengshan and manipulate the needle with sedating techniques.
3) Insert a 1.5 cun needle perpendicularly on ST36 足三里 Zusanli and manipulate the needle with tonifying techniques.
4) Needle retention time is at 15 minutes.
5) Then add warming techniques to each needle by burning 3 moxa cones on each needle.
6) Acupuncture treatment frequency is once a day for 5 days, then take 2 days off.

Cupping Treatment

Instructions:

1) The patient lies prone.
2) Bleed two or three Ashi points 阿是穴on the popliteal fossa.
3) Place acupuncture cups to drain about 2 ml blood.
4) Treatment frequency is once every other day, three times, and take 2 days off.

Formula action: Move Qi, tonify liver and kidney.

Treatment frequency: Daily.

Treatment duration: 30 minutes.

Tui Na Treatments

1) The patient lies prone with the affected leg flexed at 170 degrees and a small cushion placed under the ipsilateral ankle; the therapist stands beside the affected knee.
2) Applying thumb kneading techniques: places a thumb on the medial head of the gastrocnemius muscle and applies plucking techniques six to eight times.

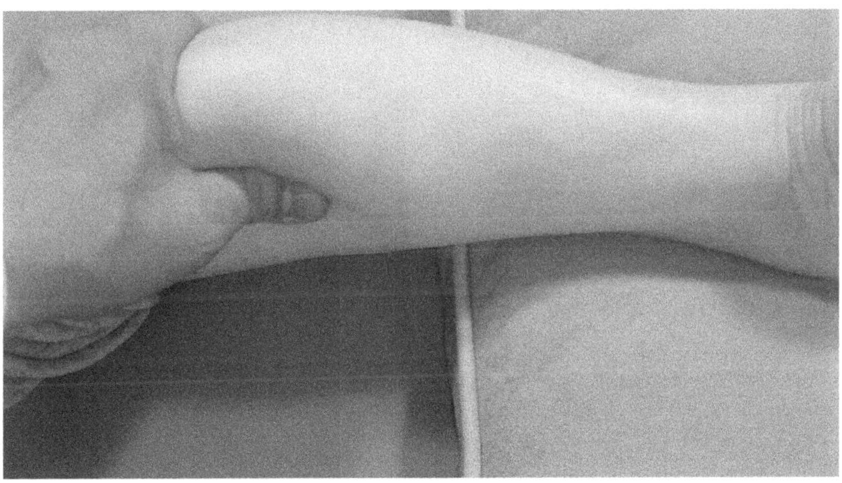

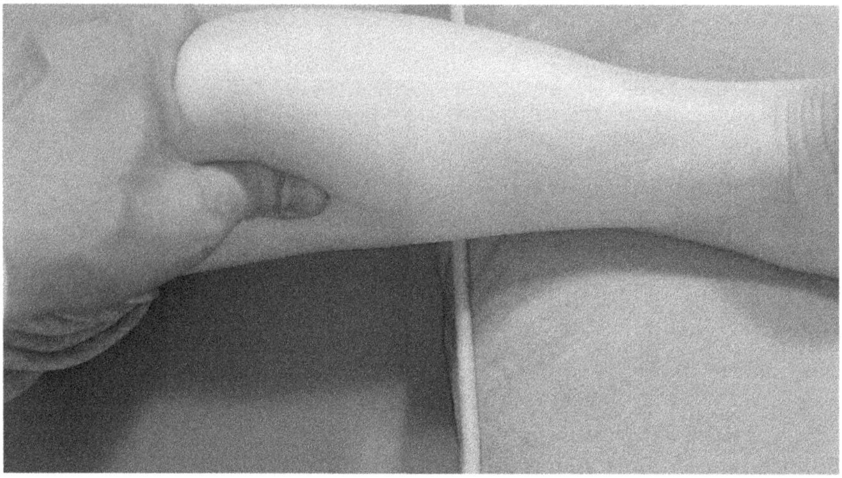

3) Applying hypothenar kneading techniques: applies kneading techniques with a hypothenar on the calf from the knee to the Achilles tendon six to eight times.

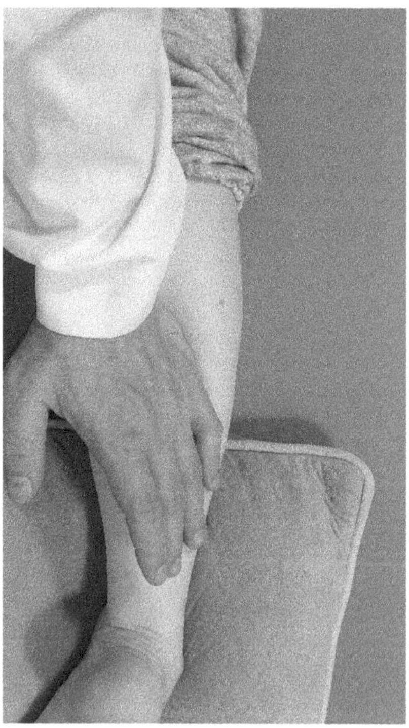

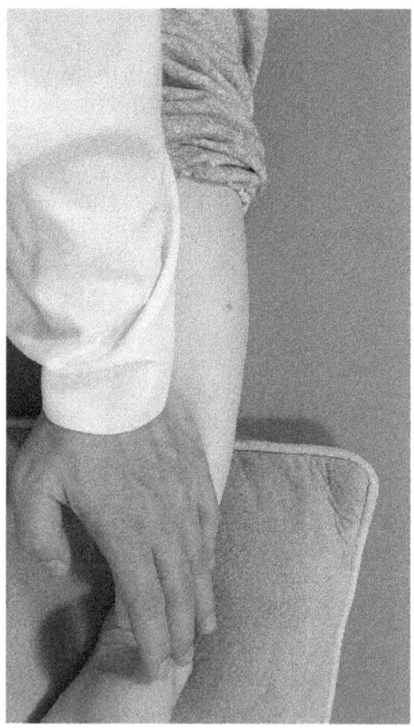

4) Applying grasping techniques: applies both hands and all fingers, grasping the gastrocnemius muscle for about 3 minutes with the appropriate pressure.

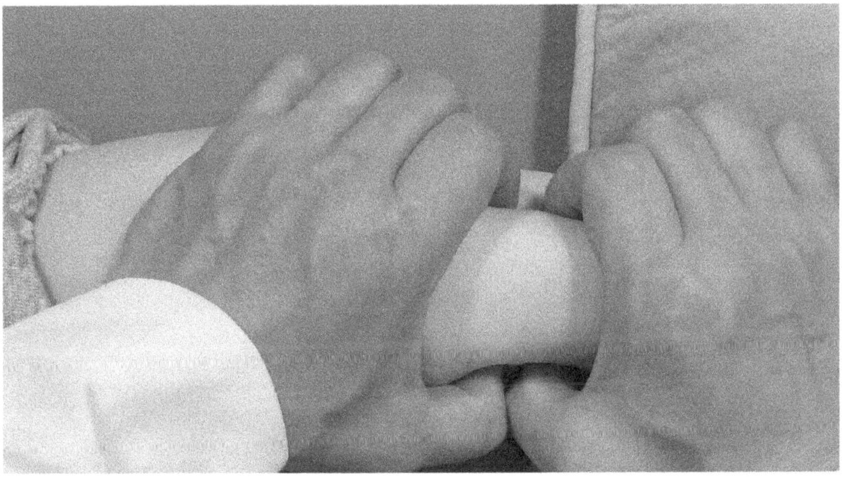

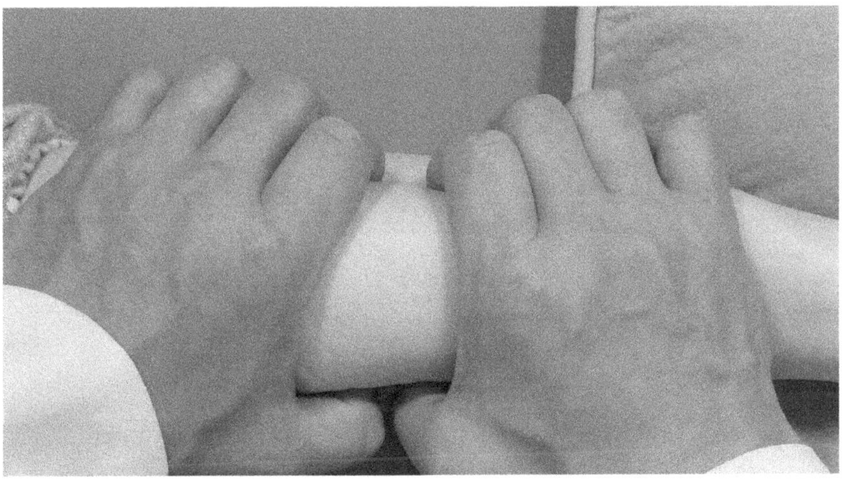

5) Applying chafing techniques: applies chafing techniques with a palm in the direction between the knee and the foot for 3 minutes.
6) Applying knocking techniques: applies knocking techniques with the heel of a hand lightly on the gastrocnemius muscle six to eight times with the appropriate pressure.

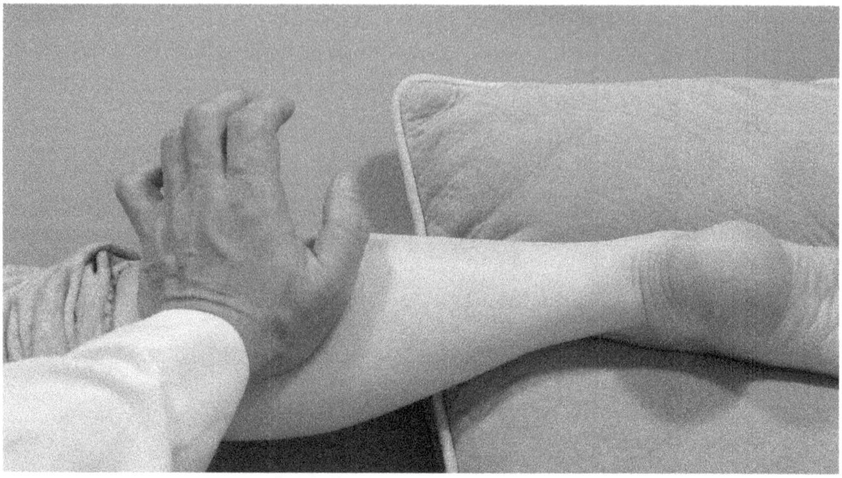

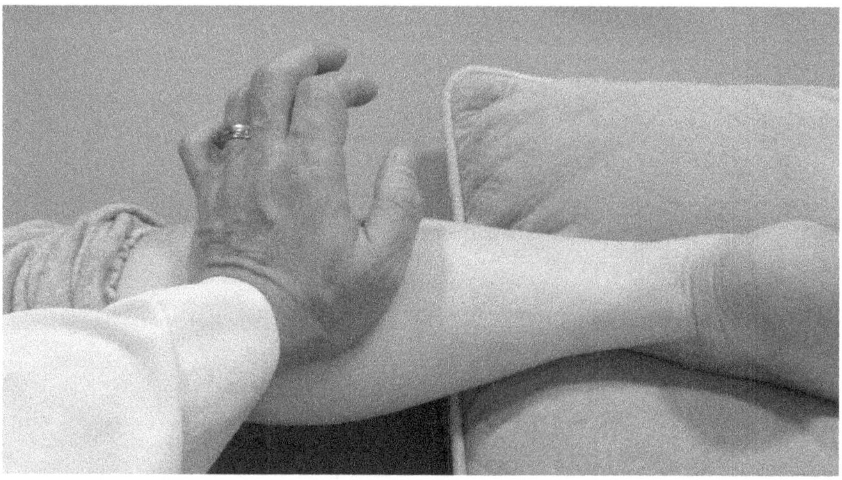

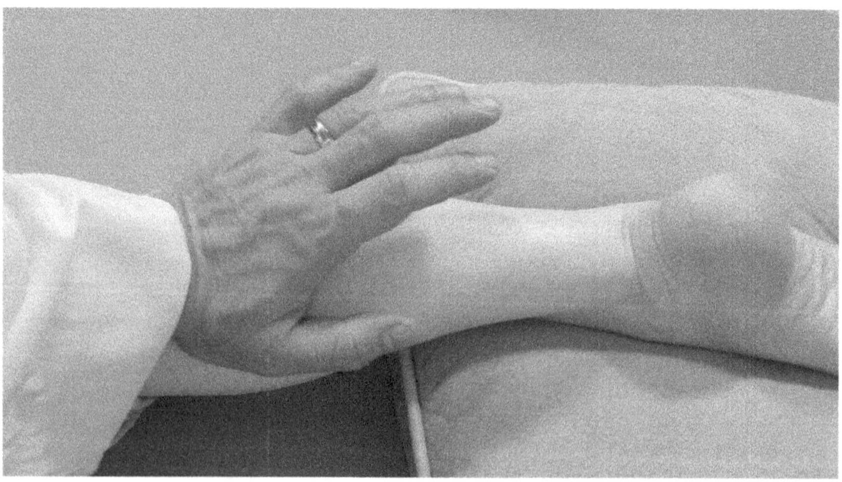

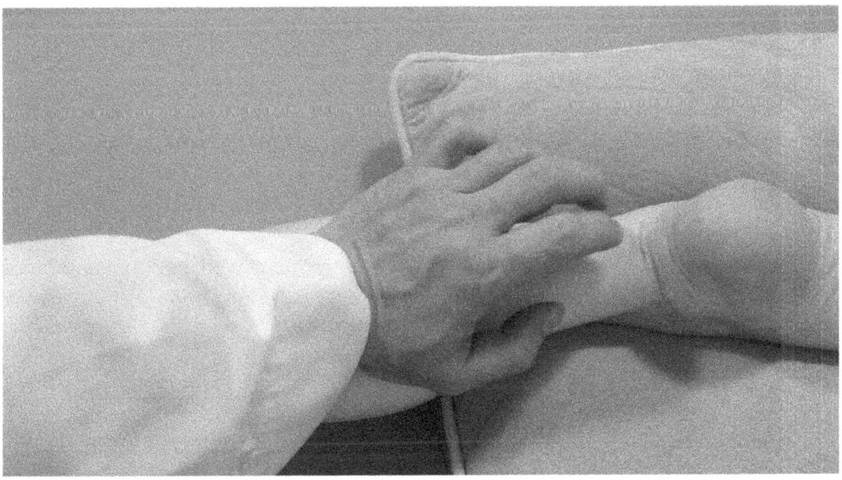

7) Treatment frequency is daily for 6 days and take 2 days off.

Patient Advisory

- For acute injury, it is recommended to apply an ice pack to the affected gastrocnemius muscle for 15–20 minutes, three or four times daily, for the first 2 days. Caution: never apply ice directly to the skin.
- Place a cushion under the calf during leg extension to raise the leg when supine.
- An elastic bandage should be considered for limiting effusion.
- A walking cane is recommended to assist walking.
- After the acute stage, stretching is helpful to assist healing:
  a) Stand facing a wall with the hands on the wall at eye level. Or stand behind a chair and place the hands on the backrest of the chair.

b) Put the leg that is going to stretch one step backward. Ensure that the knee is fully extended, the heel is on the floor, and the toes are pointing forward.
c) Flex the front knee until a stretching sensation in the back-leg gastrocnemius muscle is felt.
d) Maintain the stretch for 15 to 30 seconds.
e) Repeat on each leg for three repetitions.

- To prevent future injury, warming up the muscles before vigorous exercise can reduce the chance of injury.
- Gradually increase the intensity and duration of training.
- Discontinue the exercise when there is pain or discomfort.
- After the acute stage, a heating pad should be applied to the affected leg.
- Receive treatments from Chinese medicine periodically to promote Qi and blood flow, remove blockage, and reinforce the functions of liver and kidney.
- Drink water before, during, and after play; warm water is recommended in Chinese medicine.

Dietary suggestions:

- Avoid foods that may interfere with or decelerate healing; examples are cold food such as ice cream, tofu, cold or icy water, and raw food such as salads; also avoid deep-fried and fatty food such as chips.

  Acute stage:
- Foods that benefit Qi and blood circulation, reduce swelling, and stop pain. Examples are cooked food, chives, grapes, fish soup, beans, eggs, yams, vinegar, and brown sugar.

  Intermediate stage:
- Foods that are rich in protein, such as eggs, almonds, oats, milk, and broccoli.

  Later stage:
- Foods for kidney and liver, such as fish, shrimp, oysters, nuts, goji berries, and black beans.

PLANTAR FASCIITIS 足底筋膜炎

INTRODUCTION

Plantar fasciitis is commonly named "jogger's heel." It is one of the most common causes of heel pain in runners, dancers, or athletes in sports involving jumping; it is also a commonly seen condition in pregnancy, in obesity, and in professions that have to stand for extended periods on each workday.

Approximately 1 million office-based physician visits and hospital outpatient departments resulted in a diagnosis and treatment of plantar fasciitis during 1995–2000. Approximately 62% of all visits were to primary care practitioners, and 31% were to orthopedic surgeons (Riddle and Schapert 2004).

Anatomy

The plantar fascia is a thick fibrous connective tissue that runs between the medial calcaneal tubercle and the proximal phalanges of the toes.

The plantar fascia is intimately related to the Achilles tendon in younger people. The continuous fascial connection from the Achilles tendon to the origin of the plantar fascia at the calcaneal tubercle decreases with age to the point that it may disappear in the elderly.[18]

A tensile force in the Achilles tendon will increase tensile strain in the plantar fascia. This relationship has been applied to the treatment for plantar fasciitis for patients practicing "standing calf stretch."

Western Medicine Etiology and Pathology

The plantar fascia supports the muscles and arch of the foot. When the foot is loaded by the body weight, the plantar fascia acts as a shock-absorbing tie-rope to relieve the tensile load so as to stabilize and control foot movements. If the load on the rope is more than its tensile strain, the fascia becomes irritated and inflamed.

Risks that increase irritation of the plantar fascia:

- Obesity.
- Age around 40 to 60 years old.
- Flat feet or high arches.
- Female.
- Tight Achilles tendons.
- Careers that involve long periods of standing, such as restaurant waiter.
- Sports involving running and jumping.
- Dance such as ballet dancing and aerobic dance.

Chinese Medicine Etiology and Pathology

According to traditional Chinese medicine, plantar fasciitis is considered to be the result of "steaming bone disorder" from the deficiencies of kidney Qi, Yin and liver, Yin and blood. The wood of liver lacks nourishment from the water of kidney, liver blood and Yin are then deficient, kidney fails in governing the bone, and liver fails in storing blood and governing tendons; bone and tendons then degenerate, and wind and damp invasion blocks meridians, resulting in pain.

Many factors can cause depletion of the kidneys:

a) Age over 40 years.
b) Indulgence, such as too much sexual activity and overwork.
c) Malnutrition. The kidney system in Chinese medicine is similar to the endocrine system in Western medicine. Substances that stress the adrenal glands, such as caffeine, can drain kidney energy.
d) Emotional conditions such as fear, sadness, nervous tension and anger, worry.
e) Physical traumatic injury, especially lumbar region injury.

Manifestations

- Sharp or dull pain at the plantar aspect of the foot near the heel.
- It is usually worse in the first few steps in the morning or after sitting for a long period of time, and it is diminished after some walking.
- Some patients feel a burning or ache.
- The pain may extend toward the edges of the heel; it is more medial than lateral.
- The pain tends to be lower with activity, but it is worse again after intense activity on the feet.
- It usually develops gradually.

Physical Examination

1) Begin by obtaining the patient's medical history:
    - General health conditions, such as obesity or pregnancy.
    - History of acute injury to the plantar fascia may not be noticeable.
    - Careers that require long periods of time in a standing position.
    - Sports injury from jumping.
2) Inspection:
    - There is usually no bruising or swelling.
    - It is usually unilateral, but it could be bilateral.
3) Palpation:
    - Tight Achilles tendons.
    - Pressing or palpation of the plantar medial aspect of the heel may reproduce sharp or dull pain.
    - Pressure and stretching will increase the pain.
4) Range of motion:
    - Passive dorsiflexion of the ankle and toes may reproduce the pain.

Differential Diagnosis

Fractures of the Calcaneus

- Sudden pain.
- Inability to bear weight on the affected foot.
- Bruising of the heel and ankle.
- Swelling of the heel and ankle.

Fat Pad Atrophy

Unlike plantar fasciitis, the pain of fat pad atrophy is absent in the morning but worsens with extended activity involving standing and walking during the day.

S1 Radiculopathy

- The pain does not only present at the heel; it is also in the lumbar region and radiates along the L5S1 dermatomal distribution.
- Straight leg raise test is positive.

Tarsal Tunnel Syndrome

- The pain is typically felt on the medial side of the ankle and on the plantar aspect.
- The pain may radiate to the calf and toes.[19]
- Tingling similar to an electric shock.
- Burning.
- Numbness.
- The pain may be aggravated by extended use of the foot such as prolonged standing, walking, hiking, or dancing.[20]
- Some patients may experience more pain at night or when resting.[21]

TREATMENTS

Treatment principle:
Acute stage: Move blood, open meridians.
Chronic stage: Tonify kidney and liver.

Herbal Treatments

Acute Stage

Herbal formula:
Fu Yuan Huo Xue Tang 复原活血湯.

Formula ingredients:
Chai Hu 柴胡 15 g, Tian Hua Fen 天花粉 9 g, Dang Gui 當歸 9 g, Hong Hua 紅花 6 g, Gan Cao 甘草 6 g, Chuan Shan Jia 穿山甲 6 g, Da Huang 大黃 30 g, Tao Ren 桃仁 15 g.

Ingredient explanation:

- Chai Hu 柴胡 releases liver Qi, raises Yang Qi, disperses wind heat and phlegm.
- Tian Hua Fen 天花粉 drains heat, generates fluids, transforms phlegm, expels pus, and reduces swelling.
- Dang Gui 當歸 moves and tonifies blood, disperses cold, and stops pain due to blood stasis.
- Hong Hua 紅花 moves blood, releases bruising, opens meridians, stops pain.
- Gan Cao 甘草 releases spasms, alleviates pain, harmonizes the other herbs, and guides the herbs to all 12 meridians.
- Chuan Shan Jia 穿山甲 moves blood, reduces swelling, expels wind and damp.
- Da Huang 大黃 moves blood and disperses bruising.
- Tao Ren 桃仁 breaks up stubborn blood stasis.

Cooking instructions:

1) Place the herbs into an herb pot. Traditionally, a clay cooker is used. Gently rinse the herbs once with a strainer. You may notice some red-colored flowers in the strainer; place them back in the herb pot.

2) Add water into the pot until it is just enough to cover the herbs, and then add one more cup of water.
3) Presoak the herbs for 15 minutes.
4) Turn on the stove to high heat and bring the herbs to a boil. Then reduce the heat to medium low to simmer but with slight bubbling. Turn off the heat when the decoction is reduced to one cup.
5) Drain the decoction out into a container.
6) Repeat steps 2–5 one more time and drain each decoction into the same container.
7) Divide these two cup decoctions into two separate cups.

Dose:

- Drink the decoction one cup each time, twice a day.

Formula actions:

- Moves blood, disperses bruising, soothes liver, and opens meridians.

Formula indications:
- Injury, blood stasis blocking meridians, hypochondriac pain.

Chronic Stage
Herbal formula:

- Zhi Bai Di Huang Wan 知柏地黄丸.

Formula ingredients:
Shao Yao 山藥 12 g, Mu Dan Pi 牡丹皮 9 g, Fu Ling 茯苓 9 g, Shan Yu Rou 山茱萸 12 g, Ze Xie 澤瀉 9 g, Huang Bai 黄柏 9 g, Shu Di Huang 熟地黄 24 g, Zhi Mu 知母 9 g.

Ingredient explanation:

- Shao Yao 山藥 tonifies kidneys, benefits the Yin and the spleen Qi.
- Mu Dan Pi 牡丹皮 clears deficient heat and cools blood, moves blood, drains pus, and reduces swelling.
- Fu Ling 茯苓 drains damp, tonifies spleen.
- Shan Yu Rou 山茱萸 nourishes liver and kidney, strengthens kidney Yang.
- Ze Xie 澤瀉 drains damp, especially damp heat in lower Jiao, clears kidney deficient heat.
- Huang Bai 黄柏 clears deficient kidney Yin heat.
- Shu Di Huang 熟地黄 nourishes blood; nourishes liver and kidney Yin.
- Zhi Mu 知母 clears deficient heat and drains Fire, nourishes Yin, moistens dryness, generates body fluids.

Directions:

- The dose depends on the brand name. Read the instructions carefully and follow the suggestions. Usually, taking the pills with warm water is suggested.

Formula actions:

- Nourishes Yin, clears heat, promotes urination.

Indications:

- Yin deficiency with deficient fire, steaming bone syndrome, damp heat in lower Jiao.

Acupuncture Treatments

Acupuncture point formula:

KD1 涌泉 Yongquan, KD2 然谷 Rangu, KD3 太溪 Taixi, SP6 Sanyinjiao, UB23 肾俞 Shenshu, UB57 承山 Chengshan, LIV8 曲泉 Ququan, and Ashi acupoints 阿是穴 on the affected side.

Acupuncture instructions:

1) Insert a 1 cun needle perpendicularly on KD1 涌泉 Yongquan, KD2 然谷 Rangu, KD3 太溪 Taixi, and manipulate the needle with tonifying techniques.
2) Insert a 1.5 cun needle on UB57 承山 Chengshan, Ashi acupoints 阿是穴, and manipulate the needle with sedating techniques.
3) Insert a 1.5 cun needle on SP6 三陰交 Sanyinjiao, UB23 腎俞 Shenshu, LIV8 曲泉 Ququan, and manipulate the needle with tonifying techniques.
4) Needle retention time is at 15 minutes.
5) Withdraw the needles except from the Ashi acupoints 阿是穴.
6) Then, add warming techniques to needles of Ashi acupoints 阿是穴 by burning 3 moxa cones on each needle.
7) Acupuncture treatment frequency is once a day for 5 days, then take 2 days off.

Formula action: Moves Qi, tonifies liver and kidney.

Treatment frequency: Daily.

Treatment duration: 30 minutes.

Tui Na Treatments

Tui Na techniques:

- Pressing, grasping, rolling, rubbing, pushing

Tui Na procedures:

1) The patient lies prone, and the therapist stands near the affected foot.
2) The therapist presses and kneads with the thumbs on the medial side of the heel for 5 minutes.
3) Grasps the Achilles tendon with all fingers for 5 minutes.

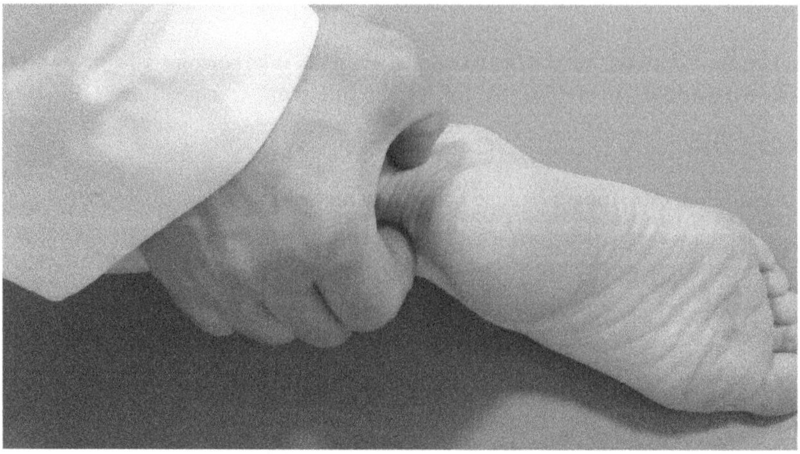

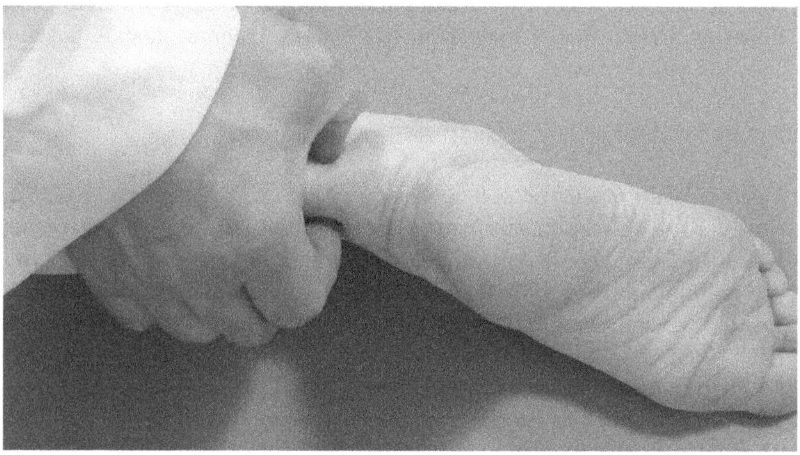

4) Grasps both sides of the foot, focusing on the heel, for 5 minutes.

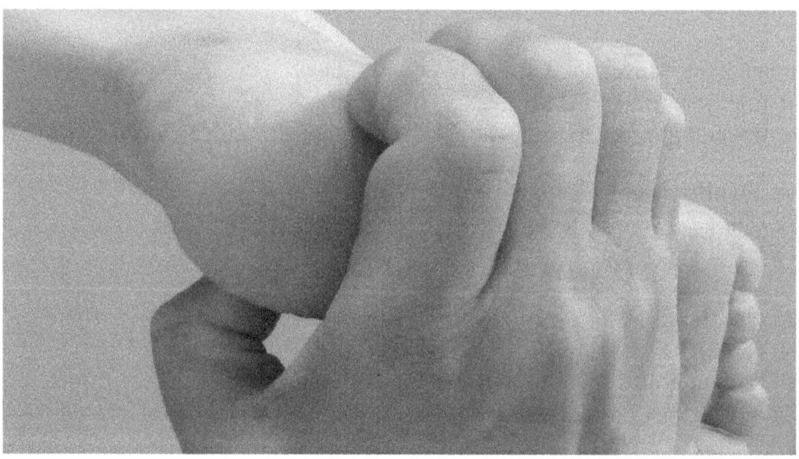

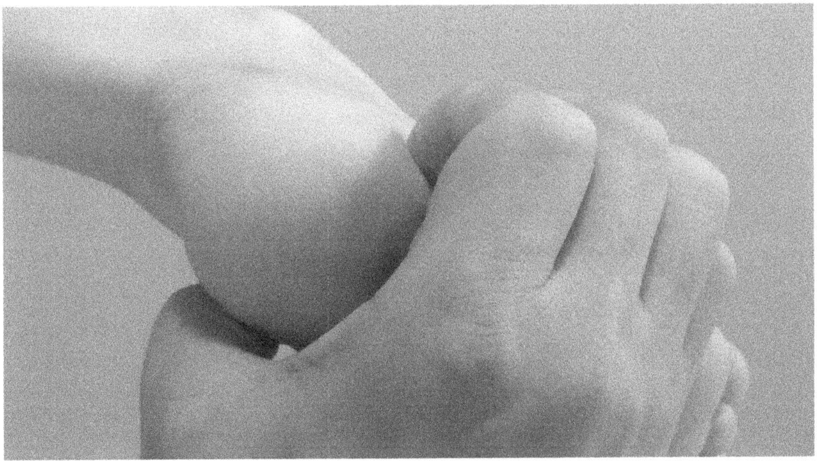

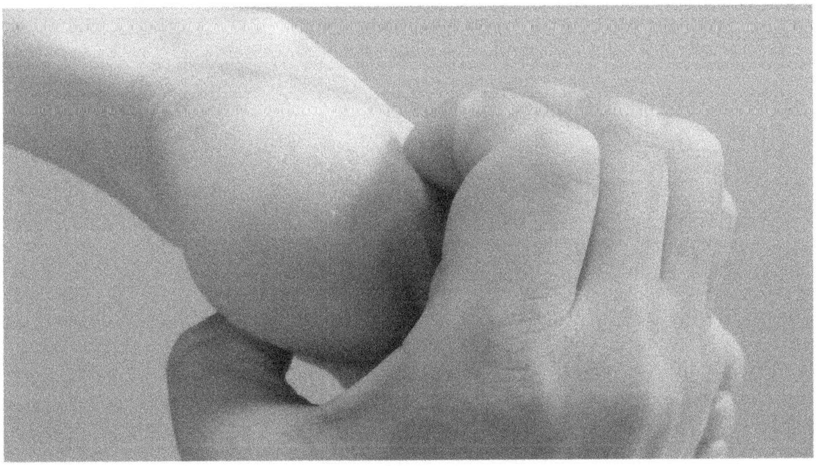

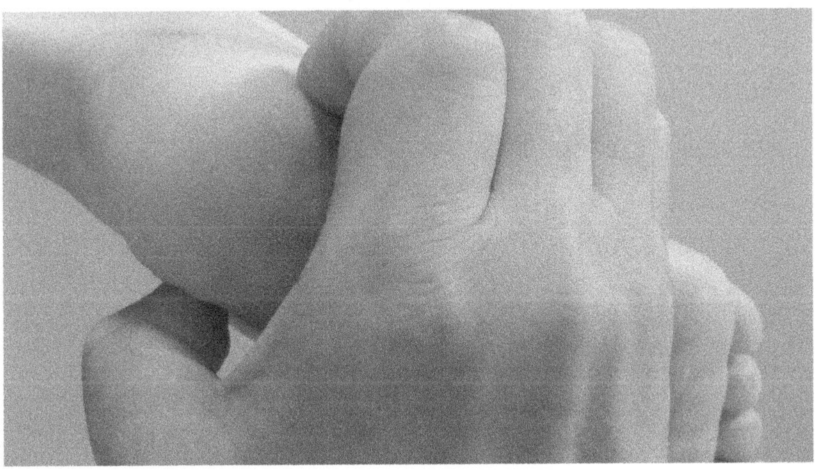

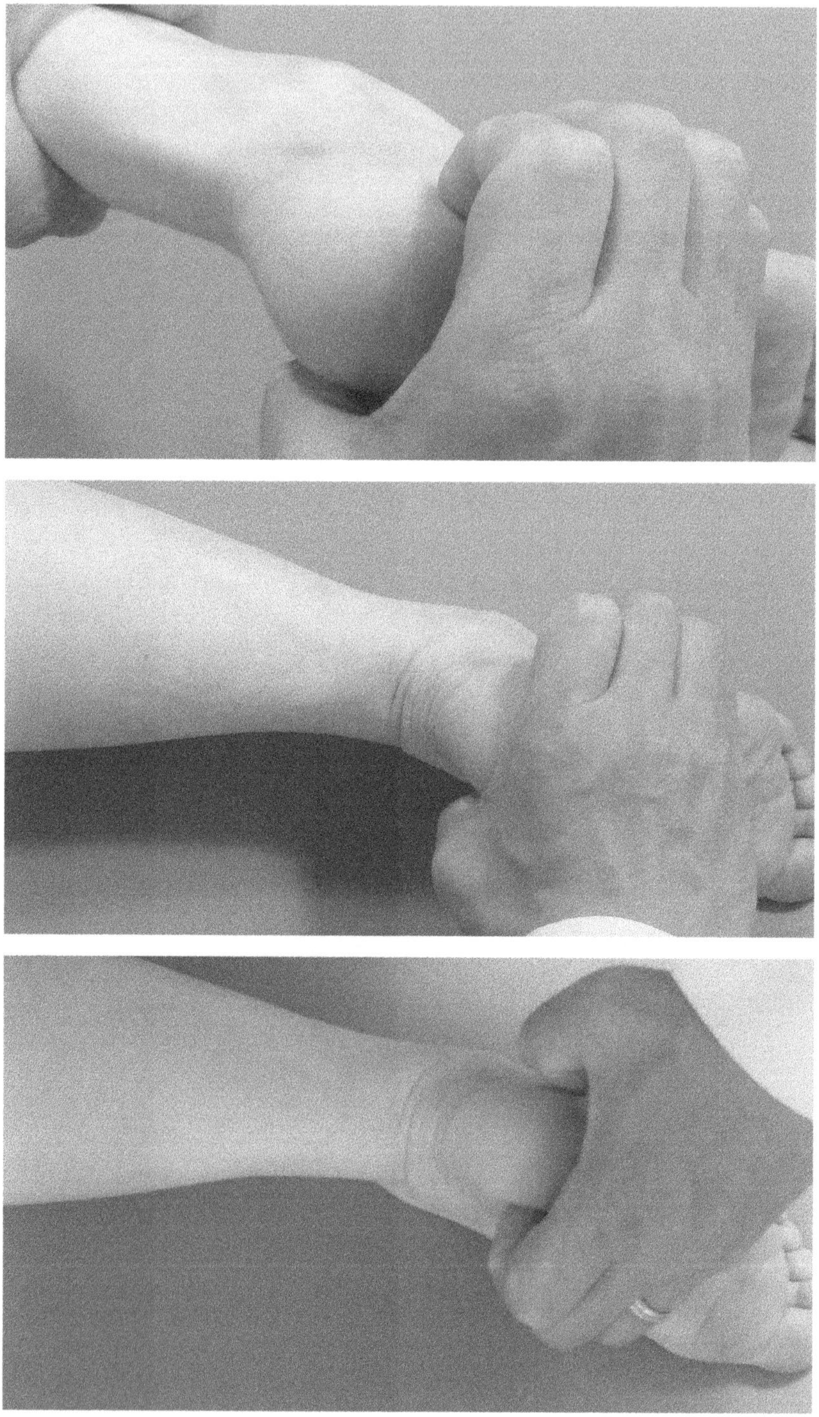

5) Grasps the gastrocnemius muscle for 5 minutes.

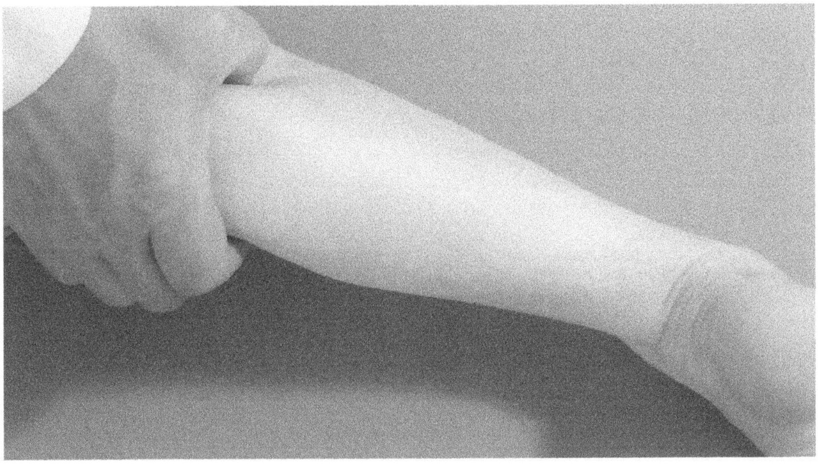

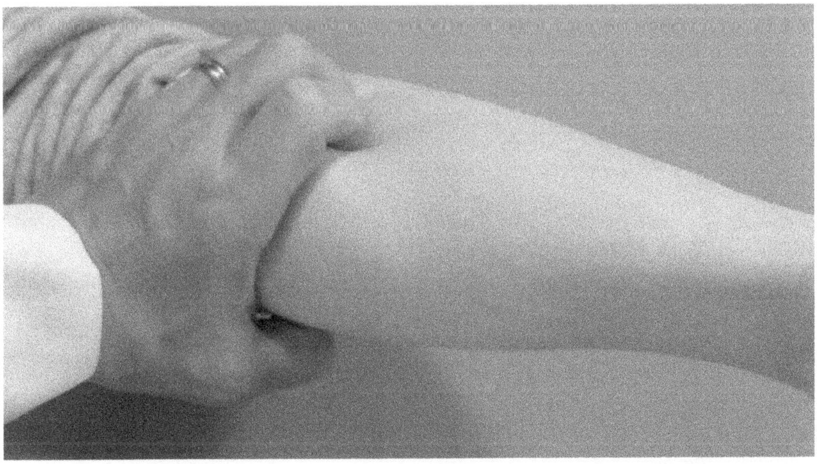

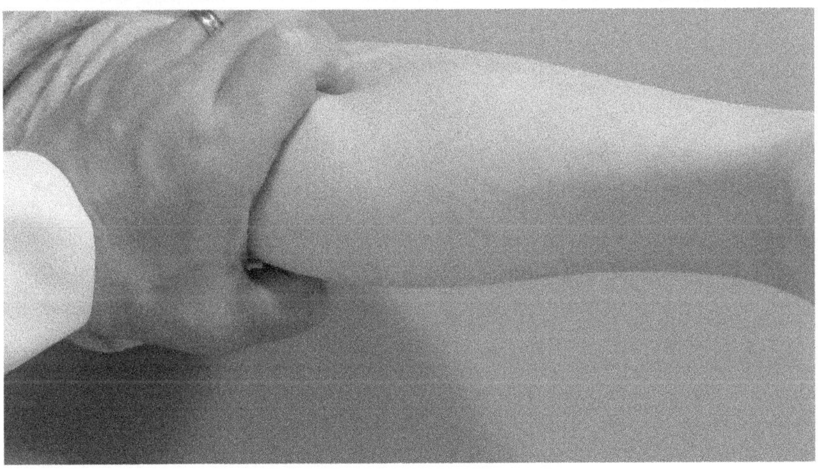

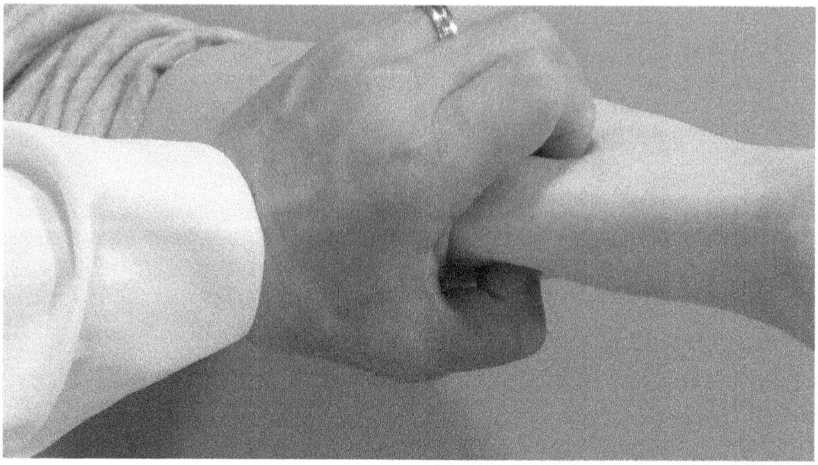

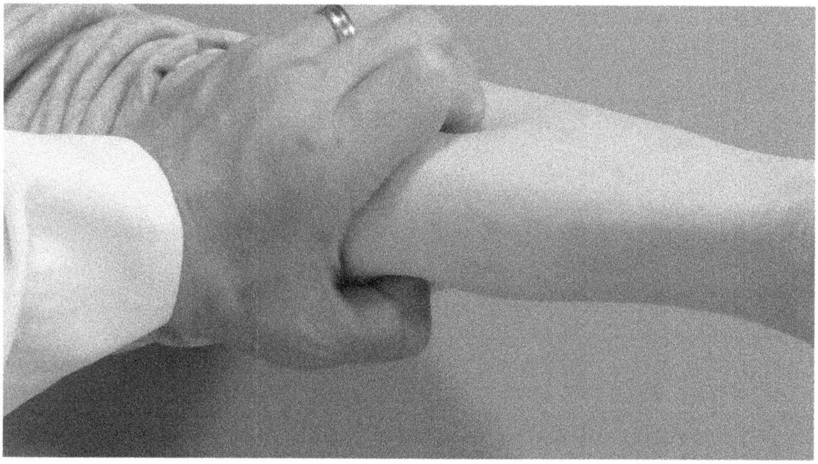

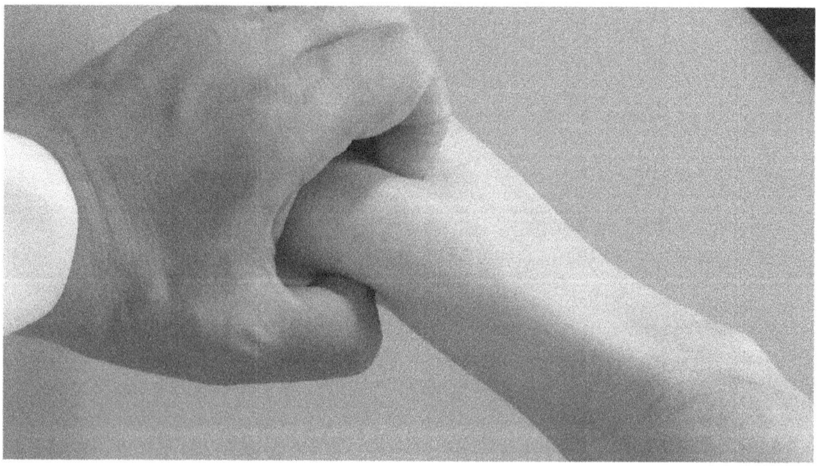

6) Rolls with the backs of the fingers on the plantar aspect of the affected foot for 5 minutes.

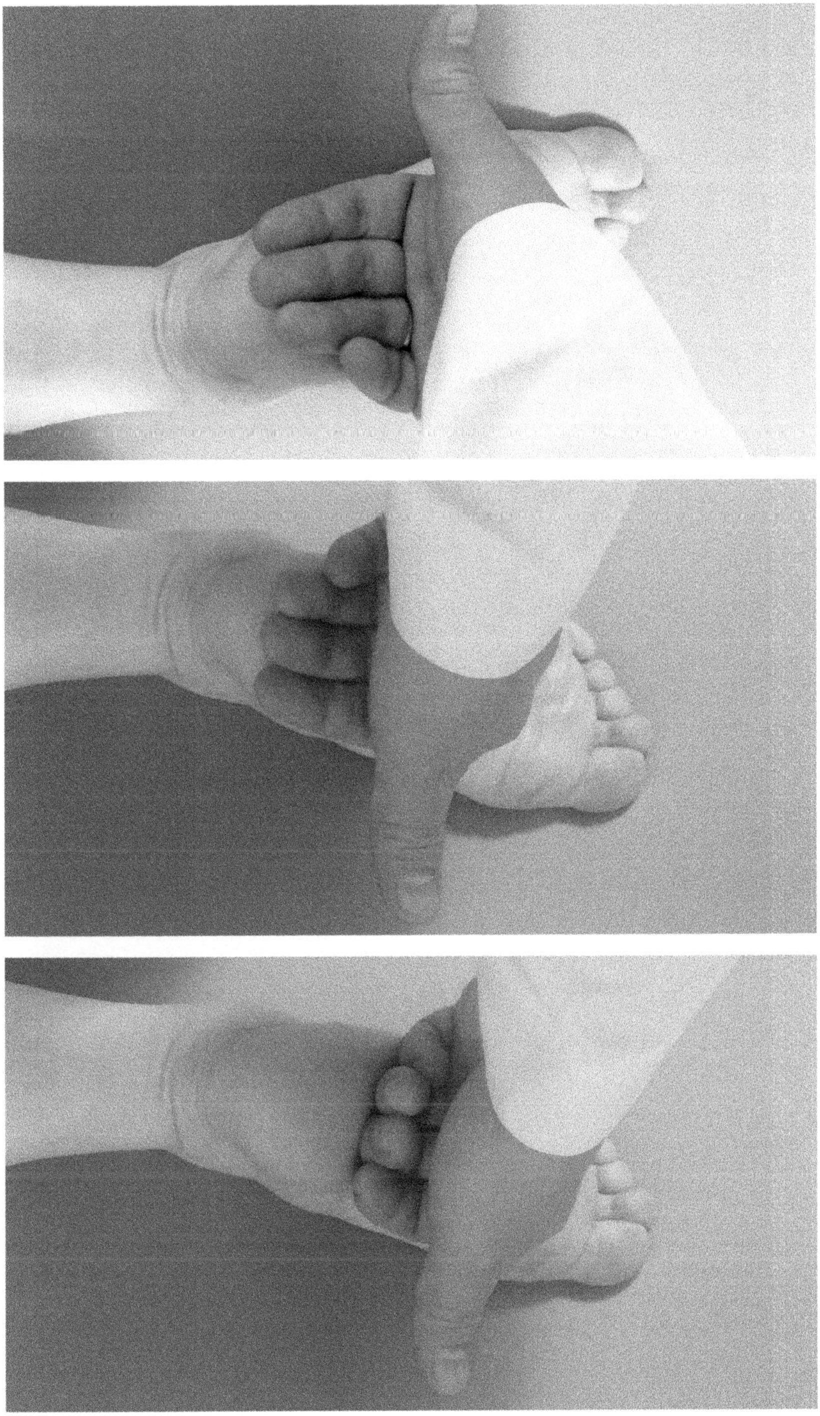

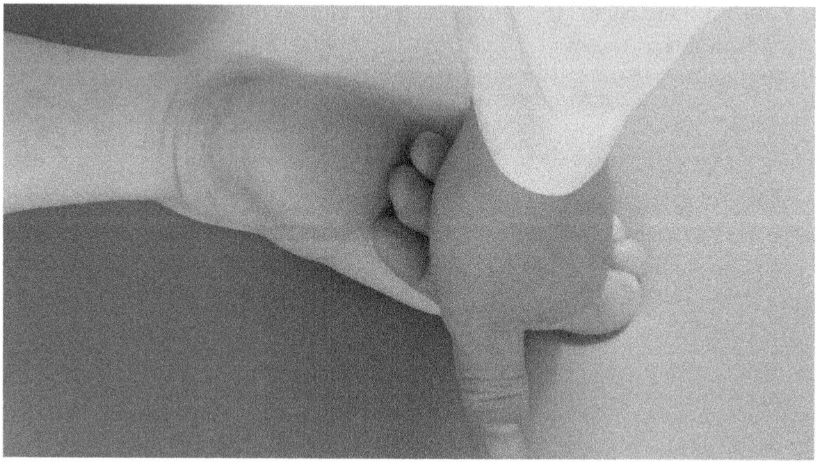

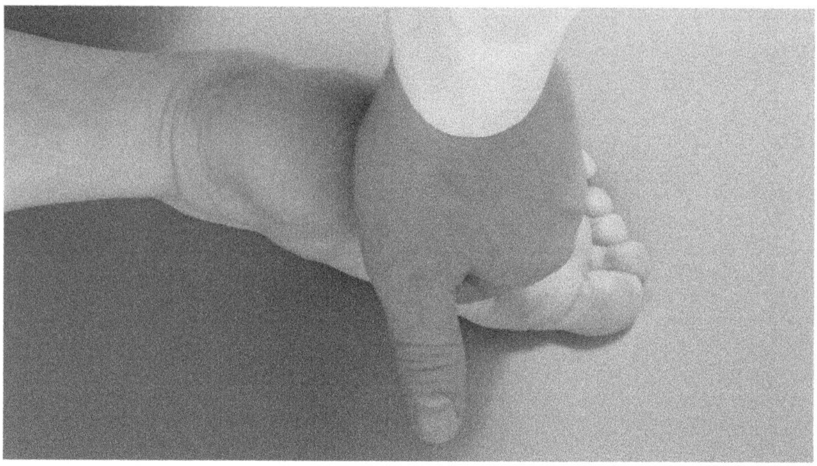

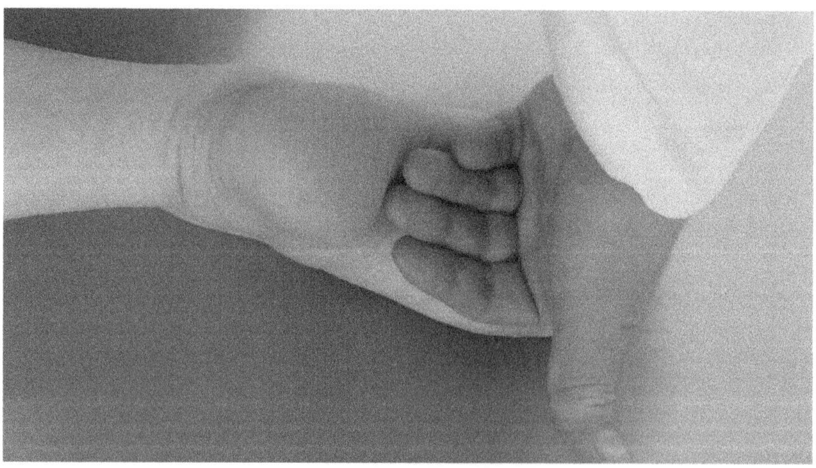

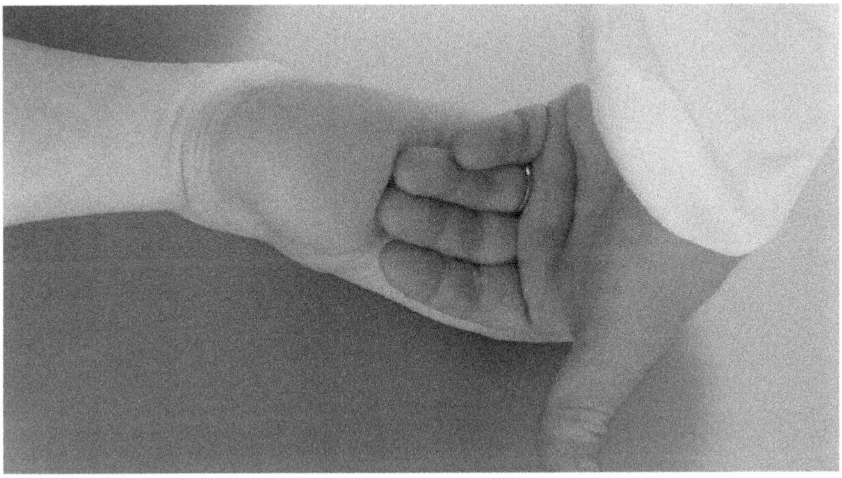

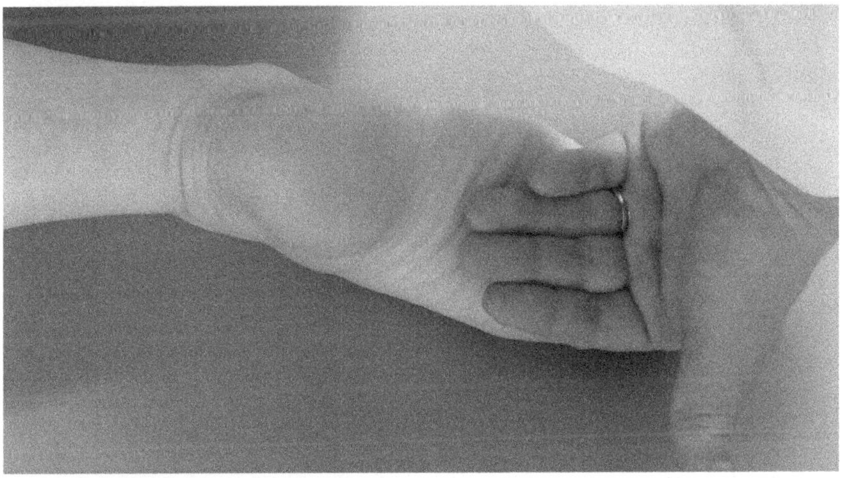

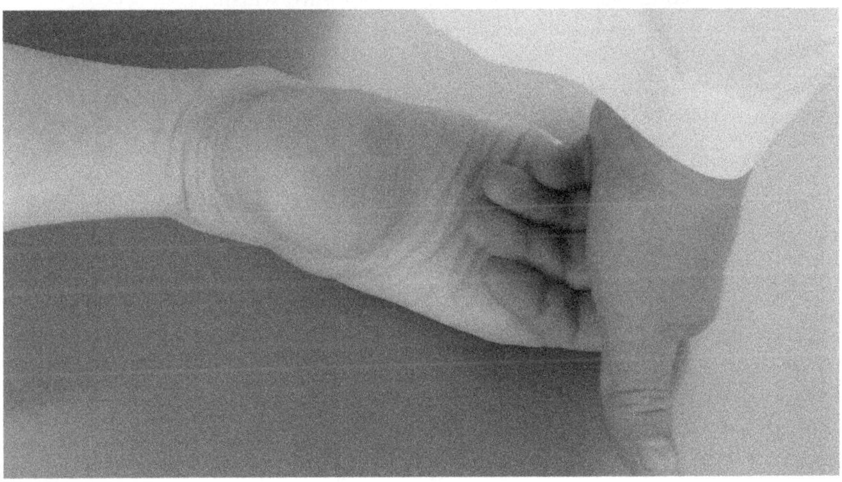

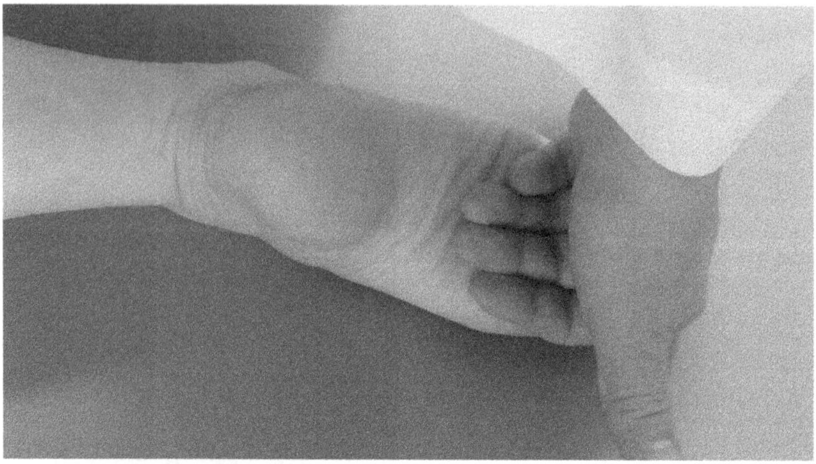

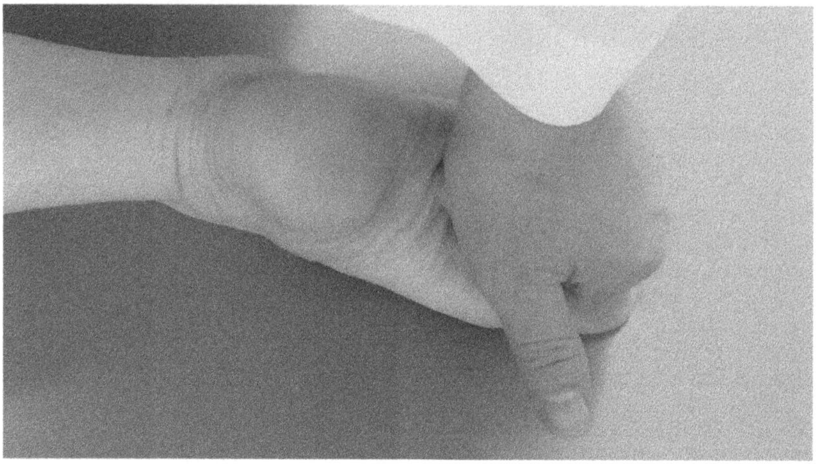

7) Rubs the plantar aspect of the affected foot until warm.

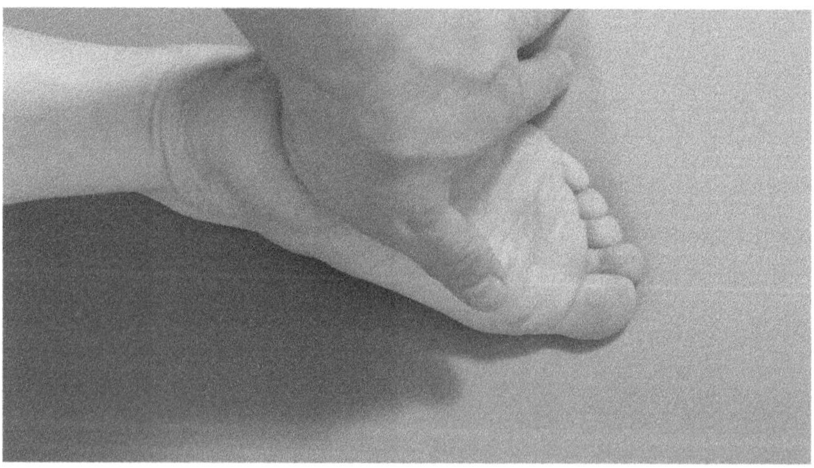

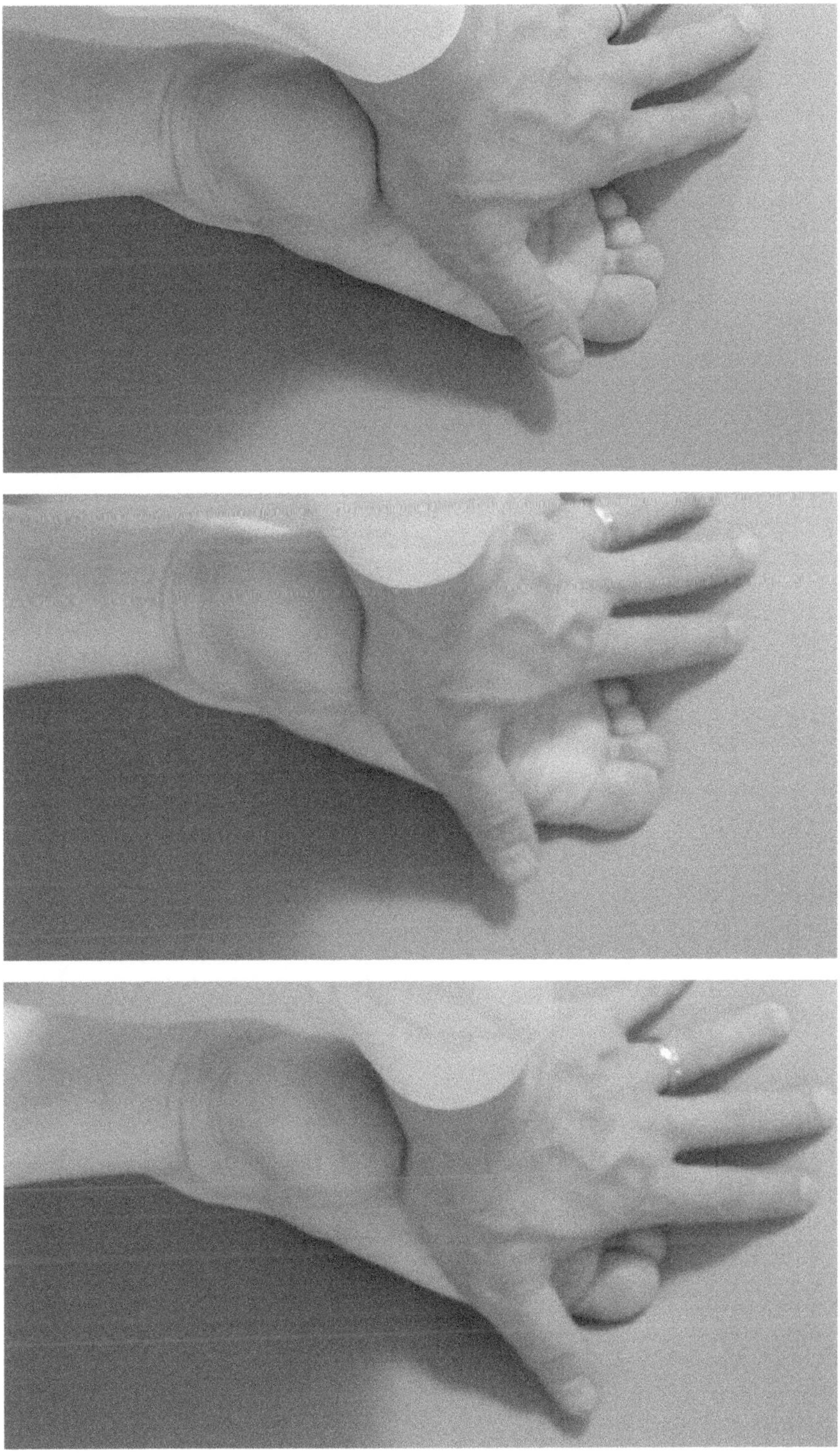

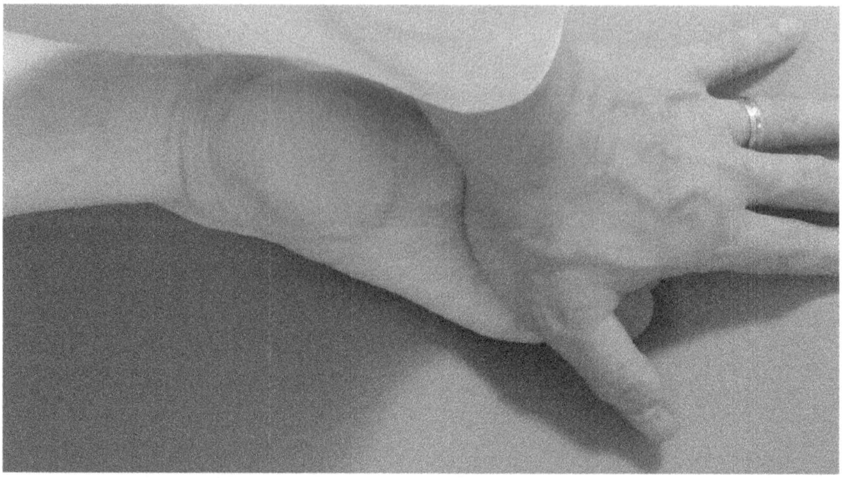

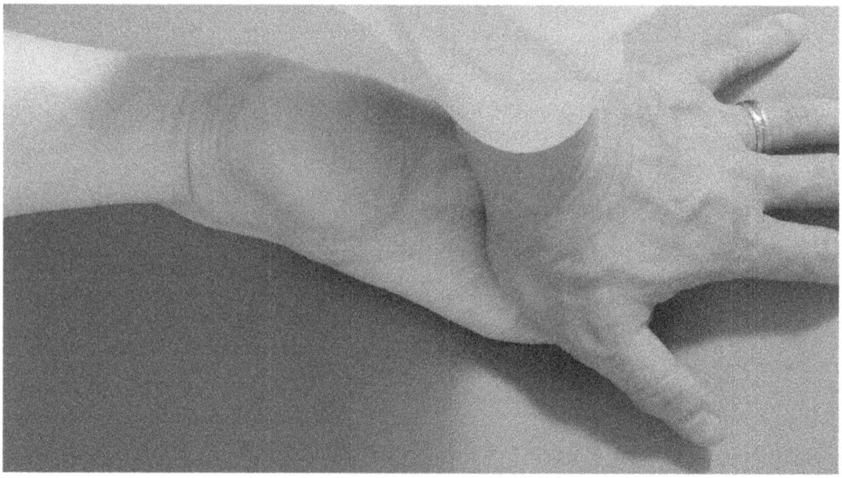

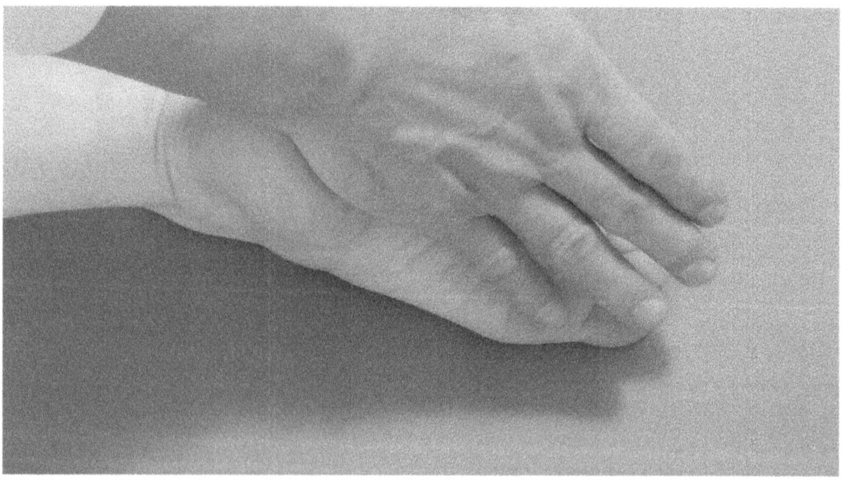

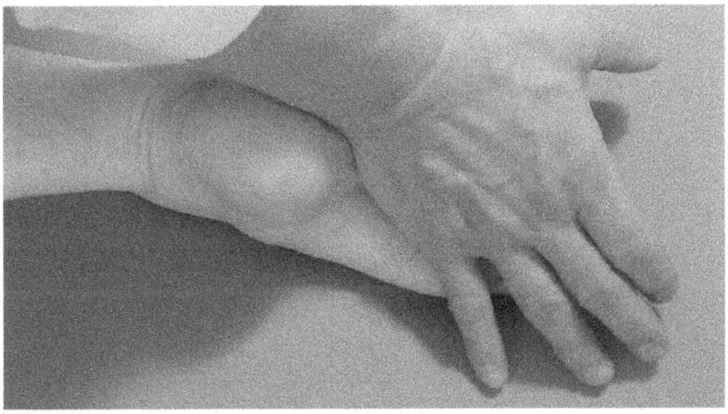

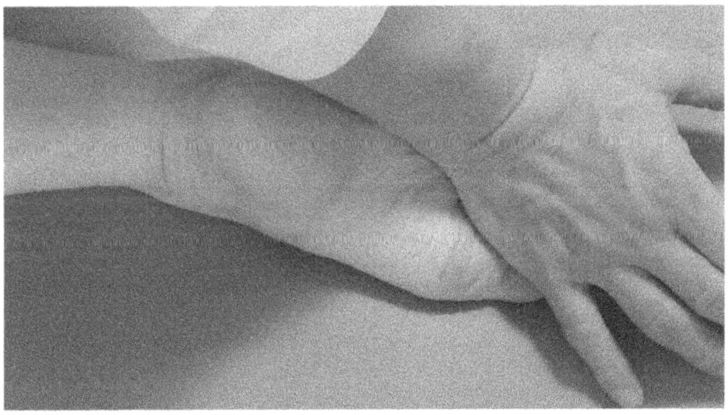

8) Both hands push the gastrocnemius muscle until warm.

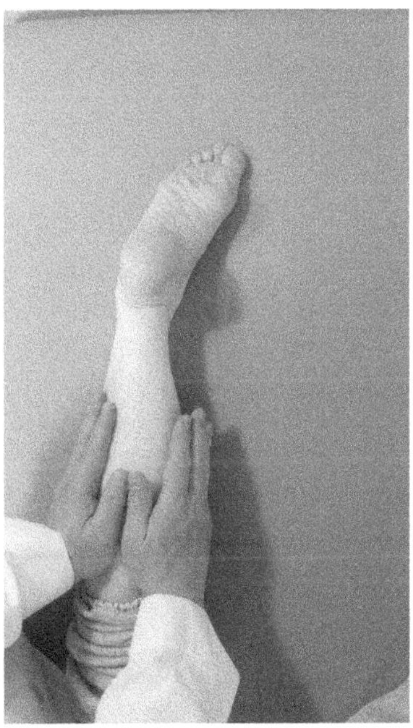

Patient Advisory

- Lose weight if overweight.
- Avoid high-impact activities such as running, hiking, jumping, and dancing; may add other low-impact exercises into routine activities, such as swimming or bicycling.
- Icing the affected heel:
  1) Freeze a bottle of water overnight.
  2) While sitting on a stool and wearing a sock, place the water bottle on the floor.
  3) Roll the bottle of water forward and backward with the affected foot.
  4) Practice the rolling for 5–10 minutes each time, two or three times a day.
  5) Never put ice directly on the skin.
- Standing calf stretching:
  1) Stand with upright position facing a wall at such a distance that both hands can reach the wall.
  2) Put the hands on the wall with arms at shoulder height.
  3) Step one leg backward far enough to feel the tightness on the calf, point the toes forward, and lift the big toe. Make sure the knee is extended.
  4) Bring the other leg forward toward the wall and flex the knee at a 90 degree angle with the toes pointing forward.
  5) Be sure to keep the head upright and the back straight, and breathe naturally.
  6) Remain in the stretched position for 30 seconds.
  7) Return to the standing upright position and repeat with the other leg.

Dietary recommendation:

- Foods that tonify kidney and liver, such as goji berries, lamb, sea cucumber, shrimp, walnut, chive, and oyster.

SPRAINED ANKLE 踝關節扭傷

Introduction

The ankle bears the full weight of the body. The bones articulate together by ligaments, but they allow some movements of the foot in flexion, extension, and rotation.[22] The injury happens when the movements pass the yield point of the ligament's range of motion, such as an excessive external rotation, inversion, or eversion of the foot. Sprains most often happen on the lateral side of the ankle; pain, bruising, and swelling may present not immediately but a few hours later. Without proper treatment and rehabilitation, the ankle may not heal well and may not be fully recovered, resulting in recurrent sprains.

Anatomy

The ankle is a synovial joint comprised of:

- Tibia
- Fibula
- Talus
- Calcaneus

The ankle consists of three distinct joints:

- The talocrural joint is formed by the tibia, fibula, and talus. This joint is responsible for flexion and extension of the foot.
- The talocalcaneal joint, also known as the subtalar joint, is formed by the talus and the calcaneus. This joint is responsible for side rotation of the foot.
- The inferior tibiofibular joint is between the tibia and the fibula. It is a syndesmosis joint, a fibrous joint between the fibular notch of the tibia and the distal epiphysis of the tibia. It is reinforced by the anterior and posterior tibiofibular ligaments and the inferior transverse and interosseous ligament.[23]

The ankle is supported by many ligaments; the lateral ligament complex is a key structure because most ankle sprains are located at the lateral side.

- *The anterior talofibular ligament* (ATFL) connects the talus with the fibula.
- *The calcaneofibular ligament* (CFL) connects the fibula with the calcaneus.[24]
- *The posterior talofibular ligament* (PTFL) originates from the malleolar fossa and inserts almost horizontally into the posterolateral talus.[25]

The medial (deltoid) ligaments are much stronger than the lateral ligament; therefore, injury is much less frequent.[26]

Western Medicine Etiology and Pathology

- Ankle sprains consist of high ankle sprains and low ankle sprains. Clinically, most ankle sprains are low ankle sprains.
- Inversion sprain is the most frequently seen clinically.[27]
- The anterior talofibular ligament (ATFL) and the calcaneofibular ligament (CFL) are the most commonly injured ligaments with inversion ankle sprain.[28] The posterior talofibular ligament (PTFL) is rarely injured except in association with a complete dislocation of the talus.
- Sports involving jumping, cutting, rolling, and twisting of the foot are more susceptible to the injury; examples are basketball, volleyball, tennis, and soccer.
- Walking or running on an uneven surface increases the risk of an ankle sprain.
- Prior ankle sprain confers a high risk of suffering from another sprain.
- Shoes that do not fit properly, high-heeled shoes, or the wrong type of shoe for an activity increase vulnerability to injury.

Chinese Medicine Etiology and Pathology

The concept of traditional Chinese medicine believes that maintaining free flow of Qi and blood circulation prevents pain occurrence. When there is a mild injury, Qi and blood are able to repair it. Qi and blood supply nutrients to nourish the traumatic tissues and clear the harmful products of damage. However, if the injury is too severe and it is beyond the ability of Qi and blood to remove the blockage, the flow of the Qi and blood circulation within the meridians near the affected area becomes stagnant and impaired, resulting in pain.

Manifestations

- Stabbing pain in the affected area
- A feeling or an audible popping sound at the time of injury if the tear is severe
- Tenderness to the touch
- Swelling
- Bruising
- Stiffness

Clinically, the sprained ankle is divided into three grades:

Grade 1: The ankle is sore and may be slightly swollen, but it is still able to bear weight. Usually, there is no bruising or joint instability. The ligament is not torn; it is just overstretched.

Grade 2: The ankle has pain, bruising, and swelling and is unable to bear weight. The ligament is partially torn.

Grade 3: The ankle is suffering from severe pain, bruising, and swelling. A feeling or an audible popping sound may present at the time of injury. The ligament is completely torn. The ankle joint is unstable, has lost its functions and range of motion, and is unable to bear any body weight.

Physical Examination

- Begin by obtaining the patient's medical history:
  - The history of the mechanism of injury, inversion, or eversion. A plantar flexion/inversion injury may indicate damage to the lateral ligament; a dorsiflexion/eversion injury may indicate damage to the medial ligament. Also, how the spraining occurred, whether from jumping, running, or hiking, or from other accidents.
  - History of ankle sprain. For example, if the same ankle was sprained in the past and was not properly rehabilitated, this may give us an idea as to whether the ankle was unstable and vulnerable to this injury.
- Inspection:
  - Inspect movement coordination, difficulty in walking, balance issues, and difficulty in grasping objects.

- Soft tissue presentations such as bruising, edema, and skin temperature.
- Palpation to location: Anterior talofibular ligament (ATFL), calcaneofibular ligament (CFL), and posterior talofibular ligament (PTFL). Tenderness elicited along their courses may indicate such damage. Be sure to begin palpating with light pressure, as this can potentially be acutely painful for the patient.
- Gait issues such as decrease in balance when standing on the affected leg.

- Range of motion:
    - Joint flexibility includes passive range of motion, with or without pain and contracture, and active range of motion, which may be limited or may have disappeared. The affected ankle may show decreased motion in the joint.[29]
    - Sensation of weakness of the muscles that move the ankle joint.[30]
    - Joint hypermobility or hypomobility can be checked by performing the following special tests.

Anterior Drawer Test (Supine)

The anterior drawer test assesses the stability of the anterior talofibular ligament.

- With the patient supine and the knee extended, place the ankle in approximately 20 degrees of plantar flexion.
- Stabilize the anterior aspect of the distal tibia with one hand, hold the posterior aspect of the calcaneus with the palm of the other hand, and attempt to bring the calcaneus and talus forward on the tibia.
- A positive test results if the talus translates forward. The foot should be able to translate slightly forward on the ankle before reaching a relatively firm endpoint provided by the anterior talofibular ligament. If the firm endpoint is absent, this indicates that there is a moderate to severe injury to the anterior talofibular ligament, or a grade 2 or 3 injury. Usually, a tear in the ligament allows the talus to slide forward on the tibia, because the anterior talofibular ligament is the only ligament preventing displacement of the talus anteriorly. Therefore, test for anterior instability between the tibia and the talus using the anterior drawer test.[31]
- Positive test results are often graded on a "0 to 3 scale," with 0 indicating no laxity and 3 indicating gross laxity.
- When performing this test, always compare the affected ankle with the uninjured ankle.

Inversion Talar Tilt Test (Supine or Sitting)

This is a forced inversion test that can assess the stability of the anterior talofibular ligament (ATFL), calcaneofibular ligament (CFL), and posterior talofibular ligament (PTFL).

- The patient is supine or sitting with the legs over the edge of the examination table. The examiner stands or sits in front of the patient.
- Stabilize the anterior aspect of the distal tibia with one hand, and hold the posterior aspect of the calcaneus with the palm of the other hand.
- To test the anterior talofibular ligament (ATFL), place the ankle in a position of approximately 20 degrees of plantar flexion, and provide an inversion stress by rolling the calcaneus medially. The test is positive if the pain is reproduced at the course of the ATFL.
- To test the calcaneofibular ligament (CFL), place the ankle in a neutral position of plantar flexion and dorsiflexion, and provide an inversion stress by rolling the calcaneus medially. The test is positive if the pain is reproduced at the course of the CFL.
- To test the posterior talofibular ligament (PTFL), place the ankle in the position of maximal dorsiflexion, and provide an inversion stress by rolling the calcaneus medially. The test is positive if the pain is reproduced at the course of the PTFL.
- When performing this test, always compare the affected ankle with the contralateral ankle.

Differential Diagnosis

Ankle Fracture

- The location of pain is at the fracture area.
- Severe pain may be elicited by a percussion on the calcaneus toward the tibia.
- Severe pain occurs immediately.
- Ankle swelling.

Ankle Dislocation

- A deformed look to the ankle.
- Severe pain occurs immediately.
- Ankle swelling.

Damage to the Medial Ligament

- Location is on the medial aspect of the ankle.

Treatment

Treatment principle:

- Move blood, disperse bruising, reduce swelling, and stop pain.

Herbal Treatments

Early Stage

Herbal formula:

- Shu Jin Huo Xue Tang 舒筋活血湯.

Formula ingredients:

- Qiang Huo 羌活 6 g, Fang Feng 防風 9 g, Jing Jie 荊芥 6 g, Du Huo 獨活 9 g, Dang Gui 當歸 12 g, Xu Duan 續斷 12 g, Qing Pi 青皮 5 g, Niu Xi 牛膝 9 g, Wu Jia Pi 五加皮 9 g, Du Zhong 杜仲 9 g, Hong Hua 紅花 6 g, Zhi Ke 枳殼 6 g.

Ingredient explanation:

- Qiang Huo 羌活 expels wind-cold-dampness, unblocks painful obstruction, alleviates or stops pain, and guides Qi to the Tai Yang and Du channels.
- Fang Feng 防風 releases the exterior, expels external wind, expels wind-dampness.
- Jing Jie 荊芥 stops bleeding, dispels wind, relieves muscle spasm.
- Du Huo 獨活 expels wind-dampness, stops pain.
- Dang Gui 當歸 moves and tonifies blood, disperses cold, and stops pain due to blood stasis.
- Xu Duan 續斷 nourishes liver and kidneys, strengthens bones and tendons, alleviates pain.
- Qing Pi 青皮 soothes liver Qi and breaks up stagnant Qi, dries dampness, and transforms phlegm.
- Niu Xi 牛膝 expels wind-dampness, tonifies liver and kidneys, directs herbs downward.
- Wu Jia Pi 五加皮 expels wind and damp, nourishes and warms liver and kidneys, strengthens bones and tendons.
- Du Zhong 杜仲 tonifies the kidneys and liver, strengthens the tendons and bones, and aids the smooth flow of Qi and blood. When Xu Duan 續斷 works together with Du Zhong 杜仲, they are good partners to strengthen the lumbar region by tonifying liver and kidneys and secure the Chong and Ren meridians to release the low back pain and weakness of the knee joint, resulting in easier walking.
- Hong Hua 紅花 moves blood, releases bruising, opens meridians, stops pain.
- Zhi Ke 枳殼 removes stagnant food, promotes Qi movement, reduces distention and pressure.

Cooking instructions:

1) Place the herbs into an herb pot. Traditionally, a clay cooker is used. Gently rinse the herbs once with a strainer. You may notice some red-colored flowers in the strainer; place them back in the herb pot.
2) Add water into the pot until it is just enough to cover the herbs, and then add one more cup of water.
3) Presoak the herbs for 15 minutes.
4) Turn on the stove to high heat and bring the herbs to a boil. Then reduce the heat to medium low to simmer but with slight bubbling. Turn off the heat when the decoction is reduced to one cup.
5) Drain the decoction out into a container.
6) Repeat steps 2–5 twice and drain each decoction into the same container.
7) Divide these three cup decoctions into three separate cups,

Dose:

- Drink the decoction one cup each time, three times a day.

External formula:

- Huo Xue Xiao Zhong San 活血消腫散.

Formula ingredients:

Da Huang 大黃 20 g, Ce Bai Ye 側柏葉 20 g, Ze Lan 澤蘭 10 g, Huang Bai 黃柏 10 g, Fang Feng 防風 10 g, Ru Xiang 乳香 10 g.

Ingredient explanation:

- Da Huang 大黃 moves blood and disperses bruising.
- Ce Bai Ye 側柏葉 cools blood, stops bleeding, clears damp heat, and dispels wind-dampness.
- Ze Lan 澤蘭 moves blood, promotes urination, and reduces swelling.
- Huang Bai 黃柏 clears deficient kidney Yin heat.
- Fang Feng 防風 releases the exterior, expels external wind, expels wind-dampness.
- Ru Xiang 乳香 moves blood, stops pain.

Formula instructions:

1) Grind the herbs into granules.
2) Spread some Vaseline on a piece of gauze.
3) Sprinkle the herbal powders onto about 0.2 cm of Vaseline.
4) Apply the gauze to the site of injury.

Intermediate and Later Stages

Herbal formula:

- Zhi Shang Er Fang 肢傷二方.

Formula ingredients:

Dang Gui 當歸 12 g, Chi Shao 赤芍 12 g, Xu Duan 續斷 12 g, Wei Ling Xian 威靈仙 12 g, Yi Yi Ren 薏苡仁 30 g, Sang Ji Sheng 桑寄生 30 g, Gu Sui Bu 骨碎補 12 g, Wu Jia Pi 五加皮 12 g.

Ingredient explanation:

- Dang Gui 當歸 moves and tonifies blood, disperses cold, and stops pain due to blood stasis.
- Chi Shao 赤芍 moves blood, reduces swelling, stops pain.
- Xu Duan 續斷 nourishes liver and kidneys, strengthens bones and tendons, alleviates pain.
- Wei Ling Xian 威靈仙 expels wind-dampness, stops pain.
- Yi Yi Ren 薏苡仁 promotes urination, leaches out dampness, strengthens the spleen, and clears damp heat.
- Sang Ji Sheng 桑寄生 expels wind-dampness, tonifies liver and kidney Yin.
- Gu Sui Bu 骨碎補 tonifies kidney Yang, strengthens tendons and bones, and moves blood, especially due to trauma.
- Wu Jia Pi 五加皮 expels wind and damp, nourishes and warms liver and kidney, strengthens bones and tendons.

Cooking instructions:

1) Place the herbs into an herb pot. Traditionally, a clay cooker is used. Gently rinse the herbs once with a strainer.
2) Add water into the pot until it is just enough to cover the herbs, and then add one more cup of water.
3) Presoak the herbs for 15 minutes.
4) Turn on the stove to high heat and bring the herbs to a boil. Then reduce the heat to medium low to simmer but with slight bubbling. Turn off the heat when the decoction is reduced to one cup.
5) Drain the decoction out into a container.
6) Repeat steps 2–5 one more time and drain each decoction into the same container.
7) Divide these two cup decoctions into two separate cups.

Dose:

- Drink the decoction one cup each time, twice a day.

External Formula

- Zheng Fu Tang Yao 正骨湯藥.

Formula ingredients:

Dang Gui 當歸 12 g, Qiang Huo 羌活 12 g, Hong Hua 紅花 12 g, Bai Zhi 白芷 12 g, Ru Xiang 乳香 12 g, Mo Yao 没藥 12 g, Gu Shui Bu 骨碎補 12 g, Xu Duan 續斷 12 g, Fang Feng 防風 12 g, Mu Gua 木瓜 12 g, Hua Jiao 花椒 12 g, Tou Gu Cao 透骨草 12 g.

Ingredient explanation:

- Dang Gui 當歸 moves and tonifies blood, disperses cold, and stops pain due to blood stasis.
- Qiang Huo 羌活 expels wind-cold-dampness, unblocks painful obstruction, alleviates or stops pain, and guides Qi to the Tai Yang and Du channels.
- Hong Hua 紅花 moves blood, releases bruising, opens meridians, stops pain.
- Bai Zhi 白芷 expels wind, drains dampness, dispels cold, and stops pain.
- Ru Xiang 乳香moves blood and stops pain.
- Mo Yao 没藥 moves blood, reduces swelling, and stops pain.
- Gu Shui Bu 骨碎補 tonifies kidney Yang, strengthens tendons and bones, and moves blood, especially due to trauma.
- Xu Duan 續斷 nourishes liver and kidney, strengthens bones and tendons, alleviates pain.
- Fang Feng 防風 releases the exterior, expels external wind, expels wind-dampness.
- Mu Gua 木瓜 promotes Qi and blood circulation and unblocks the channels, releases spasm.
- Hua Jiao 花椒 warms middle Jiao, disperses cold, dispels damp, and stops pain.
- Tou Gu Cao 透骨草 moves blood, breaks up stubborn blood stasis, reduces swelling, clears heat, cools blood, stops pain.

Application procedures:

- Place all herbs in a fabric bag.
- Steam the bag until warm.
- Place the bag on the sprained ankle for 1 hour twice daily.

Formula action:

- Moves blood, relaxes tendons

Acupuncture Treatments

Acupuncture point formula:

SP10 血海 Xuehai, LIV8 曲泉 Ququan, SP6 三陰交 Sanyinjiao, GB34 陽陵泉 Yanglingquan, LI4 合谷 Hegu, LIV3 太衝 Taichong, KD3 太溪 Taixi, BL60 崑崙 Kunlun and Ashi acupoints 阿是穴 (these are tender points that are not included in the classic acupuncture points) on the affected ankle.

Acupuncture instructions:

1) Insert a 1.5 cun needle on SP10 血海 Xuehai, GB34 陽陵泉 Yanglingquan, and Ashi acupoints 阿是穴, and manipulate the needle with sedating techniques.
2) Insert a 1.5 cun needle on SP6 三陰交 Sanyinjiao, UB23 腎俞 Shenshu, and LIV8 曲泉 Ququan, and manipulate the needle with tonifying techniques.
3) Insert a 1 cun needle perpendicularly on KD3 太溪 Taixi, and manipulate the needle with tonifying techniques.
4) Insert a 1 cun needle perpendicularly on LI4 合谷 Hegu and LIV3 太衝 Taichong, and manipulate the needle with sedating techniques.

In addition, the following acupuncture points are added according to the damaged ligament. These three points are stimulated using twisting, lifting, and thrusting techniques:

- BL60 崑崙 Kunlun is added for posterior talofibular ligament (PTFL). A needle is inserted at a 30 degree angle toward the malleolus.
- GB40 丘墟 Qiuxu is added for anterior talofibular ligament (ATFL). A needle is inserted perpendicularly.
- BL62 申脈 Shenmai is added for calcaneofibular ligament (CFL). A needle is inserted perpendicularly.

Needle retention for each session is 30 minutes.

Acupuncture treatment frequency is once a day for 5 days, then take 2 days off.

Bloodletting instructions:

1) The patient lies supine with the leg extended; the ankle is in a neutral position. The therapist wears gloves and stands beside the affected ankle.
2) Prepare the bloodletting region with the regular Clean-Needle-Technique procedures.

3) The therapist holds a medical lancet with the fingers, and using spot-pricking bleeding technique, quickly pricks the bleeding ligament to the depth of 0.05 cun.
4) Elicit 10 drops of blood.
5) Absorb the blood with sterile cotton balls.
6) Dispose of the used medical lancet, the cotton balls, and the used gloves in the biohazard trash container.
7) Place a band-aid on the bloodletting region.
8) Wash hands.

Cupping instructions:

1) The patient lies supine with the leg extended; the ankle is in a neutral position. The therapist stands beside the affected ankle.
2) Pick up a cotton ball with long medical tweezers.
3) Wet the cotton ball with rubbing alcohol.
4) Hold a glass cup with one hand and hold the tweezers with the other hand.
5) Set the cotton ball on fire.
6) Insert the flaming cotton ball into the glass cup for 3 seconds or until the glass cup feels warm. The glass cup should still be cool enough to the hand that is holding it.
7) As soon as the cotton ball has been removed from the glass cup, place the cup onto the skin over the affected ankle, release the hand, and let go. The glass cup should stay on the skin.
8) Extinguish the flaming cotton ball immediately and drop it into a fire-resistant cup containing water. Cover the cup with a lid.
9) Cupping retaining time should not be longer than 10 minutes
10) To remove a glass cup, press the skin beside its rim to let air seep in slowly.

Tui Na Treatments

Tui Na techniques:

- Grasping, kneading, pressing, bending, rotating, rubbing

Tui Na procedures:

1) The patient lies supine with the affected leg slightly flexed at approximately 30 degrees and the ankle in a neutral position; the therapist stands near to the affected ankle
2) Grasp UB60 崑崙 Kunlun and KD3 太溪 Taixi with a thumb and an index finger for approximately 3 minutes.

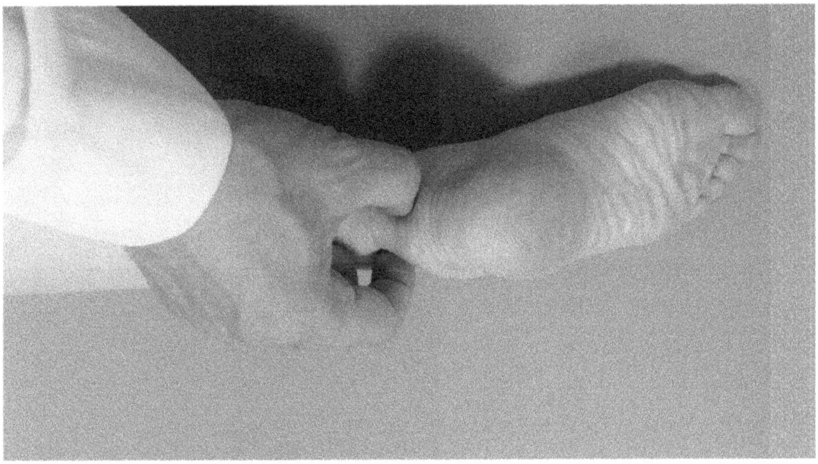

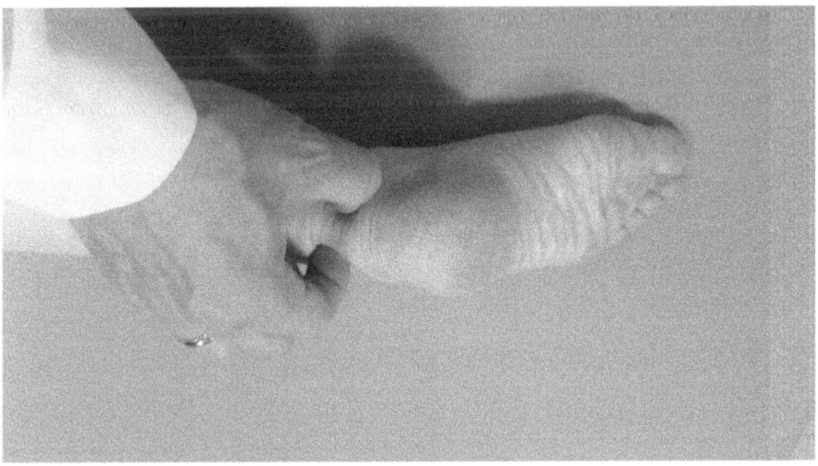

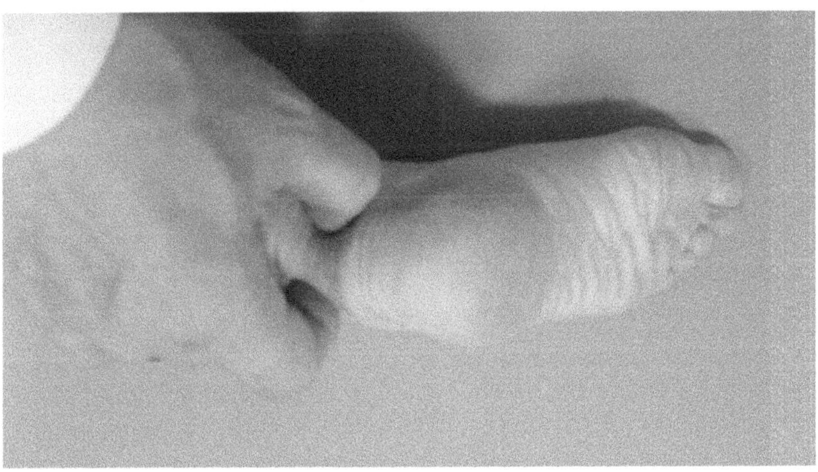

3) Knead UB60 崑崙 Kunlun, GB40 丘墟 Qiuxu, and UB62 申脈 Shenmai for 5 minutes.

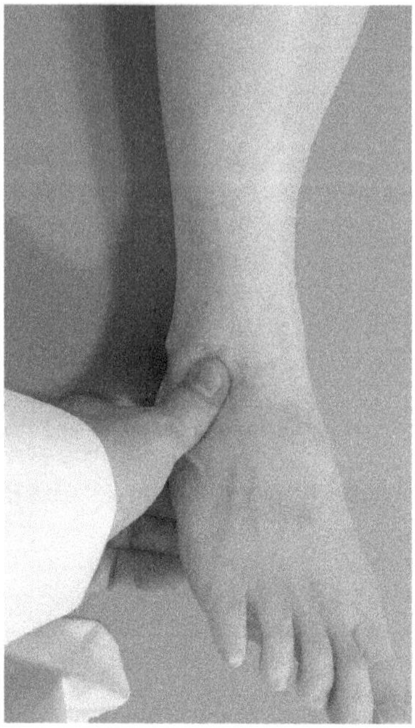

4) Repeat the last step with pressing and kneading on the acupuncture points.
5) Press with thumb tip on Ankle Anterior Blood Point, which is located at ST41 解谿 Jiexi, for 5–7 minutes or until hot.

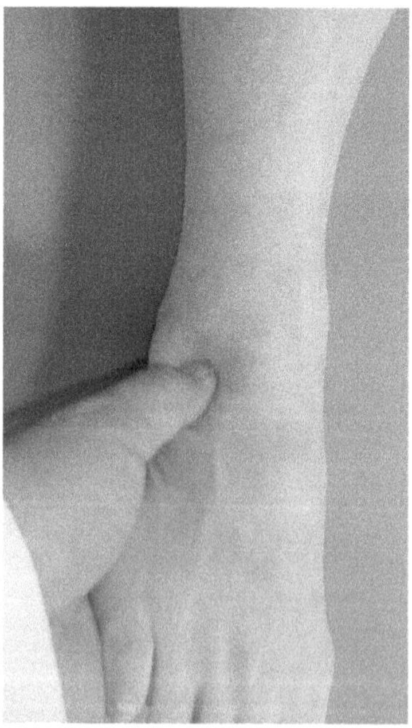

6) Press with thumb tip on Ankle Bilateral Blood Points, which are located at GB40 丘墟 Qiuxu and SP5 商丘 Shangqiu, for 5–7 minutes or until hot.

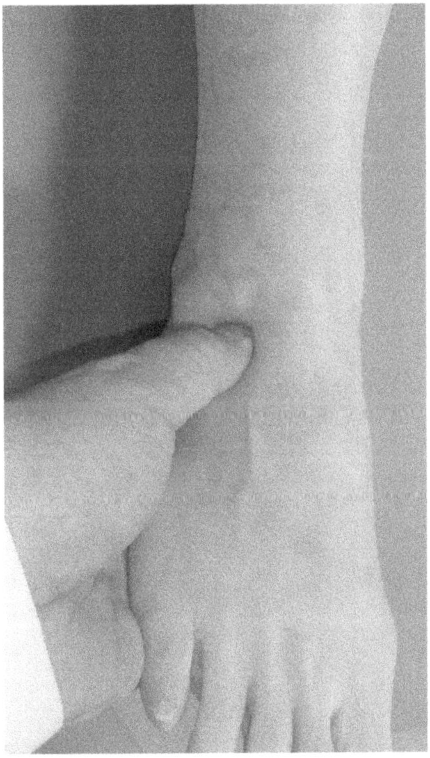

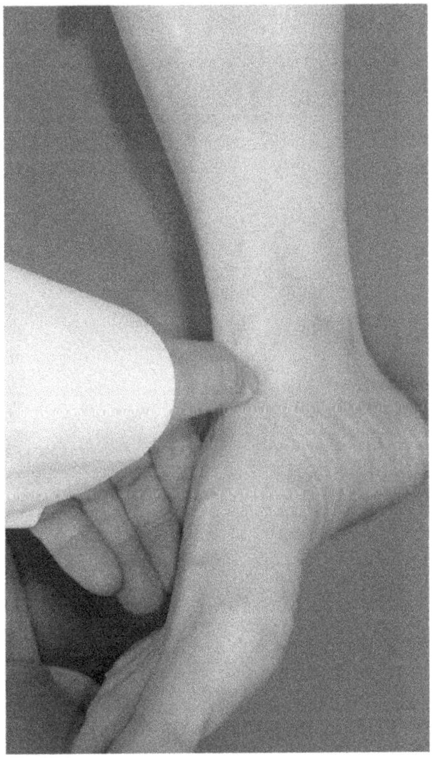

7) Place both thumbs on the affected area. Kneads the affected area with both thumbs.

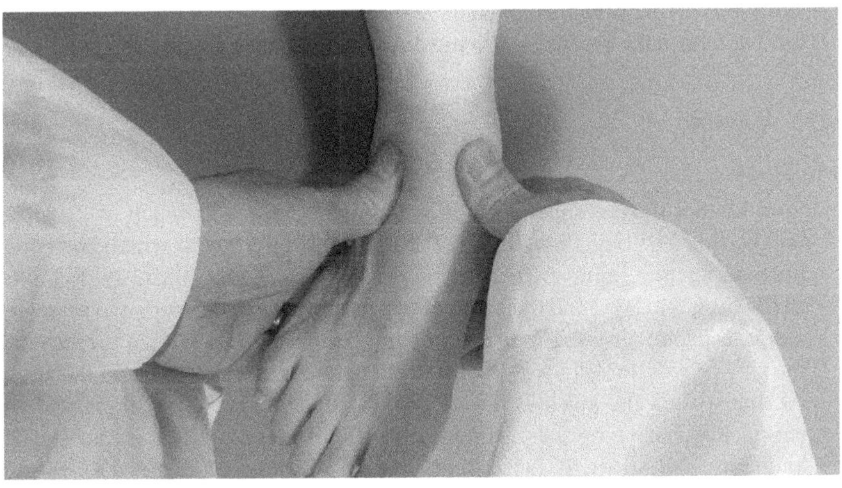

8) Dorsiflex and plantar flex the affected ankle 10 times.
9) Rotate the ankle by holding the distal tibia with one hand and holding the anterior foot with the other hand. Rotate with caution; it should start at slow speed and with small angles and increase gradually. Repeat about 10–15 times.

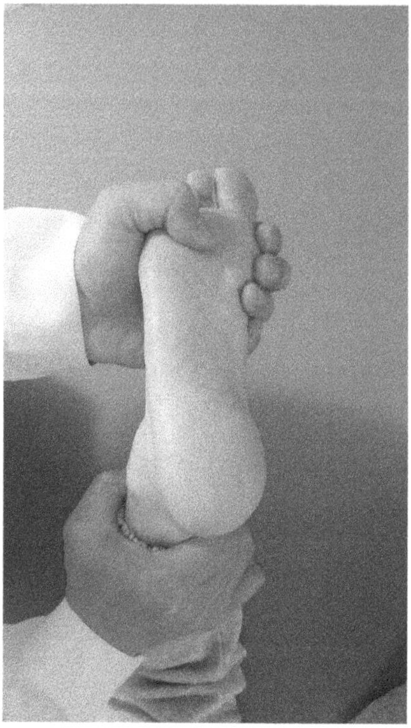

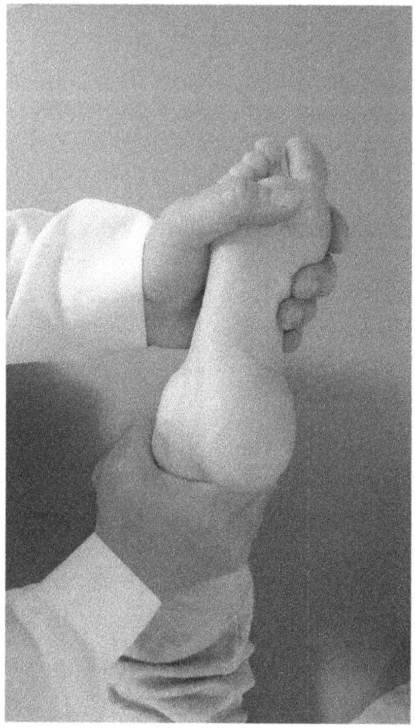

10) Apply rubbing techniques on the medial and lateral areas of the foot until red.
11) Immobilize the joint with a bandage.
12) Tui Na treatment begins on the third or fifth day after injury.

Patient Advisory

Acute Stage

- Stop further movements by the affected ankle; rest it as much as possible for the first 2–3 days. Swelling is the natural "splint" with which the body prevents further injury by restricting further movements of the ankle joint.
- In the first 2–3 days, and the sooner the better, put an ice pack on the affected ankle, especially on the injured ligament. Be sure not to put ice directly on the skin; cover the skin with a thin towel between the ice pack and the skin.
- At times when the ankle is not receiving an acupuncture treatment and an ice pack is not being used, immobilize the affected ankle with an elastic bandage such as an ACE bandage to stop bleeding, reduce swelling, and

limit the ankle movement. During the immobilization with the elastic bandage, be sure to set the ankle joint at the neutral position with slight eversion when the injury is inversion or with slight inversion when the injury is eversion. Usually, the duration of immobilization is approximately 3 weeks for grade 1 and grade 2. Immobilization for grade 3 takes 4–6 weeks.

- In the first 2–3 days, and the sooner the better, raise the affected ankle higher than the heart when seated or lying. This can greatly prevent further effusion.
- Do not let the affected ankle bear weight in the first few days until the swelling and pain improve.

Recovery Stage

- To assist recovery from an ankle sprain, the sooner movement training is started, the better the prognosis. To prevent future ankle sprain, the ankle needs to recover fully, and the ankle functions need to be fully regained.
- In grade 1 ankle sprain, range of motion exercises and muscle strengthening exercises start on the second or the third day of injury. In grade 2 ankle sprain, the exercise begins on the third or the fifth day of injury. For grade 3 ankle sprain, the day to begin exercise depends on the severity of the ligament injury and the complications of the injury, such as a fracture.
- The exercise begins with the movements of the toes during immobilization. The ankle joint exercise begins after immobilization.
- Continue to receive treatment for pain management. Chinese medicine is intended not only to provide treatment for pain management but also to tonify kidney, liver, and spleen, which means to increase the strength inside the body, to increase the strength and flexibility of the ligaments, and to promote circulation and accelerate healing as well.

Prevention

Treatment is important, but the prevention of future sprain is more crucial.

- In the beginning when returning to exercise, be careful when running or walking on an uneven surface, such as running on hiking trails.
- Always warm up before exercise.
- Always wear shoes that fit the feet well and are made for the activity.
- Practice stability and balance exercises to strengthen the leg and foot muscles.
- Do not continue the activity if there is pain or injury to the ankle.

Dietary Suggestions

Patients should be advised to avoid a Fa Wu diet. Fa Wu foods may cause more inflammation and bring old injuries back. Some examples of Fa Wu foods are shellfish such as crabs and shrimps, fish, fried foods, lamb, duck, chocolate, garlic, and green onions. This concept applies to both acute and chronic strains.

NOTES

1. www.contact.org.uk/advice-and-support/hemiplegia-support/what-is-hemiplegia/
2. https://www.ncbi.nlm.nih.gov/books/NBK534818/
3. www.ncbi.nlm.nih.gov/books/NBK534818/
4. www.acquiredbraininjury.com/abi_manual/the-cerebral-cortex
   www.headway.org.uk/about-brain-injury/further-information/about-the-brain/
5. www.webmd.com/stroke/news/20140212/could-the-weather-affect-your-stroke-risk#1
6. www.sciencedirect.com/topics/medicine-and-dentistry/yin-deficiency
7. https://physioworks.com.au/injuries-conditions-1/calf-muscle-tears
8. www.ncbi.nlm.nih.gov/pmc/articles/PMC1978447/
9. www.sciencedirect.com/topics/neuroscience/gastrocnemius-muscle
10. https://www.ncbi.nlm.nih.gov/books/NBK534766/
11. www.ncbi.nlm.nih.gov/pmc/articles/PMC3845475/
    https://sma.org.au/resources-advice/injury-fact-sheets/gastrocnemius-calf-strain/
12. www.ncbi.nlm.nih.gov/books/NBK534766/
13. www.ncbi.nlm.nih.gov/books/NBK534766/
14. https://physioworks.com.au/injuries-conditions-1/calf-muscle-tears
15. www.ncbi.nlm.nih.gov/books/NBK534766/
16. www.ncbi.nlm.nih.gov/pmc/articles/PMC2697334/
17. www.medicalnewstoday.com/articles/322809.php#other-calf-injuries
18. www.ncbi.nlm.nih.gov/pmc/articles/PMC3842207/
19. www.foothealthfacts.org/conditions/tarsal-tunnel-syndrome
20. www.foothealthfacts.org/conditions/tarsal-tunnel-syndrome
21. www.healthline.com/health/tarsal-tunnel-syndrome#symptoms
22. http://anatomy.lexmedicus.com.au/pathologies/ankle
23. http://anatomy.lexmedicus.com.au/pathologies/ankle
24. http://anatomy.lexmedicus.com.au/pathologies/ankle
25. www.ncbi.nlm.nih.gov/pmc/articles/PMC2855022/
26. https://www.physio-pedia.com/Lateral_Ligament_Injury_of_the_Ankle
27. www.ncbi.nlm.nih.gov/pmc/articles/PMC2855022/
28. www.ncbi.nlm.nih.gov/pmc/articles/PMC2855022/
29. www.runnersworld.com/advanced/a20813217/ask-the-coaches-ankle-injury-changes-gait/
30. www.runnersworld.com/advanced/a20813217/ask-the-coaches-ankle-injury-changes-gait/
31. https://books.google.com/books?id=hIq32ObcDZIC&pg=PA361&lpg=PA361&dq=Forced+Inversion+test&source=bl&ots=fQjkIQQ8XD&sig=ACfU3U3JzN8dBKwCZFG4rEhUTYZNGU4ogg&hl=en&sa=X&ved=2ahUKEwjj5M7o-IXmAhXTrJ4KHZZUA4oQ6AEwEHoECA0QAQ#v=onepage&q=Forced%20Inversion%20test&f=false

Bibliography

Alexander, Christopher E. "Lumbosacral Facet Syndrome." StatPearls [Internet]. U.S. National Library of Medicine, January 20, 2020. https://www.ncbi.nlm.nih.gov/books/NBK441906.

Alvarez, Jaime A., and Russell H. Hardy. "Lumbar Spine Stenosis: A Common Cause of Back and Leg Pain." *American Family Physician*, April 15, 1998. https://www.aafp.org/afp/1998/0415/p1825.html.

Bell, David Robert, Mason D. Smith, Anthony P. Pennuto, Mikel R. Stiffler, and Matthew E. Olson. "Jump-Landing Mechanics after Anterior Cruciate Ligament Reconstruction: A Landing Error Scoring System Study." *Journal of Athletic Training*. National Athletic Trainers Association, 2014. https://www.ncbi.nlm.nih.gov/pmc/articles/PMC4151830/.

Bhagia, Sarjoo. "What Is the Role of Blood Supply in the Pathophysiology of Meniscal Injury?" Latest Medical News, Clinical Trials, Guidelines—Today on Medscape, November 12, 2019. https://www.medscape.com/answers/308054-158697/what-is-the-role-of-blood-supply-in-the-pathophysiology-of-meniscal-injury.

Bongers, P.M., et al. "Back Pain and Exposure to Whole Body Vibration in Helicopter Pilots." *Ergonomics*, vol. 33 (1990): 1007–26. https://www.nature.com/articles/s41598-019-42597-w#ref-CR22.

Brouhard, Rod. "The Anatomy of the Coccyx." Verywell Health. Verywell Health, February 13, 2020. https://www.verywellhealth.com/coccyx-anatomy-4587599.

Chen, B., J. Tao, Y. Lin, R. Lin, W. Liu, and L. Chen. "Electro-Acupuncture Exerts Beneficial Effects against Cerebral Ischemia and Promotes the Proliferation of Neural Progenitor Cells in the Cortical Peri-infarct Area through the Wnt/β-catenin Signaling Pathway." *International Journal of Molecular Medicine*, vol., 36 (2015): 1215–22. doi: 10.3892/ijmm.2015.2334. https://www.ncbi.nlm.nih.gov/pmc/articles/PMC5713240/.

Choi, Yong-Soo. "Pathophysiology of Degenerative Disc Disease." *Asian Spine Journal, Korean Society of Spine Surgery*, June 2009, www.ncbi.nlm.nih.gov/pmc/articles/PMC2852042/.

Degowin, R.L., and E.L. Degowin, *DeGowin & DeGowin's Diagnostic Examination*, 6th edn, McGraw Hill, 1994, p. 735.

Delgado, G.J., C.B. Chung, N. Lektrakul, P. Azocar, M.J. Botte, D. Coria, E. Bosch, and D. Resnick. "Tennis Leg: Clinical US Study of 141 Patients and Anatomic Investigation of Four Cadavers with MR Imaging and US." *Radiology*, vol. 224 (2002): 112–119.

Friedman, Benjamin W., Mikaela Chilstrom, Polly E. Bijur, and E John Gallagher. "Diagnostic Testing and Treatment of Low Back Pain in United States Emergency Departments: A National Perspective." *Spine*. U.S. National Library of Medicine, November 15, 2010. https://www.ncbi.nlm.nih.gov/pmc/articles/PMC2982879/.

Garner, Matthew R., Elizabeth Gausden, Marschall B. Berkes, Joseph T. Nguyen, and Dean G. Lorich. "Extensor Mechanism Injuries of the Knee: Demographic Characteristics and Comorbidities from a Review of 726 Patient Records." *The Journal of Bone and Joint Surgery. American Volume*. U.S. National Library of Medicine, October 7, 2015. https://www.ncbi.nlm.nih.gov/pubmed/26446967.

Gerdle, Björn, Bijar Ghafouri, Malin Ernberg, and Britt Larsson. "Chronic Musculoskeletal Pain: Review of Mechanisms and Biochemical Biomarkers as Assessed by the Microdialysis Technique." *Journal of Pain Research*, Dove Medical Press, June 12, 2014, www.ncbi.nlm.nih.gov/pmc/articles/PMC4062547/.

Gliatis, John, Konstantinos Anagnostou, Pantelis Tsoumpos, Evdokia Billis, Maria Papandreou, and Spyridon Plessas. "Complex Knee Injuries Treated in Acute Phase: Long-Term Results Using Ligament Augmentation and Reconstruction System Artificial Ligament." *World Journal of Orthopedics*, Baishideng Publishing Group Inc., March 18, 2018, www.ncbi.nlm.nih.gov/pmc/articles/PMC5859197/.

Goldsmith, Laura T., and Gerson Weiss. "Relaxin in Human Pregnancy." *Annals of the New York Academy of Sciences*. U.S. National Library of Medicine, April 2009. https://www.ncbi.nlm.nih.gov/pmc/articles/PMC3856209.

Großschädl, Franziska, W. Freidl, É. Rásky, N. Burkert, J. Muckenhuber, and W. Stronegger. "A 35-Year Trend Analysis for Back Pain in Austria: The Role of Obesity." *PLoS ONE*, vol. 9, no. 9, October 2014, doi:10.1371/journal.pone.0107436.

Haefeli, Mathias, and Achim Elfering. "Pain Assessment." *European Spine Journal: Official Publication of the European Spine Society, the European Spinal Deformity Society, and the European Section of the Cervical Spine Research Society*. Springer-Verlag, January 2006. https://www.ncbi.nlm.nih.gov/pmc/articles/PMC3454549/.

Harvey, G.P., G.W. Woods. "Anterolateral Dislocation of the Proximal Tibiofibular Joint: Case Report and Literature Review." *Todays OR Nurse*, vol. 14 (1992): 23–27.

Hayat, Zara. "Patella Dislocation." StatPearls [Internet]. U.S. National Library of Medicine, January 21, 2020. https://www.ncbi.nlm.nih.gov/books/NBK538288.

Healthwise Staff. "Meniscus Repair." *Kaiser Permanente*. June 26, 2019. https://wa.kaiserpermanente.org/kbase/topic.jhtml?docId=uh2064.

Helms, C.A., R.C. Fritz, and G.J. Garvin. "Plantaris Muscle Injury: Evaluation with MR Imaging." *Radiology*, vol. 195 (1995): 201–203.

Hochschuler, Stephen. "Sciatica Causes." *Spine*, June 5, 2019, www.spine-health.com/conditions/sciatica/sciatica-causes.

Horan, J., and G Quin. "Proximal Tibiofibular Dislocation." *Emergency Medicine Journal*, vol. 23 (2006): e33.

Horst, Patrick K., and Robert F. LaPrade. "Anatomic Reconstruction of Chronic Symptomatic Anterolateral Proximal Tibiofibular Joint Instability." *Knee Surgery, Sports Traumatology, Arthroscopy: Official Journal of the ESSKA*. U.S. National Library of Medicine, November 18, 2010: 1452-5, https://www.ncbi.nlm.nih.gov/pubmed/20127312.

Hsu, Hunter. "Patellar Tendon Rupture." StatPearls [Internet], U.S. National Library of Medicine, January 21, 2020, www.ncbi.nlm.nih.gov/books/NBK513275/.

Husain, Aatif M., *A Practical Approach to Neurophysiologic Intraoperative Monitoring*. Demos Medical Publishing, 2008. p. 23.

Jackson, J.P. "The Facet Syndrome. Myth or Reality?" *Clinical Orthopaedics and Related Research*, vol. 279 (1992 June): 110–21. (level of evidence: 2C).

Jones, Bradley R. "The Effect of Q Angle on Vertical Jump in Female Athletes." *Goucher College Master of Education*, May 2013. https://mdsoar.org/bitstream/handle/11603/2400/MEd_Jones_actionres_Sp2013.pdf?sequence=1.

Kisner, Carolyn, Lynn Allen Colby, and John Borstad. *Therapeutic Exercise: Foundations and Techniques*. Philadelphia, PA: F.A. Davis Company, 2018.

Konno, Tomoyuki, Yoichi Aota, Hiroshi Kuniya, Tomoyuki Saito, Ning Qu, Shogo Hayashi, Shinichi Kawata, and Masahiro Itoh. "Anatomical Etiology of 'Pseudo-Sciatica' from Superior Cluneal Nerve Entrapment: A Laboratory Investigation." *Journal of Pain Research*. Dove Medical Press, November 1, 2017. https://www.ncbi.nlm.nih.gov/pmc/articles/PMC5677392/.

Kumar, M. "Epidemiology, Pathophysiology and Symptomatic Treatment of Sciatica: A Review." *International Journal of Pharmaceutical & Biological Archives* 2011. https://www.spine-health.com/conditions/sciatica/sciatica-causes.

Kuniya, H., Takuya Kawai, Kan-ichiro Kaneko, Tomoyuki Konno, and Tomoyuki Saito, "Prospective Study of Superior Cluneal Nerve Disorder as a Potential Cause of Low Back Pain and Leg Symptoms." *Journal of Orthopaedic Surgery and Research*, vol. 9, no. 139 (2014). https://doi.org/10.1186/s13018-014-0139-7.

Laprade, J., E. Culham, and B. Brouwer. "Comparison of Five Isometric Exercises in the Recruitment of the Vastus Medialis Oblique in Persons with and Without Patellofemoral Pain Syndrome." *Journal of Orthopaedic & Sports Physical Therapy*, vol. 27 (1998): 197–204.

Lirette, L. S., G. Chaiban, R. Tolba, and H. Eissa. "Coccydynia: an Overview of the Anatomy, Etiology, and Treatment of Coccyx Pain." *The Ochsner Journal*. U.S. National Library of Medicine, 2014. https://www.ncbi.nlm.nih.gov/pubmed?term=24688338.

Liu, Lanqing, Guibao Ke, and Azure Duan. *Introduction to Tui Na*. New Jersey: World Scientific, 2013.

Liu, S.Y., C.L. Hsieh, T.S. Wei, P.T. Liu, Y.J. Chang, and T.C. Li. "Acupuncture Stimulation Improves Balance Function in Stroke Patients: A Single-Blinded Controlled, Randomized Study." *American Journal of Chinese Medicine*, vol. 37 (2009): 483–494. doi: 10.1142/S0192415X09006990.

Logterman, Stephanie L., Frank B. Wydra, and Rachel M. Frank. "Posterior Cruciate Ligament: Anatomy and Biomechanics." *Current Reviews in Musculoskeletal Medicine*, Springer US, September 2018, www.ncbi.nlm.nih.gov/pmc/articles/PMC6105479.

Martins, Delio Eulalio, et al. "Changes in Human Intervertebral Disc Biochemical Composition and Bony End Plates between Middle and Old Age." *PloS One*, Public Library of Science, September 18, 2018, www.ncbi.nlm.nih.gov/pmc/articles/PMC614 4914/.

Meister, B.R., S.P. Michael, R.A. Moyer, J.D. Kelly, and C.D. Schneck. "Anatomy and Kinematics of the Lateral Collateral Ligament of the Knee." *The American Journal of Sports Medicine*, U.S. National Library of Medicine, November 1, 2000, www.ncbi. nlm.nih.gov/pubmed/11101111.

National Institute on Aging. "Osteoarthritis." National Institute on Aging. U.S. Department of Health and Human Services, May 1, 2017. https://www.nia.nih.gov/health/osteoarthritis.

Nieuwe Weme, R.A., M.P. Somford, and T. Schepers. "Proximal Tibiofibular Dislocation: A Case Report and Review of Literature." *Strategies in Trauma and Limb Reconstruction*. Springer Milan, November 2014. https://www.ncbi.nlm.nih.gov/pmc/articles/PMC 4278967.

Nunes, Jóni. "Acute Proximal Tibiofibular Joint Dislocation: A Sports Related Injury?-Two Different Cases." *Annals of Joint*. Accessed February 4, 2020. http://aoj.amegroups .com/article/view/4963/html.

Orthopedics 101—SimpleNeasyBook. 1.0 ed., WAGmob, 2013, https://www.amazon.com/Or thopedics-101-simpleNeasyBook-WAGmob-ebook/dp/B00H2JX752.

Physiopedia. "Straight Leg Raise Test." Physiopedia. Accessed February 3, 2020. https://ww w.physio-pedia.com/Straight_Leg_Raise_Test.

Physiopedia contributors. "Lumbar Facet Syndrome." Physiopedia. Accessed March 8, 2020. https://www.physio-pedia.com/Lumbar_Facet_Syndrome.

Platzer, Werner, *Color Atlas of Human Anatomy*, Vol. 1, Locomotor System, 5th edn. Thieme, 2004, p. 242.

Purnell, Mark L., Andrew I. Larson, and William Clancy. "Anterior Cruciate Ligament Insertions on the Tibia and Femur and Their Relationships to Critical Bony Landmarks Using High-Resolution Volume-Rendering Computed Tomography." *American Journal of Sports Medicine*, vol. 36, no. 11 (November 2008): 2083–2090.

Raj, Marc A. "Sacroiliac (SI) Joint Pain." StatPearls [Internet]. U.S. National Library of Medicine, May 12, 2019. https://www.ncbi.nlm.nih.gov/books/NBK470299.

Raj, Marc A. "Posterior Cruciate Ligament (PCL) Knee Injuries." StatPearls [Internet], U.S. National Library of Medicine, May 6, 2019, www.ncbi.nlm.nih.gov/books/NBK430726/.

Rao, Allison J., Brandon J. Erickson, Gregory L. Cvetanovich, Adam B. Yanke, Bernard R. Bach, and Brian J. Cole. "The Meniscus-Deficient Knee: Biomechanics, Evaluation, and Treatment Options." *Orthopaedic Journal of Sports Medicine*. SAGE Publications, October 23, 2015. https://www.ncbi.nlm.nih.gov/pmc/articles/PMC4714576.

Reddi, Danielle, and Natasha Curran. "Chronic Pain after Surgery: Pathophysiology, Risk Factors and Prevention." *Postgraduate Medical Journal*. The Fellowship of Postgraduate Medicine, April 1, 2014. https://pmj.bmj.com/content/90/1062/222.

Rea, Paul. "Pyramidal Tracts." Pyramidal Tracts—An overview. *ScienceDirect Topics*, 2015. https://www.sciencedirect.com/topics/neuroscience/pyramidal-tracts.

Riddle, Daniel L., and Susan M. Schappert. "Volume of Ambulatory Care Visits and Patterns of Care for Patients Diagnosed with Plantar Fasciitis: A National Study of Medical Doctors." *Foot & Ankle International*. U.S. National Library of Medicine, May 2004. http://www.ncbi.nlm.nih.gov/pubmed/15134610.

Ristić, Vladimir, Mirsad Maljanović, Iva Popov, Vladimir Harhaji, and Vukadin Vukadin. "Quadriceps Tendon Injuries." Original Studies. *Medicinski Pregled*, vol. LXVI, no. 3–4 (2013 March–April): 121–25. https://doi.org/10.2298/MPNS1304121R.

Robertson, Sally. "Types of Pain." *News*, August 23, 2018. https://www.news-medical.net/health/Types-of-Pain.aspx.

Saladin, Kenneth S. *Anatomy & Physiology: The Unity of Form and Function*, 5th edn. Boston: McGraw-Hill, 2009.

Sekiya, J.K., and J.E. Kuhn. "Instability of the Proximal Tibiofibular Joint." *The Journal of the American Academy of Orthopaedic Surgeons*, vol. 11 (2003): 120–128.

Shultz, S.J., P.A. Houglum, and D.H. Perrin, *Examination of Musculoskeletal Injuries*, 3rd edn, Human Kinetics, 2010, p. 453.

Smith, Lachlan J. "Changes in Human Intervertebral Disc Biochemical Composition and Bony End Plates Between Middle and Old Age." *PLoS One*, 2018. https://www.ncbi.nlm.nih.gov/pmc/articles/PMC6144914/.

Strobel, Michael, and Hans-Werner Stedtfeld. *Diagnostic Evaluation of the Knee*. Berlin: Springer-Verlag, 1990.

Strobel, M., and H.W. Stedtfeld. *Diagnostic Evaluation of the Knee*. New York: Springer-Verlag, 1990, pp. 166–182.

The Spine Hospital. "Degenerative Disc Disease." The Spine Hospital at The Neurological Institute of New York. Accessed March 8, 2020. https://www.columbiaspine.org/condition/degenerative-disc-disease/.

The Steadman Clinic. "Meniscus Injuries." thesteadmanclinic. Accessed March 5, 2020. https://www.thesteadmanclinic.com/patient-education/knee/meniscus-injuries.

Thomason, P.A., and M.A. Linson. "Isolated Dislocation of the Proximal Tibiofibular Joint." *Journal of Trauma*, vol. 26 (1986): 192–195.

Tyler, Timothy F., Holly J. Silvers, Michael B. Gerhardt, and Stephen J. Nicholas. "Groin Injuries in Sports Medicine." *Sports Health*. SAGE Publications, May 2010. https://www.ncbi.nlm.nih.gov/pmc/articles/PMC3445110.

Unschuld, Paul U., and Hermann Tessenow. *Huang Di Nei Jing Su Wen: An Annotated Translation of Huang Di's Inner Classic—Basic Questions: 2 Volumes*. University of California Press, 2011. http://www.biblio.nhat-nam.ru/Huang_Di_Nei_Jing_Su_Wen-Unschuld-Tessenow-1-2.pdf

Wallis, Christopher. "Management of Degenerative Meniscal Tears." assets.ossur, May 3, 2017. https://assets.ossur.com/library/37558.

Wikipedia. "Pectineus Muscle." Wikimedia Foundation, December 18, 2019. https://en.wikipedia.org/wiki/Pectineus_muscle.

Wilson, Erasmus, *The Anatomist's Vade Mecum: A System of Human Anatomy*, 1851, p. 260.

Woodmass, J.M., J.D. Johnson, I.T. Wu, A.J. Krych, and M.J. Stuart. "Patellar Tendon Repair With Ipsilateral Semitendinosus Autograft Augmentation." *Arthroscopy Techniques*, vol. 6, no. 6 (2017 December): e2177–e2181.

Yan, T., and C.W. Hui-Chan. "Transcutaneous Electrical Stimulation on Acupuncture Points Improves Muscle Function in Subjects after Acute Stroke: A Randomized Controlled Trial." *Journal of Rehabilitation Medicine*, vol. 41 (2009): 312–316. doi: 10.2340/16501977-0325. https://www.ncbi.nlm.nih.gov/pmc/articles/PMC5713240/.

Zhang, Junhua, et al. "Complex Traditional Chinese Medicine for Poststroke Motor Dysfunction." *Stroke*, June 25, 2009. https://doi.org/10.1161/STROKEAHA.109.555227.

Zhao, J.G., et al. "Effect of Acupuncture Treatment on Spastic States of Stroke Patients." *Journal of the Neurological Sciences*, vol. 276 (2009): 143–147. doi: 10.1016/j.jns.2008.09.018.

Zuke, William, Beatrice Go, Alexander Weber, and Brian Forsythe. "Quadriceps Tendon Rupture in an Adolescent Athlete." U.S. National Library of Medicine, September 26, 2017. https://www.ncbi.nlm.nih.gov/pmc/articles/PMC5634587.

Appendix: Proper Postures

Proper standing posture during pregnancy:

- Keep the face facing forward.
- Lift the head and neck straight upright.
- Keep the chin tucked in.
- Keep the ears in a straight line with the shoulders.
- Relax the shoulders, take deep breaths.
- Contract the lower abdominal muscles.
- Keep the pelvic tilt.
- Keep the feet shoulder-width distance apart.
- Point the feet forward or slightly outward.
- Keep the knees slightly flexed and relaxed.
- Keep the center of gravity in the middle of the feet.
- Avoid standing in one place for long periods of time. In case of work such as ironing or washing dishes, switch to sitting on a high stool periodically.
- Avoid high-heeled shoes, as they draw the bodyweight forward. Wear low-heeled shoes.

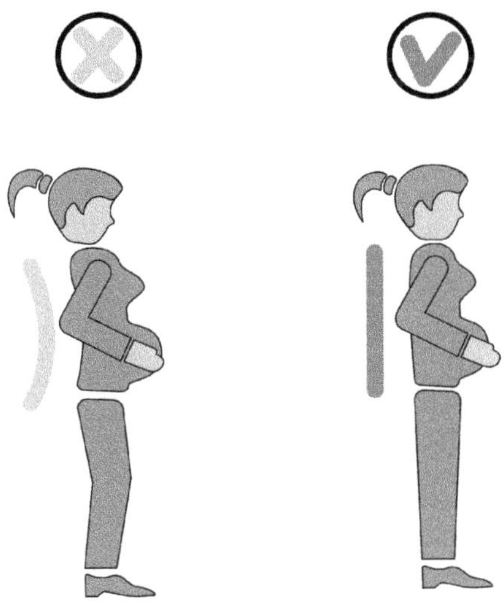

Poor standing posture during pregnancy:

- Chin drawn forward and face angled downward. Facing downward may cause tension in the neck and upper back muscles.
- Slouched abdomen. This increases stress on the lumbar and causes pressure on the urinary bladder.

- Straightened knees. This causes tension on the soft tissues of the leg.
- Bodyweight on the medial sides of the feet. This causes tension on the calves and arches of the feet.

Proper sitting posture during pregnancy:

- First, sit down on the front half of the chair, and slowly slide the buttocks backward to lean on the back of the chair. Keep the back straight, and support the lumbar with a pillow. Be sure to use the leg muscles to lower the body onto the seat.
- Rest the feet flat on the floor or on a footstool so the knees are level with the hips.

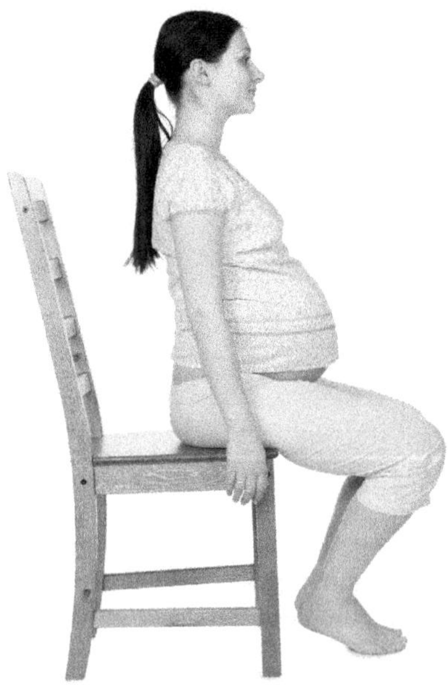

- Adjust the car seat so that the feet can reach the pedals easily if driving a car.
- Adjust the steering wheel to relax the shoulders if driving a car.
- Keep the pelvic tilt.
- Avoid slouching.
- Avoid crossing legs.
- Avoid leaning sideways.
- Avoid sitting longer than 30 minutes. Take a stretch break to relax the muscles.
- To stand up from a chair:
  1) Slide forward to sit on the front part of the chair.
  2) Lean forward, but ensure the knees are still above the feet or toes.
  3) Push the hands on the chair, and stand up without slumping the back or moving the shoulders.
  4) Ensure the body is stable and standing perfectly straight before making steps.

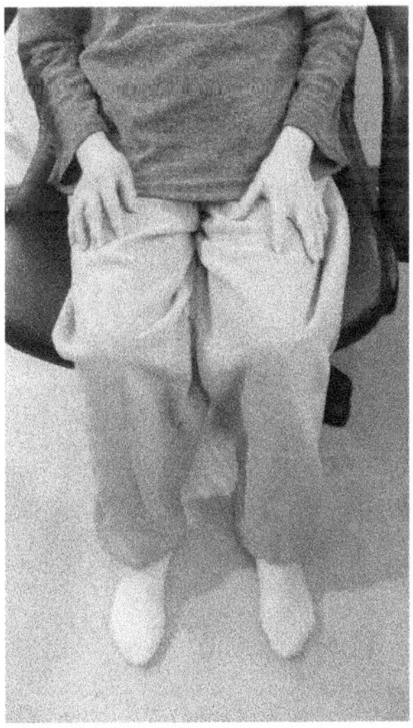
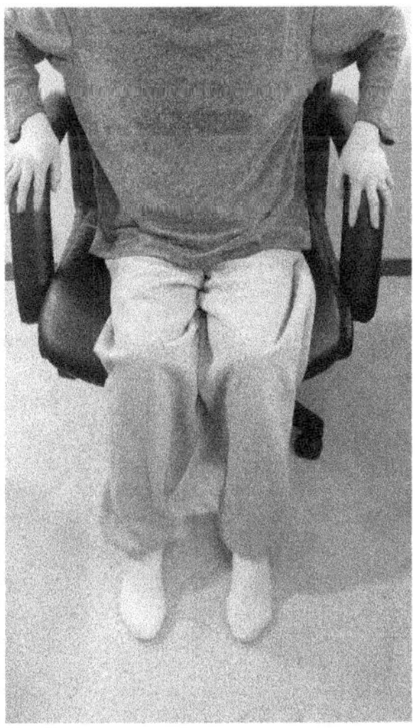
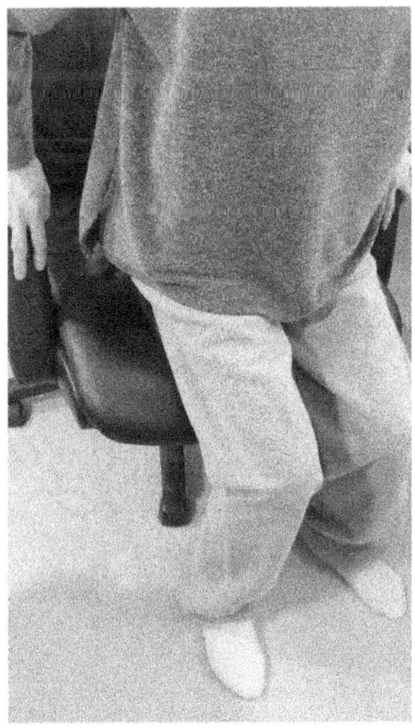
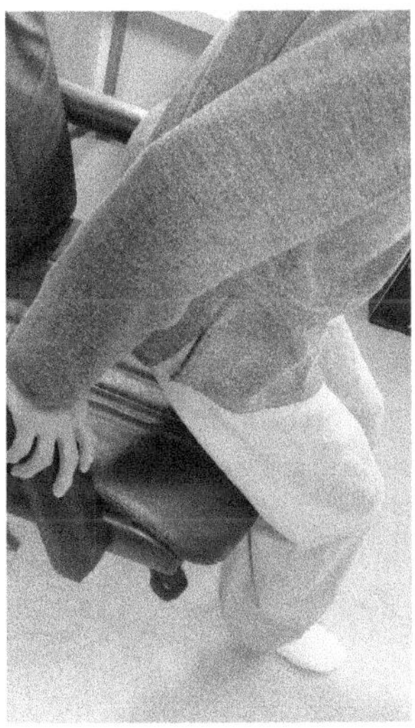

Proper lying posture during pregnancy:

- A supine position can be used for early pregnancy. Use pillows to support the head and shoulders. Place a pillow under the thighs to keep the knees slightly flexed. Let the knees roll outward naturally.
- Lying on the side is recommended for later pregnancy. Lying on the left side is better. Use pillows to support the head and neck, and place pillows behind the back for more support. The bottom leg is straightened; the upper leg is flexed with a pillow placed below the knee.

Proper way to get up from a lying position:

1. From the supine position, shift to the edge of the bed.
2. Roll to lateral lying facing the edge. Do not twist the lumbar, and ensure that the shoulders and hips roll at the same time as one unit.
3. Place the feet outside the bed.

4. Place the top hand on the bed in front of the chest, and push the hand on the bed to lift the trunk with the assistance of the lower arm and elbow pushing on the bed.
5. Avoid twisting the lumbar, only lifting up the trunk with both the hand and elbow while lowering the legs to the floor.
6. Slowly raise the body to an upright sitting position.

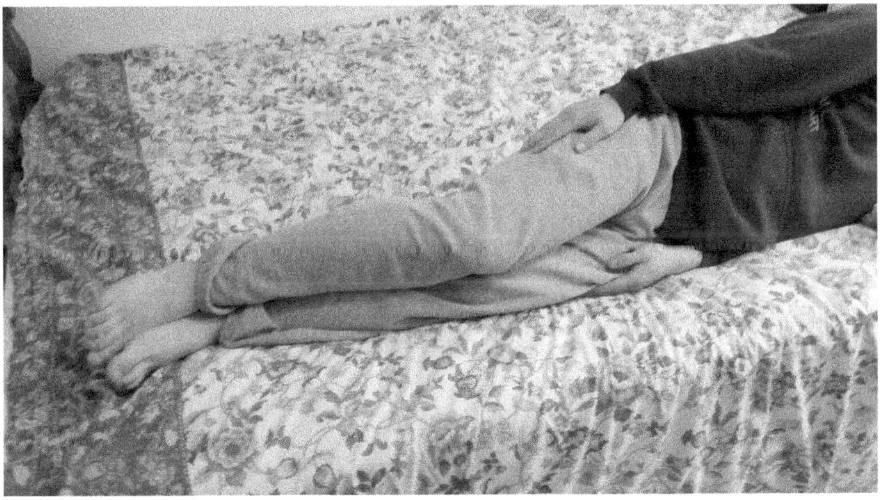

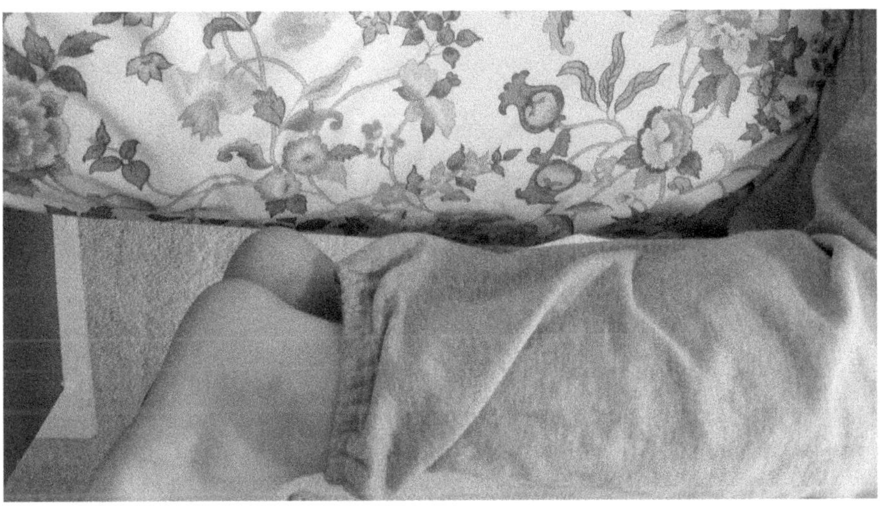

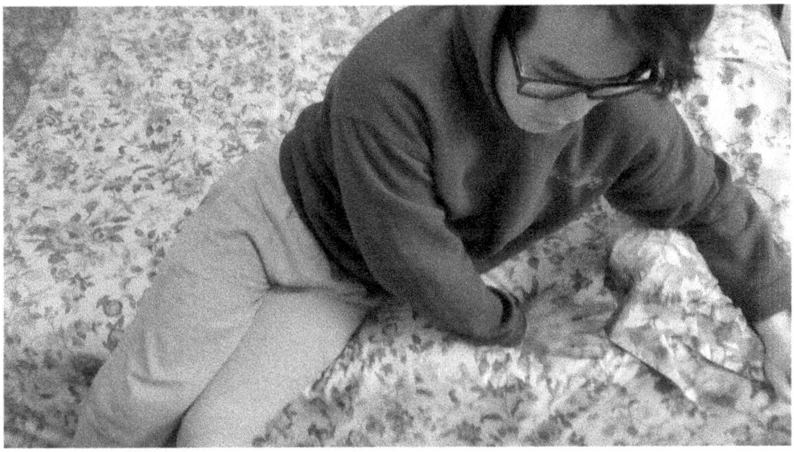

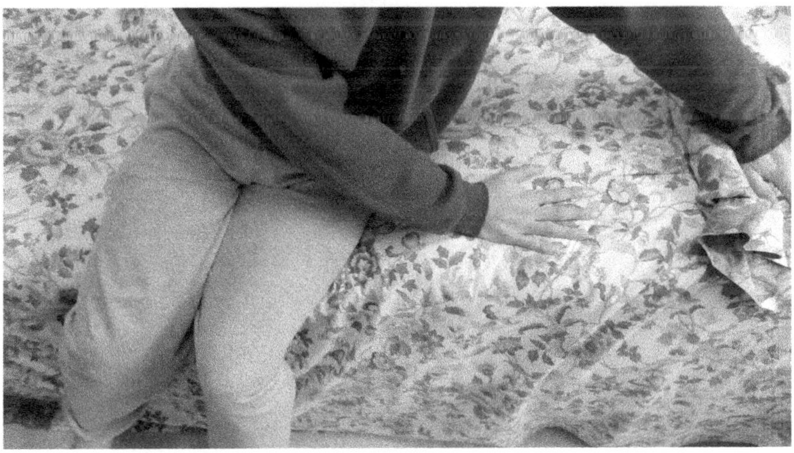

Proper way to get out of a car:

1. Roll the car seat all the way backward for maximum leg clearance.
2. Turn the whole body together with both legs facing outward. Avoid twisting the lumbar.
3. Place both feet on the floor, and slide to the side of the car seat.
4. Place one hand on the car seat, and place another hand on the wheel, door frame, or dashboard.
5. Lean forward, but ensure the knees are above the feet.
6. Push on the hands and stand up.
7. Ensure the body is stable and standing perfectly straight before making steps.

Proper way to get into a car:

1. Roll the car seat all the way backward for maximum leg clearance.
2. Sit down on the car seat while facing outward.
3. Slide to sit in the center of the car seat.
4. Place one hand on the car seat or an elbow on the seat backrest, and place another hand on the wheel, door frame, or dashboard.
5. Raise the legs and turn the whole body together with the legs toward the dashboard, collaborating with the hands.

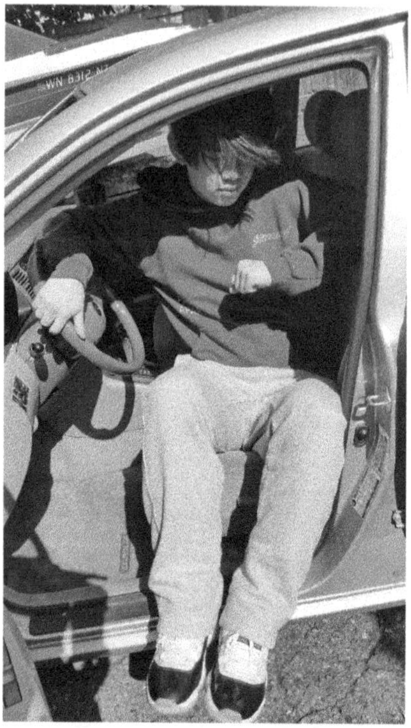

Keep good posture so that Qi and blood flow can be free:

1) Good sitting posture:
   a) Be sure to straighten the back, and avoid leaning forward. Place a cushion behind the low back to support the normal curvature of the lumbar.

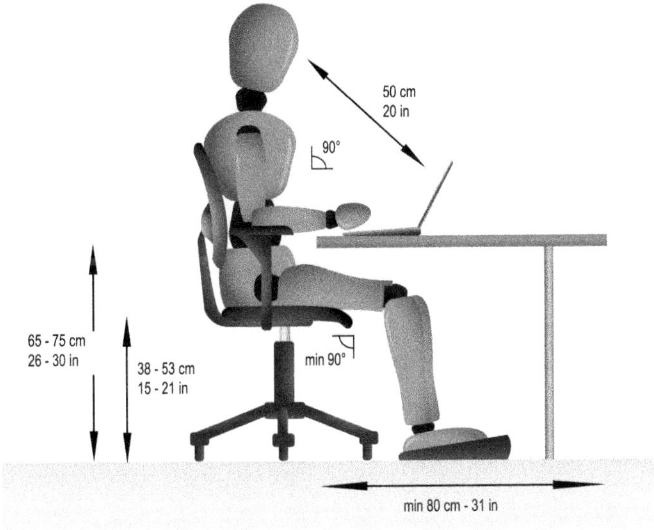

b) Flex the hips and knees at about 90 degrees.
c) Keep the legs shoulder width apart.
d) Keep the feet flat on the floor. If the feet cannot reach the floor, place the feet on a footrest.
e) Do not sit for longer than an hour. A cup of tea with a few chrysanthemum flowers and goji berries with hot water, finished within an hour, is recommended. Goji berries help eye tiredness, chrysanthemum flowers help to clear heat in the eyes, and tea helps to clear the heat inside the body. After 250cc of water, someone will need to get up and walk to the bathroom. When sitting for 8 hours, they will finish drinking 8 cups of water.

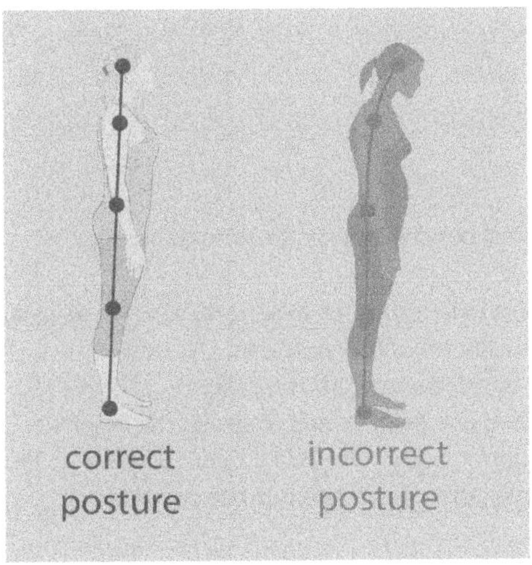

2) Good standing posture:
   a) Be sure to stand with an erect head and back. Keep the head upright on the midline between shoulders.
   b) Stand with both legs slightly apart and both knees slightly flexed.
   c) Keep the weight on the front 1/3 of the feet,
   d) Be mindful, and do not lean your body weight on either side.

3) Good sleeping posture:
   a) Sleeping on a relatively firm bed is good for sciatic pain recovery.
   b) Supine is the appropriate sleeping posture, as both sides of the body receive equal pressure.
   c) Place a cushion or pillow below the knees to alleviate lumbar pressure.
   d) In the acute stage, lift the lower legs higher by placing more pillows below the tibia.
   e) The normal lumbar curvature also can be supported by placing a thin cushion or a rolled towel.

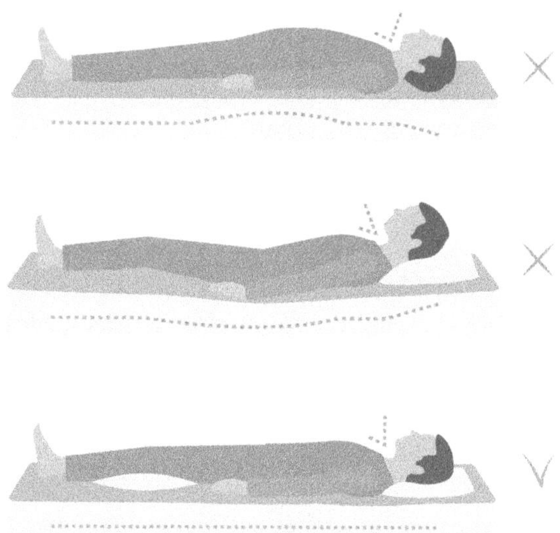

Learn how to lift and bend properly. Follow these tips:

- If an object is too heavy, big, or awkward, ask for assistance.
- Stand as close to the object as possible.
- Spread the feet apart wider than the object.
- Bend at the knees; use both hands to grasp the object.
- Straighten the upper body. DO NOT lean forward or twist sideways, so as not to lose gravity and sprain the lumbar soft tissues.

- Tighten the abdominal and hip muscles.
- Hold the object as close to the body as possible.
- Lift using the leg muscles.
- Carry the object and walk with the eyes. Slow down and be cautious when walking on an uneven surface.

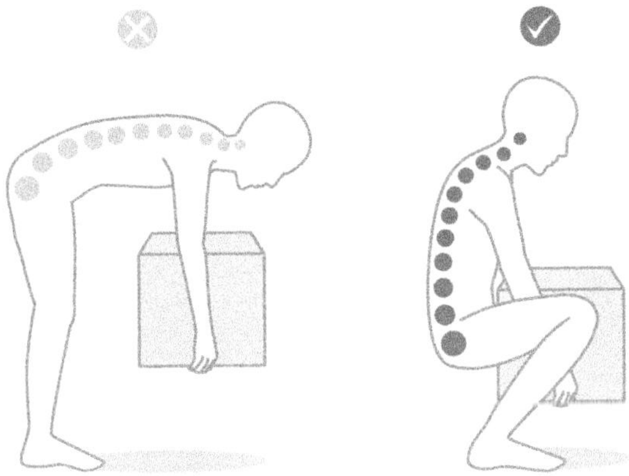

Proper lumbar and leg positions while driving:

Lumbar support:

- The lumbar support helps prevent a gap between the lumbar and the car seat. The lumbar muscles cannot relax if there is a gap.
- The lumbar support device provides warmth and prevents cold invasion from drafts.
- A good lumbar support maintains the curvature. It doesn't need to be big. It can't be too hard or too soft. Some vehicles have back support on the seat installed, and some of the back supports are adjustable.

Car seat position:

- Adjust the car seat cushion so that the thighs rest on it to just above the knee. A seat cushion can be added.
- Set the seat height as high as possible, but keep a distance of more than a fist between the head and the car roof.
- Adjust the back of the car seat to about a 100-degree angle.
- Slide the car seat until the foot is able to fully press the gas pedal and the knees are slightly flexed. Keeping the knees slightly flexed while driving will promote the Qi and blood circulation. There should be a clearance distance at about 2 to 3 fingers between the front edge of the car seat and the hollow of the knee, and there should be a clearance distance of more than a fist between the knees and the dashboard.

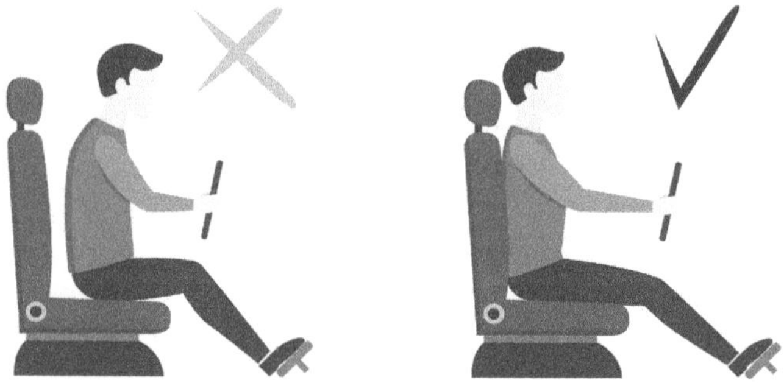

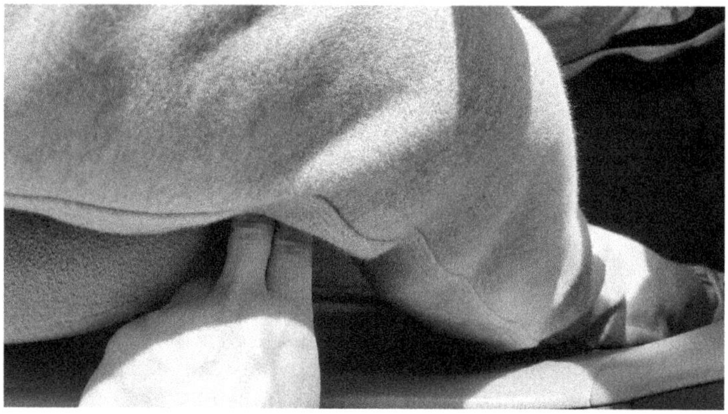

Index

E

F

G

H

I

N

O

P

www.ingramcontent.com/pod-product-compliance
Ingram Content Group UK Ltd.
Pitfield, Milton Keynes, MK11 3LW, UK
UKHW022007180926
475701UK00003B/232
9780367235857